Springer Asia Pacific Mathematics Series

Volume 7

The Springer Asia Pacific Mathematics Series promotes high-quality scientific research connected to Asia-Pacific countries. Monographs, edited volumes, lecture notes, and textbooks will cover topics related to pure and applied mathematics as well as interdisciplinary research.

***

All material will be rigorously peer-reviewed and should be related to mathematical research conducted at or with institutions in the APAC region. This series is open to subseries, which may be topical or exclusively supplied by a single research institution.

***

Sehun Chun

Moving Frames for the Numerical Solution of Partial Differential Equations in Complex Domains

Computation Using Orthonormal Basis Vectors

Springer

Sehun Chun
Underwood International College
Yonsei University
Seoul, Korea (Republic of)

ISSN 3091-2555 ISSN 3091-2563 (electronic)
Springer Asia Pacific Mathematics Series
ISBN 978-981-95-4494-3 ISBN 978-981-95-4495-0 (eBook)
https://doi.org/10.1007/978-981-95-4495-0

Mathematics Subject Classification: 65M60, 53C22, 35Q35, 35J05

This Springer imprint is published by the registered company Springer Nature Singapore Pte Ltd.
The registered company address is: 152 Beach Road, #21-01/04 Gateway East, Singapore 189721, Singapore

To Love that will not let me go

Preface

I began studying cardiac electrical signal propagation in the atrium in 2010, focusing on solving diffusion–reaction equations on a thin-walled atrial surface. I quickly realized that efficient numerical tools for solving partial differential equations (PDEs) on general curved domains were lacking. My postdoctoral advisor, Spencer J. Sherwin, recommended a paper by Bernard et al., "*High-order discontinuous Galerkin schemes on general 2D manifolds applied to shallow water equations*" (Bernard et al., 2009), which addressed the shallow water equations using tangent vectors defined on the tangent plane of a curved surface. Inspired by this approach, I began applying similar methods to more general PDEs, particularly diffusion–reaction equations for cardiac electric signal propagation on a complex domain similar to a human atrium. The original work employed non-orthogonal and non-unit tangent vectors as basis vectors.

Soon afterward, I realized that using orthonormal basis vectors offered clear advantages in both computation and analysis. I successfully implemented a numerical solution for diffusion–reaction equations and used it to simulate cardiac electrical signal propagation on a 2D atrial geometry. As I prepared to publish the results, it became necessary to give the method a name. After searching through Internet and several geometry texts, I encountered the concept of the *moving frame*, introduced by Élie Cartan, which closely resembled the framework I had developed. I therefore chose to adopt that name for the approach I believed I had devised.

Since then, my journey into the world of moving frames has continued for over a decade, much much longer than I had anticipated. The moving frame has been both fascinating and inspiring, though at times difficult, and occasionally even painful, to fully comprehend. The method appears deceptively simple, yet I spent many days wrestling with why it works and what it truly means to differentiate or align moving frames. This book is a mid-course reflection: a summary of what I have studied and applied in the numerical solution of PDEs using moving frames.

I vividly recall the first time I presented a talk on the numerical method for solving the shallow water equations with moving frames. The conference, *Partial Differential Equations on the Sphere*, was held in 2017 at the École Normale Supérieure (ENS)—the very institution where Élie Cartan had studied, earned his

Ph.D., and taught in the early twentieth century. I had explored various PDEs on curved surfaces using moving frames, but among them, the shallow water equations (SWEs) had the largest and most active community working on numerical solutions for curved domains, particularly the sphere. Together with my coauthor, Claes Eskilsson (Aalborg University), I had incorporated moving frames into the SWEs. After multiple rejections from different journals, we were finally on the verge of publishing our paper in the *Journal of Computational Physics* (Chun and Eskilsson, 2017). The publication itself was a major achievement after years of effort, but giving a talk that was both intellectually and emotionally tied to Cartan's original work marked another deeply meaningful milestone. I often wondered how Élie Cartan might have responded to this work if he were alive today. Would he dismiss it? Offer new insights? Critique my interpretations? I will likely never know—and perhaps I do not need to. Still, presenting that talk at the ENS gave me a strange and unforgettable sense of connection across a vast divide of space and time.

There is a wealth of literature on the method of moving frames. Nearly every book on differential geometry includes a section or subsection on the topic. Most of these are mathematical texts that assume some background in differential geometry, though more accessible introductions can also be found in the physics literature. However, I believe this book differs in kind from the existing works on moving frames. I personally struggled to grasp the intuitive aspects and practical applications of moving frames in scientific computing and applied mathematics. For that reason, I have aimed to share my insights and intuition with minimal mathematical abstraction. Moreover, inspired by Tristan Needham's *Visual Differential Geometry and Forms* (Needham, 2021), I have presented the concepts and mathematical ideas through many visualizations. Years of presentations and discussions with non-mathematicians have also shaped the content and tone of this book.

The moving frames used in this book are closely related to orthonormal basis vectors aligned with the local geometry. They represent only a small fraction of Cartan's original formulation of moving frames, which was developed in the broader context of symmetry and equivalence. Therefore, the moving frames adopted here may not be directly related to Cartan's full and modern theoretical framework (Olver and Valiquette, 2008; Sharpe, 1997; Arnold et al., 2010).

This book outlines how moving frames, constructed at discrete three-dimensional grid points, can be employed to solve various PDEs on complex domains. It is intended for undergraduate and graduate students, as well as researchers in scientific computing, particularly those interested in the numerical solution of PDEs. The primary discretization techniques used are spectral/hp methods (Karniadakis and Sherwin, 2013) and discontinuous Galerkin schemes (Hesthaven and Warburton, 2008). While familiarity with these numerical methods may aid in understanding the implementation of moving frames, it is not strictly necessary, except Chap. 5. C++ code, implemented in the open-source spectral/hp element library Nektar++ (Cantwell et al., 2015), is included in Chap. 5 as pseudocode. The complex meshes—such as the sphere, pseudosphere, and atrium—shown in this section were generated using the open-source three-dimensional finite element mesh generator `Gmsh` (Geuzaine and Remacle, 2009). For higher accuracy and compatibility with high-order numerical schemes, high-order meshes can be generated using `NekMesh` (Green et al., 2024).

To further enhance the usability and practical value of this book, MATLAB scripts are also provided for (1) constructing moving frames, (2) computing the metric tensor and Christoffel symbols, (3) computing connections, and (4) applying differential operators on the sphere and pseudosphere.

Seoul, Republic of Korea Sehun Chun
May, 2025

Acknowledgments I am deeply grateful to Linyu Peng (Keio University) for his insightful consultations on moving frames, particularly from diverse mathematical perspectives. I also thank Jae-Hung Jung (POSTECH) for his inspiration and encouragement, both of which played a vital role in the completion of this book. I am indebted to Misun Min (Argonne National Laboratory) and Paul Fischer (University of Illinois Urbana–Champaign) for their valuable collaboration and inspirational advice throughout this journey. I would also like to express my deepest appreciation to Janusz Pudykiewicz (Environment and Climate Change Canada), who supported me during the most challenging periods of my exploration into moving frames. Special thanks go to Hae-Jeong Park (Yonsei University College of Medicine) for his guidance and encouragement, particularly regarding neuroscience.

I am also grateful to the Nektar++ development team for helping me understand the core codebase and enabling the implementation of my own moving frame routines within the library. In particular, I thank Spencer J. Sherwin (Imperial College London), Mike R. Kirby (University of Utah), Chris Cantwell (Imperial College London), and David Moxey (King's College London) for their support and generosity.

My academic journey would not have been possible without the mentorship of my Ph.D. advisor, Jan S. Hesthaven (EPFL), and my postdoctoral supervisor, Spencer J. Sherwin (Imperial College London), whose influence has been truly foundational. I will never forget the warm welcome and lasting friendships I experienced at the African Institute for Mathematical Sciences (AIMS) in Cape Town, South Africa. That period shaped my character and enriched my research capabilities through inspiring interactions with many voluntary visitors at AIMS and in-house researchers like Barry Green (AIMS), Jeff Sanders (AIMS), Paul Taylor (NIMH), Raouf Ghomrasni (Howard University), Lafras Uys (Stellenbosch University), and Barry Bassett (University of Cape Town).

I am sincerely thankful to my former postdoctoral fellow, Abdoulkadri Chama, and to my current and former students—Youngwoo Kim (KT), Hoeyoung Kim (Seoul National University), and Woosung Choi (Yonsei University)—for reading multiple chapters and providing helpful comments. I would like to express my heartfelt thanks to Semin Park for creating many of the descriptive illustrations featured in this book, and to Soohyun Hwang for meticulous proofreading and English editing.

This book would not exist without the unwavering love and support of my family—my vibrant young daughter, Sian; my beloved wife, Jihyun; and my gracious mother, Hyesin.

Finally, I gratefully acknowledge financial support from the National Research Foundation of Korea (NRF-2021R1A2C109297811).

Contents

Acronyms

C++	C Plus Plus
CPU	Central Processing Unit
DG	Discontinuous Galerkin
DG-FEM	Discontinuous Galerkin Finite Element Method
DoF	Degrees of Freedom
FEM	Finite Element Method
FFT	Fast Fourier Transform
GPU	Graphics Processing Unit
MATLAB	Matrix Laboratory (MATLAB)
MF	Moving Frames
MMF	Methods of Moving Frames
ODE	Ordinary Differential Equation
PDE	Partial Differential Equation
RMS	Root Mean Square
SDF	Signed Distance Function
SEM	Spectral Element Method
CL	Conservational Laws
SWE	Shallow Water Equations

Chapter 1
Introduction to Moving Frames for Curved and Anisotropic Domains

> *"Only those who will risk going too far can possibly find out how far one can go."*
>
> *—T. S. Eliot*

When describing physical or biological phenomena in a multidimensional space, **referencing** is essential and is typically provided by a **coordinate system**, or **frames**. Referencing involves determining the position and representing the velocity of moving objects. For Euclidean spaces, such as a plane or a cube, the Cartesian coordinate system has been the most widely adopted choice. Setting up an x- and y-axis for a plane, or equivalently x-, y-, and z-axes for a cube, provides an effective reference system for the space (see Fig. 1.1). Any point in the multidimensional space can be precisely represented using the adopted coordinate system. When an object moves within the domain, its velocity can also be represented in the coordinate system by tracking its positional changes over a given time interval.

Setting up the Cartesian coordinate system for a domain is not unique, even in a Euclidean space. The origin may be placed at different locations, and the coordinate axes can be rotated, as illustrated by the (x', y', z') system in Fig. 1.1. When a single coordinate system is used across the entire domain, it is referred to as a **global coordinate system**. The global coordinate system is conceptually straightforward and frequently encountered in daily life, for example, in mailing addresses or the Global Positioning System (GPS). If multiple observers occupy the same space and agree to use the global coordinate system, then the positioning of a moving object will be consistently interpreted by all observers.

In contrast, to provide a unique representation for each point that depends on the local context, a different referencing approach called a **local referencing system** may be considered. **Local referencing** means that the coordinate system varies with the observer's position and is therefore not universally applicable across all points. Instead, a distinct coordinate system exists at each location. When multiple observers are present in the space, each uses their own coordinate system, as shown in Fig. 1.2.

S. Chun, *Moving Frames for the Numerical Solution of Partial Differential Equations in Complex Domains*, Springer Asia Pacific Mathematics Series 7,
https://doi.org/10.1007/978-981-95-4495-0_1

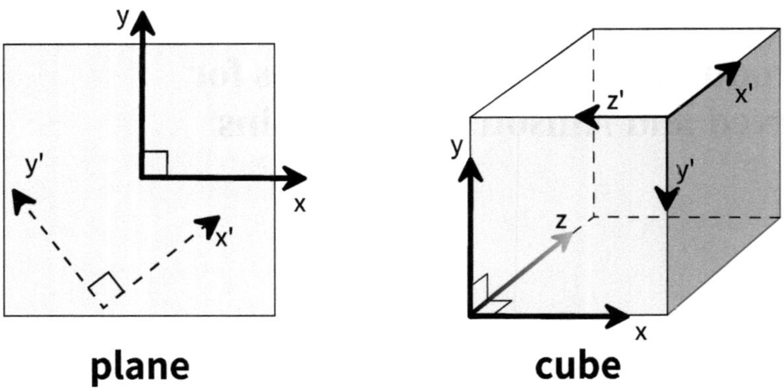

Fig. 1.1 Cartesian coordinate system for a two-dimensional plane and a three-dimensional cube. Two different reference systems (x, y) and (x', y') in the plane and (x, y, z) and (x', y', z') in the cube

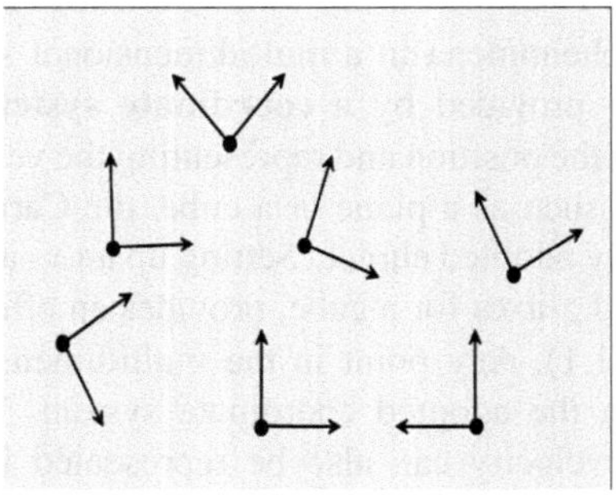

(a) Local reference frames

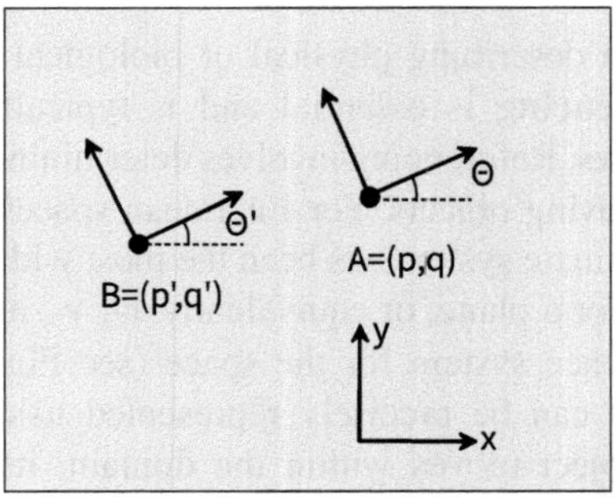

(b) Two local reference frames

Fig. 1.2 Local reference systems in which every point has its own unique coordinate system. (**a**) Each point possesses a distinct (random) reference system. (**b**) Points A and B each have a reference system oriented at angles θ and θ', respectively, with respect to the classical Cartesian coordinate system

The origin of each local coordinate system is located at the corresponding point, and the orientation and scale of the axes may vary arbitrarily, as long as they lie within the same space. To interpret a position specified in one local frame using another, a transformation accounting for both translation and frame rotation is required. For example, suppose observers A and B are observing a moving object, and they are 1 m apart, with observer A positioned 1 m to the left of observer B. If observer A states, "The object is 1 m to my left," then observer B, who is already 1 m to the right of A, should interpret the object's location as 2 m to their left.

A local frame is treated as Euclidean within an infinitesimal neighborhood—effectively a tangent space (Bishop 1975). In addition, local reference systems are all aligned in the same orientation at the same scale.

Consider two local reference frames located at points $A = (p, q)$ and $B = (p', q')$ in a global Cartesian coordinate system, where the first and second components represent distances along the x- and y- axes, respectively. In this case, (a, b) in the A-reference system corresponds to $(a - p', b - q')$ in the B-reference system. Furthermore, assume that the first axes of the local frames at A and B make angles θ and θ', respectively, with respect to the x-axis. Then, the unit tangent vectors $\mathbf{e}_1$ and $\mathbf{e}_2$ of the A-reference system can be expressed in terms of θ as:

$$\mathbf{e}_1 = (e_1^x, e_1^y, e_1^z) = (\cos\theta, \sin\theta), \tag{1.1}$$

$$\mathbf{e}_2 = (e_2^x, e_2^y, e_2^z) = (-\sin\theta, \cos\theta). \tag{1.2}$$

Now, consider a velocity vector $\mathbf{v}$. In the Cartesian coordinate system, it is written as:

$$\mathbf{v} = v^x\mathbf{x} + v^y\mathbf{y}, \tag{1.3}$$

where v^x and v^y are the components along the x- and y-axes, respectively. In the local A-reference frame, the same vector is represented as:

$$\mathbf{v} = v^1\mathbf{e}_1 + v^2\mathbf{e}_2, \tag{1.4}$$

or simply as the coordinate pair (v^1, v^2). Using the dot product, the components v^1 and v^2 can be calculated as:.

$$v^i = \mathbf{v} \cdot \mathbf{e}_i, \quad 1 \le i \le 2. \tag{1.5}$$

However, the B-reference system has the following two unit tangent vectors:

$$\mathbf{e}'_1 = (e'^x_1, e'^y_1, e'^z_1) = (\cos\theta', \sin\theta'), \tag{1.6}$$

$$\mathbf{e}'_2 = (e'^x_2, e'^y_2, e'^z_2) = (-\sin\theta', \cos\theta'), \tag{1.7}$$

Suppose that A- and B-reference systems share the origin and are linear, but not affine, transformation. Accordingly, the velocity vector $\mathbf{v}$ can be expressed in the B-reference system as:

$$\mathbf{v} = v'^1\mathbf{e}'_1 + v'^2\mathbf{e}'_2, \tag{1.8}$$

or, more concisely, as (v'^1, v'^2). Similarly, the components v'^1 and v'^2 as:

$$v'^i = \mathbf{v} \cdot \mathbf{e}'_i, \quad 1 \le i \le 2. \tag{1.9}$$

Since Eqs. (1.4) and (1.8) represent the same vector, the components v^i and v'^i are related by the following transformation:

$$\begin{bmatrix} v^1 \\ v^2 \end{bmatrix} = \begin{bmatrix} e_1^x & e_2^x \\ e_1^y & e_2^y \end{bmatrix}^{-1} \begin{bmatrix} (e')_1^x & (e')_2^x \\ (e')_1^y & (e')_2^y \end{bmatrix} \begin{bmatrix} v'^1 \\ v'^2 \end{bmatrix}. \tag{1.10}$$

Alternatively, if the coordinate systems differ only by a rotation, the relation simplifies to:

$$\begin{bmatrix} v^1 \\ v^2 \end{bmatrix} = \begin{bmatrix} \cos(\theta - \theta') & \sin(\theta - \theta') \\ -\sin(\theta - \theta') & \cos(\theta - \theta') \end{bmatrix} \begin{bmatrix} v'^1 \\ v'^2 \end{bmatrix}. \tag{1.11}$$

Equation (1.11) indicates that vector components in one local reference system should be transformed to be interpreted in another local reference system. Each local frame is treated as a local Euclidean approximation—i.e., a tangent space—where the metric tensor is approximately the identity. This simplification assumes both frames are orthonormal and differ only by a rotation.

Naturally, the use of a local reference system introduces additional complexity. One of the most fundamental operations in analyzing dynamic phenomena is differentiation, that is, quantifying the difference of two scalar or vector quantities. In a global coordinate system, a derivative of a vector in a Euclidean space can be directly computed because all positions and vector components are described within a single, consistent frame of reference. For instance, consider two velocity vectors $\mathbf{v}_1$ and $\mathbf{v}_2$ located at different spatial positions. Since vectors are independent of the origin, the difference between them can be computed as:

$$\mathbf{v}_1 - \mathbf{v}_2 = (v_1^x - v_2^x)\mathbf{x} + (v_1^y - v_2^y)\mathbf{y}. \tag{1.12}$$

In contrast, within a local reference system, one must first designate a specific coordinate system (e.g., the one associated with the point of interest). Let us call it the first local reference system. The velocity of a neighboring point is measured in its own second local reference system, and must be transformed into the first system before their difference can be computed. These extra steps are not required in a global system. Vector comparisons across distinct local frames require a consistent transformation—typically via **parallel transport**—especially in curved geometries. Here, we adopt an intuitive notion of parallel transport via angle preservation, which is valid under **Levi-Civita connections** in orthonormal frames. A formal treatment using connection 1-forms is presented in Chap. 3. Each local frame is assumed to be **flat** and **Cartesian** (i.e., locally Euclidean in the sense that the metric approximates the identity tensor at a point).

A more rigorous treatment appears in Chap. 3.

This book aims to illustrate why local reference systems can be advantageous in certain problems. In particular, it is devoted to explaining how a local coordinate system can be effectively utilized for the numerical solution of partial differential equations (PDEs) in complex domains. A complex domain refers to a non-Euclidean space (roughly speaking, not a flat space), such as a curved space or volume. Moreover, such domains often exhibit anisotropic media, where the conductivity or propagation speed varies with direction. In contrast, isotropic media allow uniform propagation in all directions. Anisotropic media frequently occur in biological tissues such as cardiac or neural fibers. Mathematically, such domains are described by **Riemann geometry**, which defines **geometry** as the combination of both the **shape** of space and the **conducting property** of the medium. Consequently, complex domains incorporate both spatial and directional heterogeneity.

One particular local coordinate system emphasized in this book is the **moving frames**, a concept originating from the early twentieth-century work of Cartan (1922, 1924, 2001a,b) and Treibergs (2019). This book adopts a numerically grounded interpretation of **moving frames**, referring to pointwise **orthonormal bases** aligned with local geometry, distinct from Cartan's full method of equivalence. In physics, moving frames are also referred to as **orthonormal basis vectors** (Frankel 2011; Arfken et al. 2012). In mathematics, the local coordinate systems discussed here are discrete moving frames Beffa et al. (2013), defined at a finite number of points (chapter 7 of Mansfield 2010; Burke 1985). While moving frames may encompass general Euclidean local systems, including oblique basis vectors, this book restricts attention to **orthonormal moving frames**, that is, those composed of orthogonal unit vectors. Orthonormality ensures numerical tractability and eliminates the need for explicit use of the metric tensor in frame-based derivatives. Before delving into the applications of moving frames, it is worth clarifying the term itself. The phrase **moving frames** is a translation of the French term **repére mobile** from Cartan's original work, and may initially be a source of confusion.

1.1 What Are Moving Frames? A Localized Geometric Perspective

The concept of **moving frames** may not be immediately intuitive (Bishop 1975). The term **frames** refers to a **reference coordinate system** used to represent the position of a scalar or the direction and magnitude of a vector. Moving frames possess several essential characteristics (Olver 1995). One of the most significant is their **locality**—different frames can be assigned to different points in space. Any scalar or vector in the vicinity of a point P is located or measured using the local frame defined at P.

This notion is reminiscent of the Lagrangian approach in classical field theory, where the observer follows particles to observe how quantities evolve. In that approach, the coordinate system is attached to the moving particle, resulting in a moving frame of reference. In contrast, **moving frames** in this book refer to spatially distributed but temporally fixed orthonormal frames defined at discrete points. This differs from body-fixed or time-varying frames used in mechanics. Paradoxically, despite their name, moving frames do not "move" with the object. Their origins remain fixed at grid points, although the orientations or definitions of the frames may vary.

1.1.1 Directional Freedom and Rotational Adaptivity

In an isotropic two-dimensional medium, moving frames can be constructed at each grid point. However, the direction of the frames may remain unfixed, as illustrated in Fig. 1.2a. This directional freedom means the orientation of frames may vary as a particle travels or a wave propagates—akin to a compass lying flat on a table and rotating in response to a nearby magnetic field. In this sense, the frames may be seen as **rotating** but not translating. Adapting the local frame dynamically along a path—such as when following a particle—relates to the idea of using the most relevant local reference frame for that region.

1.1.2 Local Flatness and Tangent Approximations

Although the global domain may be curved, each tangent space is locally Euclidean, and moving frames span these local flat spaces. Local coordinate systems may locally resemble Cartesian systems in flat spaces, as illustrated in Fig. 1.3. However, this assumption only holds in a sufficiently small neighborhood around a point—on an infinitesimal scale.

Consider a circle. Viewed from afar, it is clearly curved, but when zooming in on a small arc, it appears nearly flat (Fig. 1.4). This is the concept of being **locally Euclidean**. At a point P, one may verify this by drawing a tangent line: if the curve resembles this line more closely as we zoom in, then the curve is locally Euclidean at P. For a surface in $\mathbb{R}^3$, a similar idea holds using tangent planes. Conversely, points

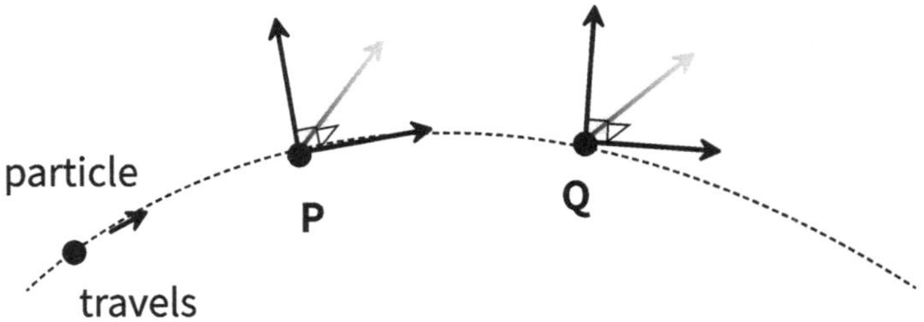

Fig. 1.3 Moving frames at points P and Q for a particle traveling along a curved path. The frames at P and Q are not necessarily identical, even at neighboring points

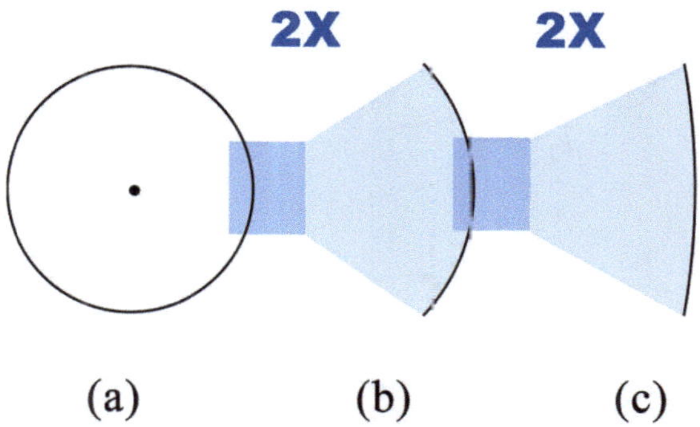

Fig. 1.4 Intuitive meaning of *local Euclidean* for a circle. A smaller segment of a curved line appears increasingly flat (Euclidean) when viewed locally. (**a**) The original circle; (**b**) the surface magnified by a factor of two; (**c**) the surface further magnified by an additional factor of two

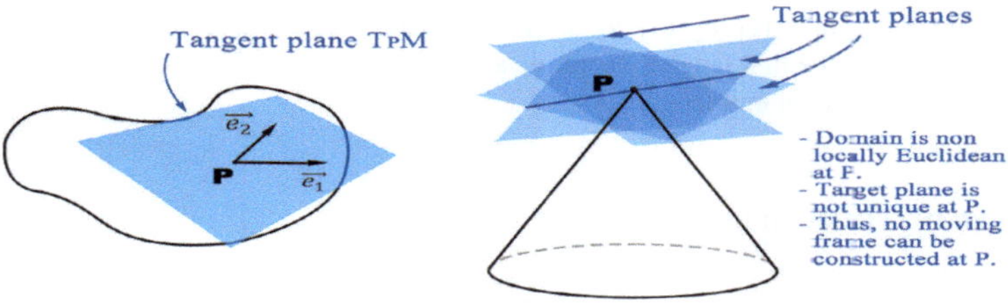

Fig. 1.5 Examples of curved surfaces, some of which are locally Euclidean when a unique tangent plane exists at every point on the surface. The cone (right) is not locally Euclidean at its peak, since an infinite number of tangent planes exist there (see do Carmo 1976; Lee 2018)

like cusps, where no such tangent exists, represent non-locally Euclidean regions (Fig. 1.5).

1.1.3 Tangent Bases and Coordinate Representation

On 2D surfaces, moving frames typically consist of two orthonormal tangent vectors. In 3D, they form a **trihedron**–three mutually orthogonal unit vectors (Fig. 1.6). These vectors typically lie on the tangent plane of a surface and are referred to as **tangent vectors**. In this book, moving frames are denoted by:

$$\mathbf{e}_i = \{\mathbf{e}_1, \mathbf{e}_2, \mathbf{e}_3\} : \text{moving frames.} \tag{1.13}$$

Although general moving frames may be non-orthonormal, this book focuses on the orthonormal case, unless stated otherwise. Consider a vector $\mathbf{v}$ expressed in terms of basis vectors $\mathbf{e}_i$:

$$\mathbf{v} = \sum_i v^i \mathbf{e}_i, \tag{1.14}$$

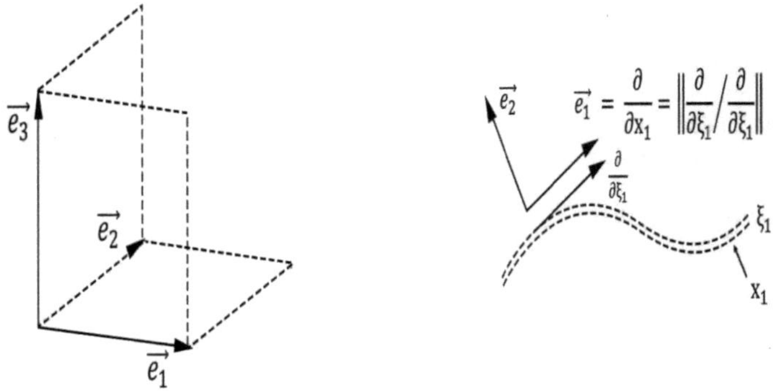

Fig. 1.6 Moving frames $(\mathbf{e}_1, \mathbf{e}_2, \mathbf{e}_3)$ as a trihedron (left). The frames can be oriented along traveling paths, with the first tangent vector $\mathbf{e}_1$ aligned with a curved coordinate direction of the domain while maintaining unit length

which can be written in matrix notation as:

$$\mathbf{v} = \mathbf{E}(\mathbf{v}^T\hat{\mathbf{E}}) = \mathbf{E}\hat{\mathbf{v}} \tag{1.15}$$

where $\hat{\mathbf{v}}$ is an $n \times 1$ column vector of coefficients and $\mathbf{E} = [\mathbf{e}_1\ \mathbf{e}_2\ \cdots\ \mathbf{e}_n]$. Now consider a new basis $\hat{\mathbf{E}}'$ related to the original basis via a transformation matrix $\mathbf{A}$:

$$\mathbf{E}' = \mathbf{E}\mathbf{A} \quad \longleftrightarrow \quad \begin{bmatrix} \mathbf{e}'_1 \\ \mathbf{e}'_2 \\ \vdots \\ \mathbf{e}'_n \end{bmatrix}^T = \begin{bmatrix} \mathbf{e}_1 \\ \mathbf{e}_2 \\ \vdots \\ \mathbf{e}_n \end{bmatrix}^T \mathbf{A}. \tag{1.16}$$

When the translation of basis vectors is ignored, the transformation matrix modifies only the rotation and elongation of the original basis vectors $\mathbf{e}_i$. If $\mathbf{A}$ is an orthogonal matrix (i.e., represents a pure rotation), then $\mathbf{A}^{-1} = \mathbf{A}^T$. Otherwise, the inverse and transpose generally differ. The corresponding transformed coefficients $\hat{\mathbf{v}}'$ are thus given by:

$$\hat{\mathbf{v}}' = \hat{\mathbf{v}}\mathbf{A}^{-1} = \hat{\mathbf{v}}\mathbf{A}^T. \tag{1.17}$$

For the original coordinate system x_i and the transformed coordinate system x'_i, the transformation matrix is expressed as:

$$\mathbf{A} = \left[\frac{\partial x'_i}{\partial x_j}\right]_{1\le i,j\le n} = \begin{bmatrix} \partial x'_1/\partial x_1 & \partial x'_1/\partial x_2 & \cdots & \partial x'_1/\partial x_n \\ \partial x'_2/\partial x_1 & \partial x'_2/\partial x_2 & \cdots & \partial x'_2/\partial x_n \\ \vdots & \vdots & \vdots & \vdots \\ \partial x'_n/\partial x_1 & \partial x'_n/\partial x_2 & \cdots & \partial x'_n/\partial x_n \end{bmatrix}. \tag{1.18}$$

Since the vector components transform with the inverse of the basis transformation, the vector **v** is said to be **contravariant**. Contravariant components transform with the inverse Jacobian $(\partial x_i/\partial x'_j)$, while covariant components transform with the Jacobian $(\partial x'_j/\partial x_i)$. Common examples include tangent vectors and velocity vectors. In three dimensions, the transformation of contravariant components is given by:

$$(v')^i = \sum_{j=1}^{3} v^j \left(\frac{\partial x'_i}{\partial x_j}\right) = v^1 \frac{\partial x'_i}{\partial x_1} + v^2 \frac{\partial x'_i}{\partial x_2} + v^3 \frac{\partial x'_i}{\partial x_3}, \quad 1 \le i \le 3. \tag{1.19}$$

By contrast, a **covariant** vector in three dimensions transforms as:

$$v'_i = \sum_{j=1}^{3} v_j \left(\frac{\partial x'_j}{\partial x_i}\right) = v_1 \frac{\partial x'_1}{\partial x_i} + v_2 \frac{\partial x'_2}{\partial x_i} + v_3 \frac{\partial x'_3}{\partial x_i}, \quad 1 \le i \le 3. \tag{1.20}$$

One notable example of a covariant vector is the gradient of a scalar field f, denoted by ∇f:

$$\nabla f = v_1 \mathbf{e}^1 + v_2 \mathbf{e}^2 + v_3 \mathbf{e}^3. \tag{1.21}$$

Comparing Eqs. (1.19) and (1.20) highlights the difference in transformation rules between covariant and contravariant components. Here, the basis vectors $\mathbf{e}^i$ are covariant and form the **dual basis**, which is related to the tangent (contravariant) vectors $\mathbf{e}_i$ via:

$$\mathbf{e}^1 = \frac{\mathbf{e}_2 \times \mathbf{e}_3}{\mathbf{e}_1 \cdot (\mathbf{e}_2 \times \mathbf{e}_3)}, \tag{1.22}$$

$$\mathbf{e}^2 = \frac{\mathbf{e}_3 \times \mathbf{e}_1}{\mathbf{e}_2 \cdot (\mathbf{e}_3 \times \mathbf{e}_1)}, \tag{1.23}$$

$$\mathbf{e}^3 = \frac{\mathbf{e}_1 \times \mathbf{e}_2}{\mathbf{e}_3 \cdot (\mathbf{e}_1 \times \mathbf{e}_2)}. \tag{1.24}$$

In $\mathbb{R}^3$, the dual basis can be expressed via cross products as shown. In general manifolds, the dual basis is defined via the condition $\mathbf{e}^i(\mathbf{e}_j) = \delta^i_j$. In the special case where the moving frames are orthonormal, i.e., mutually orthogonal with unit magnitude, we have $\mathbf{e}^i = \mathbf{e}_i$. Accordingly, throughout this book, we consistently use $\mathbf{e}_i$ to denote contravariant basis vectors.

1.1.3.1 Visualizing Frames in the Tangent Space

Consider a smooth curved surface. We assume that two tangent vectors, $\mathbf{e}_1$ and $\mathbf{e}_2$, lie on the tangent plane of the surface, as illustrated in Fig. 1.7. The surface normal vector is a unit vector oriented in the direction orthogonal to the surface. Hence, it is perpendicular to the tangent plane and consequently to both $\mathbf{e}_1$ and $\mathbf{e}_2$. Thus, the third tangent vector, $\mathbf{e}_3$, is necessarily aligned with the surface normal, which lies outside the tangent space.

The magnitude of the tangent vectors $\mathbf{e}_i$ (for $1 \le i \le 3$) need not be unitary. However, for the remainder of this book, we assume that all moving frames are of unit length unless stated otherwise. For each tangent vector $\mathbf{e}_i$, suppose that there exists a curved axis ξ^i such that its tangent vector points in the same direction as $\mathbf{e}_i$. Let the tangent vector to the axis ξ^i be denoted by $\partial/\partial\xi^i$. Since the axis ξ^i is aligned in the direction of $\mathbf{e}_i$, the tangent vector $\partial/\partial\xi^i$ is also aligned in the same direction. However, its magnitude is generally not unitary. Therefore, the unit tangent vector $\mathbf{e}_i$ can be obtained as:

$$\mathbf{e}_i = \frac{\partial}{\partial\xi^i} \Bigg/ \left\| \frac{\partial}{\partial\xi^i} \right\|, \tag{1.25}$$

where each ξ^i represents a coordinate axis whose tangent vector $\partial/\partial\xi$ is normalized to define $\mathbf{e}_i$, ensuring unit magnitude in the direction of motion.

Furthermore, the magnitude of $\mathbf{e}_i$ can be adjusted by modifying the conducting properties of the domain. This reflects the core idea of Riemannian geometry: that propagation shaped by geometry can be altered by the medium's conductivity. For instance, consider the Euclidean axis $(x^i)'$ defined as $(x^i)' = 2x^i$ for the classical

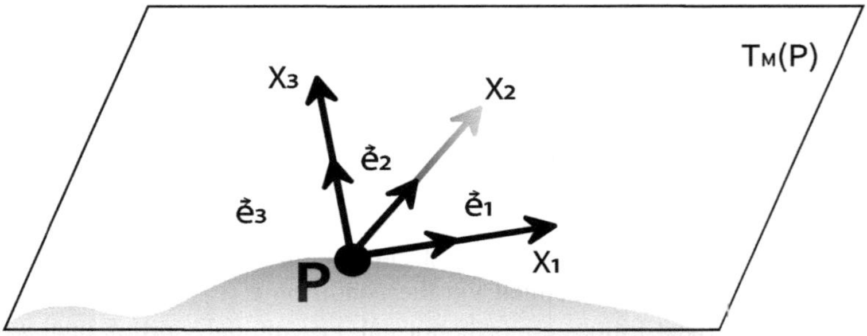

Fig. 1.7 Moving frame and its axes on the tangent plane at a point P of the domain, denoted by $T_M(P)$. The unit tangent vectors $\mathbf{e}_1$ and $\mathbf{e}_2$ lie in the tangent plane, while $\mathbf{e}_3$, aligned with the surface normal vector, is orthogonal to them. The tangent vectors $\mathbf{e}_1$ and $\mathbf{e}_2$ may be aligned with the local coordinates x_1 and x_2, respectively

Cartesian coordinate axis x^i for $1 \leq i \leq 3$. Then, the point located at $x^i = 1$ corresponds to $(x^i)' = 2$. The associated tangent vectors satisfy:

$$\frac{\partial}{\partial x^i} = 0.5 \frac{\partial}{\partial (x^i)'}, \tag{1.26}$$

which means that the tangent vector of the axis x^i is 0.5 times smaller than that of the axis $(x^i)'$. The Euclidean axis x^i is thus aligned with ξ^i but possesses a unit tangent vector:

$$\frac{\partial}{\partial x^i} = \frac{\partial}{\partial \xi^i} \Big/ \left\| \frac{\partial}{\partial \xi^i} \right\|. \tag{1.27}$$

In orthonormal coordinates, the moving frame $\mathbf{e}_i$ coincides with the unit tangent vector along the coordinate axis x^i. In general, it serves as a normalized vector aligned with a local coordinate direction. Regardless of the magnitude of the tangent vector of the curved axis, throughout this book, we assume moving frames are orthonormal–orthogonal unit vectors spanning the local tangent space–unless otherwise specified:

$$\left\| \frac{\partial}{\partial x_i} \right\| = \|\mathbf{e}_i\| = 1, \qquad 1 \leq i \leq 3. \tag{1.28}$$

As a result, a trihedron of unit vectors $\mathbf{e}_i$ is established at point P, as shown in Fig. 1.7. The Euclidean axes x_1 and x_2 lie on the tangent plane, denoted by $T_M(P)$, and thus $\mathbf{e}_1$ and $\mathbf{e}_2$ also lie within $T_M(P)$. The third tangent vector, $\mathbf{e}_3$, aligns with the surface normal vector. In this context, the unit tangent vectors are referred to as **local unit basis vectors**.

When constructed orthonormally and tangent to the domain, moving frames serve as both local coordinate axes and local bases for tensorial computations. In this book, the term **moving frames** primarily refers to the set of orthonormal tangent vectors $\mathbf{e}_i(P)$ at a point P. For example, any vector at P may be expressed in terms of these basis vectors as:

$$\mathbf{v}(P) = v^1 \mathbf{e}_1(P) + v^2 \mathbf{e}_2(P) + v^3 \mathbf{e}_3(P) \qquad \text{at} \quad P. \tag{1.29}$$

The moving frame's role as a local axis becomes apparent when the location of an object is measured relative to the moving frame at a specific point P. The coordinates of a velocity vector—or its components v^i—can be described using the local axes $x_i(P)$, such as:

$$\mathbf{v} = (v^1, v^2, v^3) = v^1 \mathbf{e}_1 + v^2 \mathbf{e}_2 + v^3 \mathbf{e}_3 \in \mathbf{R}^3. \tag{1.30}$$

This localized representation is particularly useful for certain applications, such as measuring distances between distinct points. Nevertheless, the global Cartesian coordinate system (x, y, z) is generally preferred for positional representations, such as $\mathbf{v}(x_p, y_p, z_p)$.

1.2 Encoding Geometry and Material Structure with Moving Frames

Suppose that moving frames are constructed at every point of a domain to serve as **local reference frames** or **local unit basis vectors**. Throughout this book, moving frames are generally assumed to be orthonormal, although they may be oblique or non-unit in length in certain cases, as illustrated in Fig. 1.8. While oblique moving frames are not commonly employed, they can be useful for representing geometric deformations relative to a previously defined orthonormal domain.

The solution to PDEs strongly depends on the geometry of the domain, referring to the spatial region of interest where discrete points are defined. Geometry, in this context, generally includes the shape and conductivity of the domain. Even with the same governing equations, the solution may vary significantly depending on geometric variations. For example, consider a diffusion-reaction equation on a two-dimensional domain. The resulting solution may differ dramatically between geometries such as a plane, a sphere, or a pseudosphere. Moreover, changes in conductivity—such as the diffusion coefficient—profoundly influence the propagation of solutions. Even when the domain shape is fixed, varying the conductive material alters the propagation pattern.

Moving frames provide a means to represent the geometry underlying such PDEs (Fels and Olver 1998; Beffa et al. 2013). When constructed at every point in the domain, they encapsulate both the shape and medium properties of the domain. In mathematical terms, moving frames are a tool for expressing **Riemannian geometry**. The directional ordering of moving frames reflects local geometric features of the domain, which are encoded in the connection and ultimately influence the **Riemannian curvature** (Kenmotsu 2003).

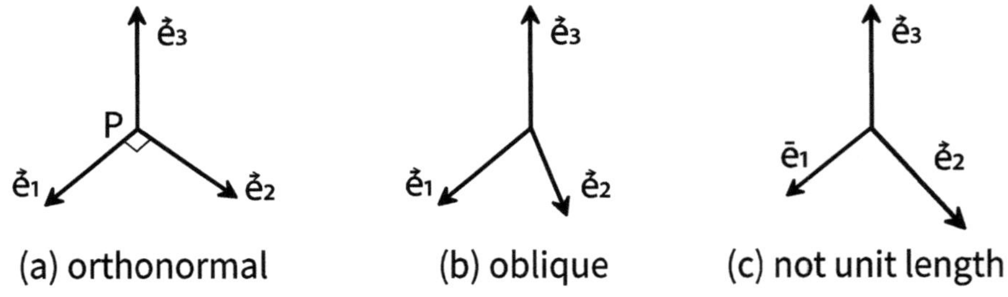

Fig. 1.8 Diverse configurations of moving frames: (**a**) Orthonormal—mutually orthogonal and of unit length, (**b**) Oblique—not all vectors are orthogonal, (**c**) Non-unit length—some vectors are not of unit length

1.2.1 Describing Curvature via Tangent Variation

The representation of the shape using moving frames is intuitive. Consider three discrete points P_i for $1 \leq i \leq 3$ on a curved line Ω. Let $T_{P_i}M$ denote the tangent vector at point P_i. The geometry of Ω can be locally approximated by the collection of tangent vectors $T_{P_i}M$. The approximation becomes less accurate as one moves away from P_i, especially in regions of high curvature (Sternberg 2012). Nevertheless, at each P_i, the tangent vector offers an exact representation. While each tangent plane $T_{P_i}M$ provides a linear approximation of the surface at point P_i, the geometry of Ω can be approximated by interpolating between these local frames under smooth sampling.

A similar argument applies to curved surfaces. Let P_1, P_2, P_3 be points on a curved surface Ω, and let $T_{P_i}M$ denote the tangent plane at each point (Fig. 1.9). With two tangent vectors $\mathbf{e}_1$ and $\mathbf{e}_2$ defined in $T_{P_i}M$, the overall geometry of the surface can be reconstructed by combining these tangent planes. The finer the sampling (i.e., smaller spacing between grid points), the better the approximation.

Additionally, the curvature of a domain can be characterized by the relative angles between neighboring moving frames. Consider a curved line C lying on a tangent plane, as shown in Fig. 1.10. Let P_1 and P_2 be two nearby points on C, located at $\mathbf{x}_{P_1}$ and $\mathbf{x}_{P_2}$, respectively. Define $\mathbf{e}_1(\mathbf{x}_{P_1})$ and $\mathbf{e}_1(\mathbf{x}_{P_2})$ as tangent vectors to the curve at these points. The angle θ_{12} between these vectors indicates the local curvature. For smoother curves, such as C, θ_{12} is relatively small; for more highly curved lines, like C_1, the angle θ'_{12} is larger:

$$\theta_{12} < \theta'_{12} \quad \longrightarrow \quad C_1 \text{ curve has a higher curvature at } P'_2. \tag{1.31}$$

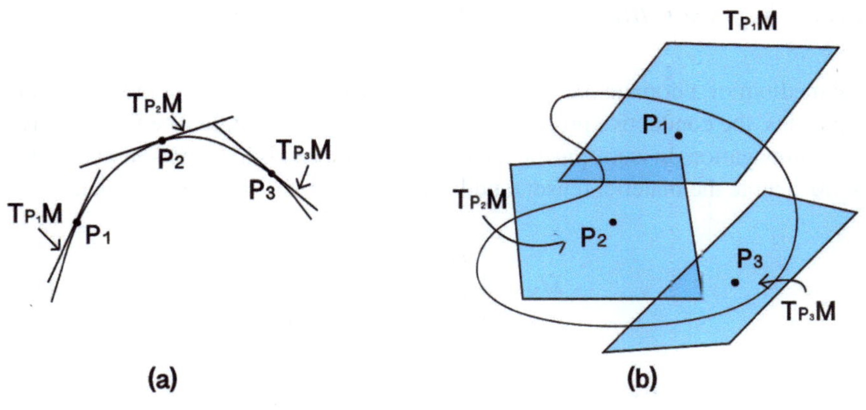

Fig. 1.9 A collection of tangent vectors and planes provides an effective approximation of curved geometries. A set of tangent lines approximates a curved line (left), while a set of tangent planes approximates a curved surface (right)

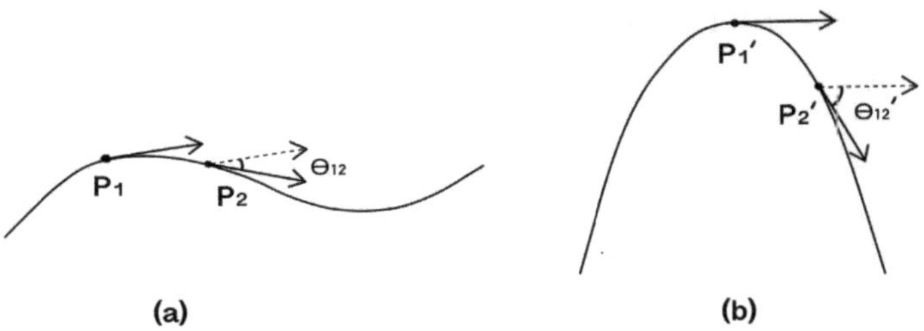

Fig. 1.10 The angle between neighboring moving frames captures curvature. Along a curved line, the frames at P_1 and P_2 form an angle θ_{12}, which reflects the change in direction of the curve. Greater curvature corresponds to a larger θ_{12}. Change of angle θ_{12} in a (**a**) smooth and (**b**) steeper curve

Since the tangent vector corresponds to the first derivative of the curve, and their difference approximates the second derivative, the angles θ and θ' effectively measure the shape of the medium such as curvature and such angular variations can be described by **connection**. To formalize how vector orientation varies across space, we introduce the notion of a **connection**, which quantifies angular deviation between transported and local frames. The connection components depend on the choice of frame. In orthonormal moving frames, these are captured by connection 1-forms, which quantify the infinitesimal rotation of the basis vectors. Here, we interpret **connection** informally as the angular variation between neighboring frames. A rigorous formulation using covariant derivatives and connection 1-forms is presented in Chap. 3.

1.2.2 Representing Anisotropy Through Frame Scaling

The medium of a domain can also be effectively represented using moving frames. In physics, the conductive properties of a medium directly influence the velocity of wave propagation. In isotropic media, propagation speed is independent of direction. Let $\widehat{M}$ denote the material tensor, and let $\mathbf{v} = (v_x, v_y, v_z)$ be a vector. Then:

$$\mathbf{v}_{new} = \widehat{M}\mathbf{v} = \begin{bmatrix} m_{11} & m_{12} & m_{13} \\ m_{21} & m_{22} & m_{23} \\ m_{31} & m_{32} & m_{33} \end{bmatrix} \begin{bmatrix} v_x \\ v_y \\ v_z \end{bmatrix}. \tag{1.32}$$

In the isotropic case:

$$\widehat{M} = \begin{bmatrix} m & 0 & 0 \\ 0 & m & 0 \\ 0 & 0 & m \end{bmatrix} = m\widehat{I}, \quad m > 0 \quad \text{in an isotropic medium}, \tag{1.33}$$

where $\widehat{I}$ is the identity matrix, and the constant m is any arbitrary positive number. In this case, waves propagate uniformly in all directions. Moving frames can be constructed as orthonormal local reference frames, though the first tangent vector need not be aligned with the propagation direction.

For an **anisotropic** medium, anisotropy can be expressed by aligning the moving frames along the anisotropic direction, with the length of each frame vector corresponding to the strength of anisotropy. A fully anisotropic material tensor can be represented as a full matrix (Abraham et al. 1988):

$$\widehat{M} = \begin{bmatrix} m_{11} & m_{12} & m_{13} \\ m_{21} & m_{22} & m_{23} \\ m_{31} & m_{32} & m_{33} \end{bmatrix}, \tag{1.34}$$

where, in an anisotropic medium, $\widehat{M}$ is typically symmetric and positive-definite, but individual components m_{ij} may be negative or zero depending on basis orientation. To analyze the material tensor, we can decompose it as follows:

$$\widehat{M} = Q \Lambda Q^T = \begin{bmatrix} e_1^x & e_2^x & e_3^x \\ e_1^y & e_2^y & e_3^y \\ e_1^z & e_2^z & e_3^z \end{bmatrix} \begin{bmatrix} \lambda_1 & 0 & 0 \\ 0 & \lambda_2 & 0 \\ 0 & 0 & \lambda_3 \end{bmatrix} \begin{bmatrix} e_1^x & e_2^x & e_3^x \\ e_1^y & e_2^y & e_3^y \\ e_1^z & e_2^z & e_3^z \end{bmatrix}^T, \tag{1.35}$$

where Q is an orthogonal matrix whose columns are the eigenvectors $\mathbf{v}_i$ of $\widehat{M}$, and Λ is a diagonal matrix with eigenvalues λ_i on the diagonal. If $\lambda_i \neq 1$ at a point P, then λ_i represents the strength of anisotropy. The corresponding eigenvector $\mathbf{v}_i$ is aligned with the anisotropy direction in the medium. Therefore, orthonormal eigenvectors $\mathbf{e}_i$ can serve as moving frames. This relationship can be verified by evaluating the product of the material tensor and a velocity vector:

$$\widehat{M}\mathbf{v} = Q \Lambda Q^T \mathbf{v} = \begin{bmatrix} \mathbf{e}_1 & \mathbf{e}_2 & \mathbf{e}_3 \end{bmatrix} \begin{bmatrix} \lambda_1 & 0 & 0 \\ 0 & \lambda_2 & 0 \\ 0 & 0 & \lambda_3 \end{bmatrix} \begin{bmatrix} \mathbf{e}_1 \cdot \mathbf{v} \\ \mathbf{e}_2 \cdot \mathbf{v} \\ \mathbf{e}_3 \cdot \mathbf{v} \end{bmatrix}. \tag{1.36}$$

Define the resized moving frames as:

$$\mathbf{d}_i = \sqrt{\lambda_i}\mathbf{e}_i, \quad 1 \le i \le 3, \tag{1.37}$$

where $\mathbf{d}_i$ has the same direction as $\mathbf{e}_i$ but possibly with the different magnitude. Then, Eq. (1.36) becomes:

$$\widehat{M}\mathbf{v} = Q \Lambda Q^T \mathbf{v} = \begin{bmatrix} \mathbf{d}_1 & \mathbf{d}_2 & \mathbf{d}_3 \end{bmatrix} \begin{bmatrix} \mathbf{d}_1 \cdot \mathbf{v} \\ \mathbf{d}_2 \cdot \mathbf{v} \\ \mathbf{d}_3 \cdot \mathbf{v} \end{bmatrix}. \tag{1.38}$$

Equation (1.36) demonstrates that if the lengths of the tangent vectors are adjusted appropriately, anisotropy can be represented more naturally-though not eliminated-by aligning frame directions with eigenvectors of the material tensor and scaling accordingly. In other words, anisotropy can be geometrically represented by scaling and aligning frame vectors along eigenvectors of the material tensor. By scaling frame vectors according to eigenvalues of the material tensor, one can embed the anisotropic conductivity into the local reference frame for numerical modeling.

Consider a medium whose anisotropy is aligned along a direction $\mathbf{d}_{ani}$ with a non-unit magnitude, $\|\mathbf{d}_{ani}\| \neq 1$. The conductive properties of the medium can be encoded into the moving frames by performing the following steps:

1. Align a tangent vector (e.g., $\mathbf{e}_1$) with the direction of anisotropy.
2. Scale its magnitude by $\|\mathbf{d}_{ani}\|$, while keeping the magnitudes of the remaining vectors unchanged.

For instance, if the anisotropy is along the x-axis with a magnitude of 2, as illustrated in Fig. 1.11, the new tangent vector becomes:

$$\mathbf{e}_1' = 2\mathbf{e}_1. \tag{1.39}$$

This stretching implies that the associated x_1' coordinate is scaled accordingly:

$$\mathbf{e}_1' = 2\mathbf{e}_1 \iff x_1' = \frac{1}{2}x_1. \tag{1.40}$$

This implies that motion along the stretched frame $\mathbf{e}_1'$ covers half the distance in the original coordinate frame $\mathbf{e}_1$. Thus, a unit step in x_1' corresponds to two

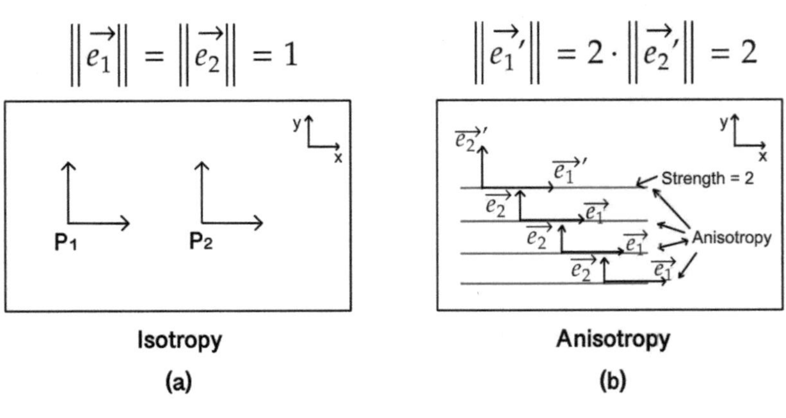

Fig. 1.11 Stretching the first tangent vector according to the strength of anisotropy along its direction. In an isotropic medium, the moving frame vectors have unit length. In the presence of anisotropy (illustrated by the solid line in the right plot), the length of $\mathbf{e}_1$ can be adjusted to represent changes in conduction velocity along the anisotropic direction relative to its orthogonal direction. (**a**) isotropic medium, (**b**) anisotropic medium

units in the x_1 axis. The local differential geometry and material anisotropy of a domain can be effectively encoded using moving frames. First, place orthonormal moving frames at every grid point, lying within the tangent plane. Next, encode the conductive properties of the medium into both the direction and magnitude of these frames. The angular differences between neighboring frames capture the **shape** of the domain, while the relative magnitudes of tangent vectors represent the **conductivity** of the medium. The remarkable utility of moving frames lies in their ability to encoding geometric and material features in a unified representation suited for scientific computing, offering a powerful and efficient tool to understand the complexities of geometric structures.

1.3 Connection: Quantifying Frame Rotation Across Space

The construction of moving frames is uniquely determined in a two-dimensional curved domain with an anisotropic medium. The first tangent vector, $\mathbf{e}_1$, is aligned with the anisotropy direction, and its length is scaled according to the strength of the anisotropy. The second tangent vector, $\mathbf{e}_2$, is orthogonal to $\mathbf{e}_1$ and also lies on the tangent plane. The third vector of the moving frame is oriented along the surface normal. The magnitudes of the tangent vectors should be defined in accordance with the anisotropy. Due to two constraints-the curvature of the domain and the presence of anisotropy-the alignment of moving frames is largely constrained by curvature and anisotropy, leaving at most a discrete ambiguity such as orientation (See Fig. 1.12a).

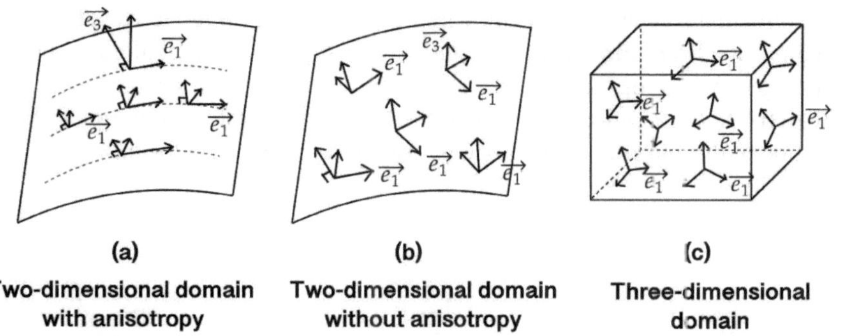

Fig. 1.12 Degrees of freedom for moving frames in two- or three-dimensional space with or without anisotropy. (**a**) In a two-dimensional domain with anisotropy, the first tangent vector aligns with the anisotropy, the second is orthogonal to it on the plane (two possible orientations), and the third is determined accordingly. (**b**) In a two-dimensional domain without anisotropy, the first tangent vector may point in any direction, the second remains orthogonal within the plane, and the third is fixed as the surface normal (two possible orientations). (**c**) In three dimensions, no restrictions exist on the alignment of moving frames

In the absence of anisotropy, there remains one degree of freedom concerning the direction of the first tangent vector. Although the two tangent vectors $\mathbf{e}_1$ and $\mathbf{e}_2$ lie on the tangent plane, their orientations are arbitrary. The only constraint is that both vectors must reside in the two-dimensional surface. (See Fig. 1.12b.)

Now consider an isotropic three-dimensional domain. If a two-dimensional curved surface exists within this domain, then moving frames constrained to the surface will be fixed accordingly. However, in the absence of both surface constraints and anisotropy, all tangent vectors $\mathbf{e}_i$ may be aligned arbitrarily. In this case, no constraints govern the orientation of the moving frames, and the magnitudes of the tangent vectors are constant but undetermined (See Fig. 1.12c).

1.3.1 From Random to Aligned: Establishing Frame Order

Consider three adjacent points P_1, P_2, and P_3 in an isotropic multidimensional domain Ω, as shown in Fig. 1.13. Let us construct moving frames in the tangent plane at each point. Denote by $\mathbf{e}_1(P_i)$ and $\mathbf{e}_2(P_i)$ the tangent vectors lying on the tangent plane $T_{P_i}M$ at each point P_i. In an isotropic medium, $\mathbf{e}_1$ at each point may be aligned in any arbitrary direction; the same holds for $\mathbf{e}_2$. In the absence of any wave or any **form of anisotropy**, the vectors $\mathbf{e}_1$ and $\mathbf{e}_2$ may point in random directions. We refer to this configuration as a **random state** or a **chaotic state without order**, with respect to the direction of the first tangent vector.

At a specific moment, a **signal** wave can induce an **organized** state. This signal may be an **electromagnetic wave** or a diffusion-reaction-type signal that triggers rotation of the first tangent vector $\mathbf{e}_1$. This signal does not alter the domain's shape but modifies only the direction or magnitude of the first tangent vector. Nonetheless, these directional changes in the moving frames reflect **geometric variations** of the domain in the **Riemannian sense**. Figure 1.5 illustrates various configurations of the first tangent vector $\mathbf{e}_1$. In Fig. 1.5a, all vectors $\mathbf{e}_1$ point in the same direction along the x-axis. In this configuration, the dashed circle is not effectively captured by the direction of the tangent vectors. By contrast, in Fig. 1.5b and c, the tangent vector $\mathbf{e}_1$

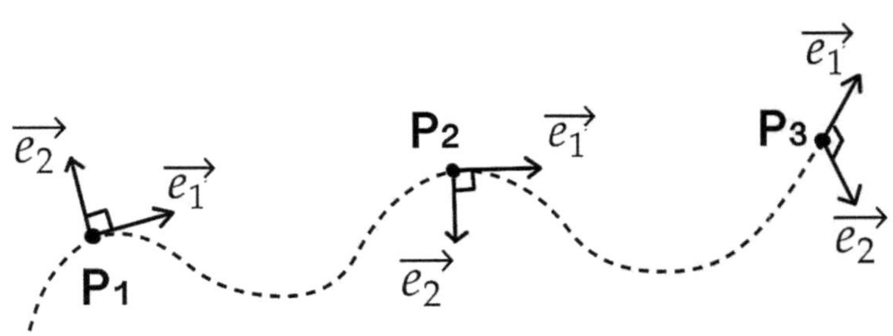

Fig. 1.13 Variations of moving frames along a curved path. Along such a path, the frames at different points may differ

points in different directions at each grid point—specifically along the radial (r) and tangential (θ) directions in a polar coordinate system, respectively. Observe that in these two cases, the moving frames accurately represent a specific geometry—the dashed circle. In this context, the **order** of moving frames refers to the **organized arrangement** of their directions, which can be mathematically expressed through their **connection**.

Now consider two points P_1 and P_2 on a curved domain Ω. To compare their tangent vectors, we use the notion of **parallel transport**, first introduced by the Italian mathematician Tullio Levi-Civita in 1917 (Needham 2021). Let the parallel transport of a vector $\mathbf{v}$ defined at P_1 to P_2 be denoted as follows Misner et al. (2017) and Needham (2021):

$$\text{Parallel transport of } \mathbf{v} \text{ from } P_1 \text{ to } P_2 \equiv \mathbf{v}_{\|}(P_2). \tag{1.41}$$

Parallel transport of a tangent vector $\mathbf{e}_1$ from P_1 to P_2 along a curved path preserves its angle relative to the path. The connection can be viewed as the infinitesimal rate of rotation of $\mathbf{e}_1$ with respect to the transport path. Thus, **parallel transport** using a Levi-Civita connection preserves inner products between vectors, and thus angles, along the transport path between P_1 and P_2.

In Fig. 1.14a, P_1 may be transported to P_2 via a straight line that preserves the angle of $\mathbf{e}_1$. In Fig. 1.14b, the vector $\mathbf{e}_1$ can be parallel transported along the radial axis connecting P_1 and P_2. In Fig. 1.5c, the circle itself defines the path along which $\mathbf{e}_1$ is parallel transported. See further the illustration in Fig. 1.15. Since tangent vectors at neighboring points on a curved surface cannot be directly compared, parallel transport is essential for vector calculus on curved domains, enabling consistent comparison of vectors at different locations. The connection thus plays a fundamental role by defining the geometric relation between these comparisons.

Suppose the length of the moving frames remains constant. In that case, their comparison is determined solely by angular differences. In contrast, comparisons

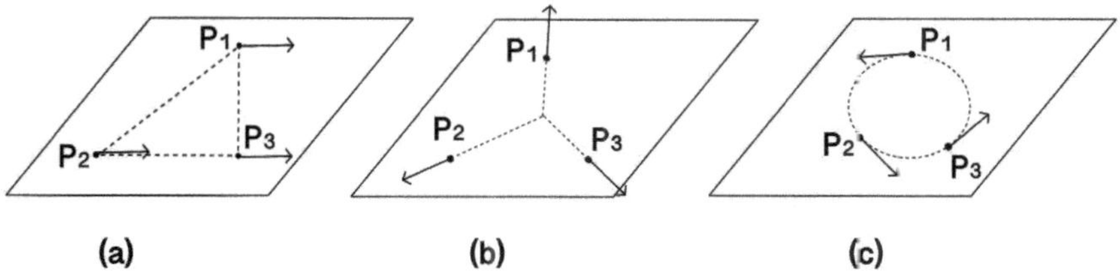

Fig. 1.14 Examples of varying moving frames at three points. The construction of moving frames may reflect different dynamics at these points: (**a**) For planar propagation from left to right, the frames align consistently, possibly matching the Cartesian coordinate system. (**b**) For propagation from the origin, the frames align with the trajectory, similar to polar or spherical coordinates; (**c**) For periodic propagation along a circle enclosing all points, the frames align with the circular trajectory through them

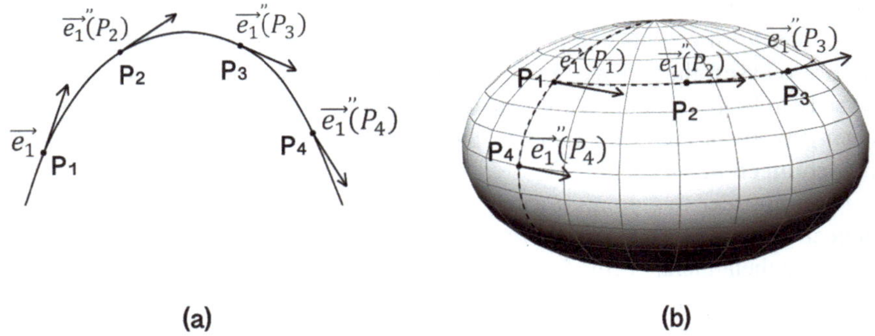

Fig. 1.15 Illustration of parallel transport. (**a**) The tangent vector $\mathbf{e}_1$ is transported from P_1 to P_4 along the curved path; in this sense, $\mathbf{e}_1$ at P_1 and at P_4 are considered the same vector. (**b**) In a curved domain, parallel transport may follow different paths (e.g., from P_1 to P_3 or from P_1 to P_4), so the transported result need not be unique

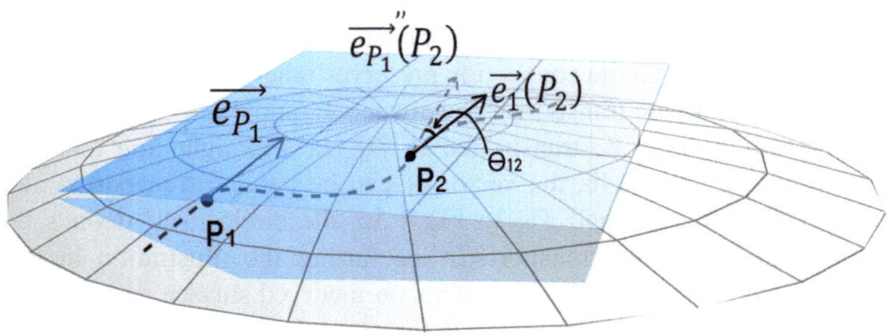

Fig. 1.16 Angular difference between two neighboring tangent vectors. Parallel transport along different paths may yield different angular differences at point P_2, denoted by θ_{12}

between general tangent vectors involve both angular and magnitude differences. For example, the moving frame vector $\mathbf{e}_1$ at P_1 can be parallel transported to P_2, resulting in an angular difference θ_{21}, as shown in Fig. 1.16. Assume P_1 and P_2 lie along the same curved path, and that $\mathbf{e}_1$ is the unit tangent vector of that path. Then, the angular difference is zero. Hence, θ_{21} becomes nontrivial only when the tangent vectors are not aligned with the path between P_1 and P_2. In the 2D orthonormal case, we interpret the angular rate of rotation of frame vectors as a scalar connection. In general, connection is a matrix-valued 1-form governing frame variation. The angle can be computed by:

$$\theta_{21} = \text{sign}(\mathbf{e}_{1\parallel}(P_2) \times \mathbf{e}_1(P_2)) \cos^{-1}\left(\mathbf{e}_{1\parallel}(P_2) \cdot \mathbf{e}_1(P_2)\right), \tag{1.42}$$

where we implicitly assume that $\mathbf{e}_i$ is a unit vector. The sign is positive if the rotation from $\mathbf{e}_{1\parallel}(P_2)$ to $\mathbf{e}_1(P_2)$ is counterclockwise and negative if clockwise. If additional grid points are placed along the dashed curve, both the angular difference and the arc length become smaller. Thus, the magnitude of the angular difference remains meaningful in the limit. Let ε denote the arc length between P_1 and P_2, defined as:

$$\varepsilon = \int_{\mathbf{x}=P_1}^{\mathbf{x}=P_2} ds. \tag{1.43}$$

In the 2D case, the **connection** can be represented as the angular rate of change of a frame vector.

$$\theta(P_1) = \lim_{\varepsilon \to 0} \frac{\theta_{21}}{\varepsilon}. \tag{1.44}$$

In general, the connection is a tensorial object defining parallel transport rules via covariant differentiation.

Equation (1.44) captures only changes in angle. Connection (e.g., angular rate of change) provides the structure from which covariant derivatives are derived. In orthonormal frames, the two may coincide operationally, which can account for both angular and magnitude differences of tangent vectors between points. Suppose P_2 lies in the direction of $\mathbf{v}$, at a distance ε from P_1, and let $\mathbf{w}_{\|}(P_1)$ be the vector $\mathbf{w}$ at P_1 obtained via parallel transport from P_2. Then, the covariant derivative is given by:

$$\nabla_{\mathbf{v}} \mathbf{w}(P_1) = \lim_{\varepsilon \to 0} \frac{\mathbf{w}_{\|}(P_1) - \mathbf{w}(P_1)}{\varepsilon}. \tag{1.45}$$

In the context of orthonormal moving frames, we define the angular difference from Eq. (1.44) as the **connection**. It quantifies the change in the direction of the first tangent vector $\mathbf{e}_1$ as one moves from P_1 to P_2. Since moving frames consist of three orthonormal vectors $\mathbf{e}_i$ for $1 \leq i \leq 3$, changes in $\mathbf{e}_1$'s direction form a vector and can be expressed in the local basis of $\mathbf{e}_i$. Although the connection itself is not a tensor–since it changes under frame transformation–the curvature derived from it is **tensorial**. In orthonormal moving frames, the angular change of frames can be encoded using **connection 1-forms**, which behave tensorially under certain conditions.

1.3.2 Wave Geometry and Connection

We are also interested in the application of the connection to the **geometry of wave propagation**. Wave propagation is commonly described by PDEs. By **geometry of wave propagation**, we refer to the wave's **trajectory pattern**. Imagine tracing the motion of individual particles to form curved lines. A collection of such lines represents a trajectory bundle, and the resulting pattern is interpreted as the trajectory pattern. As the wave propagates, the particle trajectories are traced to observe how they converge or diverge, and how much they deviate from the **ideal**

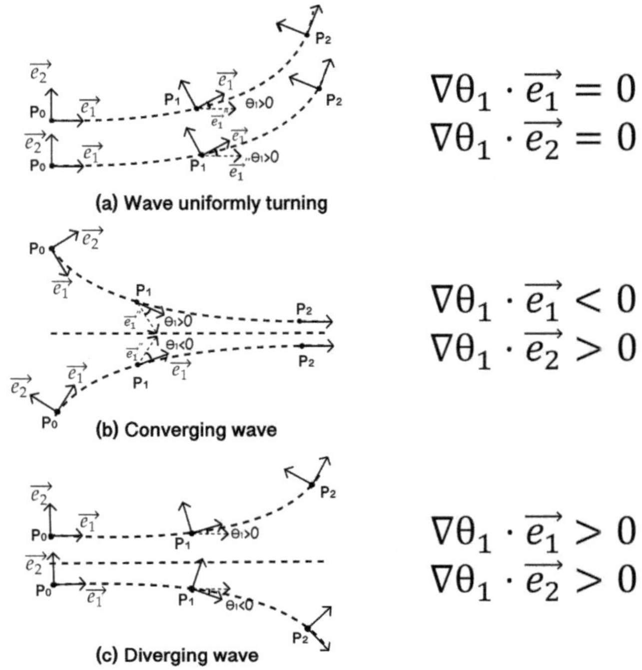

Fig. 1.17 Connection represents the geometry of a trajectory. (**a**) For uniform upward turning, there is no significant change in $\nabla\theta \cdot \mathbf{e}_i$ for the angle θ of the first tangent vector. (**b**) For a converging trajectory, $\nabla\theta \cdot \mathbf{e}_1$ is negative while $\nabla\theta \cdot \mathbf{e}_2$ is positive. (**c**) For a diverging trajectory, all components $\nabla\theta \cdot \mathbf{e}_i$ are positive

path. This trajectory pattern reflects key propagation phenomena. For example, divergence or convergence in electromagnetic waves is widely used to infer the presence of massive or energetic structures in space, such as black holes. Similarly, the divergence or convergence of cardiac electrical waves can indicate conduction blocks where electrical propagation fails.

Figure 1.17 illustrates several trajectories evolving from left to right. In Fig. 1.17a, the wave uniformly curves upward as it moves from left to right. Suppose $\mathbf{e}_1$ is aligned with the direction of wave propagation, and $\mathbf{e}_2$ is orthogonal to it in the plane. In the plane, we interpret $\mathbf{e}_1 \times \mathbf{e}_2$ as indicating orientation (into or out of the page) to help define sign conventions for angular change. Assume the x- and y-axes of the plane are aligned horizontally and vertically, respectively.

Consider $\theta(P_1)$ as defined in Eq. (1.42). Let us adopt this $\theta(P_1)$ as the definition of connection. Then, the connection of $\mathbf{e}_1$ is positive everywhere. The angle from $\mathbf{e}_{1\parallel}(P_2)$ to $\mathbf{e}_1(P_2)$ is counterclockwise, implying that $\mathbf{e}_{1\parallel}(P_2) \times \mathbf{e}_1(P_2)$ is positive.

To evaluate changes in θ in the directions of $\mathbf{e}_1$ or $\mathbf{e}_2$, we employ the following directional derivative, denoted by $\nabla_{\mathbf{e}_i}\theta(\mathbf{x})$:

$$\nabla_{\mathbf{e}_i}\theta = \lim_{h\to 0}\frac{\theta(\mathbf{x}+h\mathbf{e}_i)-\theta(\mathbf{x})}{h}. \tag{1.46}$$

The directional derivative quantifies how a scalar function changes along the direction of a frame vector. In our case, $\nabla_{\mathbf{e}_i}\theta(\mathbf{x})$ computes the change of connection along the direction of the first and second tangent vector $\mathbf{e}_i$. Connection captures angular variation of moving frames, which indirectly reflects curvature or deformation patterns of the wavefront–such as **convergence** or **divergence** of trajectories.

(I) **Uniformly curved trajectory**: For the trajectory of a constant curvature, $\theta(P)$ is constant everywhere. Thus, the changes of the connection in the direction of $\mathbf{e}_1$ and $\mathbf{e}_2$ are zero, i.e.,

$$\nabla_{\mathbf{e}_1}\theta = 0, \quad \nabla_{\mathbf{e}_2}\theta = 0. \tag{1.47}$$

(II) **Converging trajectory**: Figure 1.17b shows a wave whose trajectories converge as they propagate from left to right. Consider a center line at $y = 0$, with the upper region defined by $y > 0$ and the lower region by $y < 0$. The connection of $\mathbf{e}_1$ is positive in the upper region: $\mathbf{e}_{1\parallel}(P_2)$ turns counterclockwise toward $\mathbf{e}_1(P_2)$. Conversely, in the lower region, the connection is negative because the angle from $\mathbf{e}_{1\parallel}(P_2)$ to $\mathbf{e}_1(P_2)$ is clockwise. This variation causes the connection to increase in the direction of $\mathbf{e}_2$; hence, $\nabla_{\mathbf{e}_2}\theta$ is positive. However, the connection tends to decrease or flatten in the direction of $\mathbf{e}_1$, as the curve converges to a point or becomes parallel. Therefore, $\nabla_{\mathbf{e}_1}\theta$ is non-positive:

$$\nabla_{\mathbf{e}_1}\theta \le 0, \quad \nabla_{\mathbf{e}_2}\theta \ge 0. \tag{1.48}$$

(III) **Diverging trajectory**: A more intriguing scenario arises when the trajectory diverges—particularly when the wave encounters an obstacle or energy source. Figure 1.17c shows a positive connection in the upper trajectory region and a negative connection in the lower region. As in the converging case, the connection increases in the direction of $\mathbf{e}_2$, making $\nabla_{\mathbf{e}_2}\theta$ positive. However, unlike the converging trajectory, the curvature increases in the direction of $\mathbf{e}_1$, making $\nabla_{\mathbf{e}_1}\theta$ also positive. This reflects the **divergence of trajectory**:

$$\nabla_{\mathbf{e}_1}\theta \ge 0, \quad \nabla_{\mathbf{e}_2}\theta \ge 0. \tag{1.49}$$

In both converging and diverging cases, the derivative of the connection changes along the propagation direction $\mathbf{e}_1$. To distinguish convergence from divergence by sign alone, the orientation of $\mathbf{e}_2$ must be considered. This will be further discussed in Section 6. However, Eqs. (1.47)–(1.49) serve as the basis for the **categorization** and

quantification of trajectory convergence and divergence. The directional variation of the connection leads to curvature–captured by the **Riemann tensor** in the orthonormal basis or the **relative acceleration**–through second derivatives of frame rotation or commutators of covariant derivatives.

1.4 From Geometry to Numerics: Moving Frames in PDE Solvers

We have established that moving frames can represent the geometry of a surface domain by placing two tangent vectors on the tangent plane at each point. Anisotropy can be captured through the relative scaling of the tangent vectors in the frame. Moreover, moving frames facilitate the analysis of wave trajectories via the connection, whose derivative yields a curvature tensor—such as the Riemann curvature tensor in orthonormal bases. Thus, moving frames constitute a powerful geometrical tool for studying geometry in complex media. This book, in particular, aims to explore the application of moving frames to the numerical solution of PDEs.

While moving frames were originally introduced in the context of **continuous** geometry, their adaptation to numerical methods involves a transition to **discrete** settings. Significant progress on discrete moving frames was initiated by Beffa et al. (2013) and further developed by Mansfield et al. (2019). However, the approach adopted in this book diverges from these earlier studies. In what follows, we briefly discuss the implications of using moving frames in discrete space.

In a **discrete space**, moving frames are no longer continuous or defined at infinitely many points. Instead, they are constructed at a finite number of discrete grid points. In Élie Cartan's formulation, the derivative—or difference—between neighboring points presupposes that the distance between them is **infinitesimally** small. However, when implementing moving frames in numerical schemes, these grid points are separated by non-negligible distances. In a sufficiently fine mesh, the inter-grid spacing may be considered negligible relative to the wavelength. Nonetheless, when the spacing is appreciable, discretization errors arise in approximating functions, derivatives, and integrals (cf. Fischer et al. 2005). Furthermore, the use of moving frames at these discrete grid points introduces additional error.

Despite this, the error introduced by moving frames must be analyzed in the context of numerical accuracy and stability. In this book, we examine how the finite spacing between grid points affects the performance of schemes that utilize moving frames.

In scientific computing, the efficiency and accuracy of a numerical method are often characterized by its order of convergence—i.e., the rate at which discretization error diminishes as grid spacing is refined. Consider a curved domain where moving frames are aligned along curved trajectories. Moving frames are locally Euclidean in the sense that they provide orthonormal bases in which the metric approximates

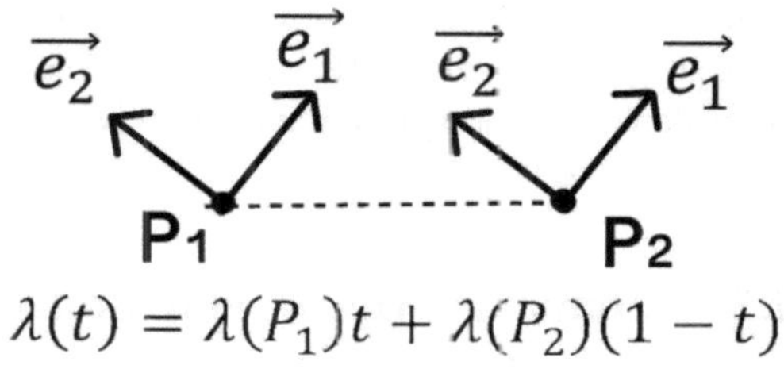

Fig. 1.18 Moving frames in a discrete space. The moving frames associated with the discrete points P_1 and P_2 can be regarded as varying continuously between the two points

the identity. However, when frames are constructed at discrete points on a curved domain, they represent linear approximations to curved geometry.

This mismatch between the Euclidean moving frame and the underlying curved geometry may degrade numerical accuracy. Specifically, except at the common point shared by the frame and the domain, quantities defined along the tangent vector only approximate those on the curved surface. For instance, in Fig. 1.18, the original point P_2 on the curved domain is approximated by $\widehat{P_2}$, which lies on the tangent vector $\mathbf{e}_1$. Using linearized frames introduces approximation error when quantities defined on curved surfaces are represented in flat local bases. This modeling error is distinct from and additive to discretization error. The most direct way to eliminate this additional error would be to use infinitesimally close grid points to compute derivatives.

However, our approach does not seek to eliminate this error outright. Rather, we demonstrate that under certain conditions, the error introduced by using moving frames is negligible relative to the discretization error inherent in numerical differentiation and integration on curved surfaces. As will be discussed in later chapters, the resulting numerical scheme is both sufficiently accurate and stable.

This book provides a comprehensive and systematic treatment of how to use moving frames for solving PDEs, particularly in complex domains such as curved surfaces or volumes containing isotropic or anisotropic media. The various features of moving frames discussed earlier are incorporated into the framework for representing PDEs in complex geometries.

The elegance of moving frames lies in their simplicity and consistency—and this extends to their application in numerical PDEs. A moving frame acts as a **local Euclidean coordinate system** at each point, that is, a coordinate system in which the metric locally approximates the identity matrix. Both vectors and gradients can be expressed **locally** in terms of the frame components. Because moving frames form a Euclidean system at each point, classical numerical methods originally developed for flat (Euclidean) space can be adapted to work in curved geometries with relative ease.

However, the use of local reference frames such as moving frames necessitates the introduction of covariant derivatives and integrals. Covariant calculus deals with the differentiation of vectors or gradients within local coordinate systems, accounting for the fact that neighboring frames may express the same vector differently. In global Euclidean coordinates—such as the Cartesian system—covariant derivatives are not required. But in curved surfaces or volumes, where global coordinates are difficult or impossible to define, local frames are the only viable reference (see do Carmo 1976; Lee 2018). Therefore, the use of moving frames is essentially equivalent to adopting covariant derivatives and integrals in differential calculus.

1.5 Summary: Why Moving Frames Matter

This chapter introduced the conceptual and mathematical foundations of **moving frames** as local orthonormal coordinate systems for describing geometry and analyzing partial differential equations (PDEs) on curved or anisotropic domains. The key takeaways are summarized as follows:

Global Cartesian coordinate systems provide uniform reference frames across space; however, local systems offer greater flexibility for modeling complex geometries. A **local reference system** assigns a distinct frame to each spatial point, requiring careful transformation rules for vector comparison and differentiation. Moving frames consist of local basis vectors—typically denoted by $\mathbf{e}_1, \mathbf{e}_2, \mathbf{e}_3$—attached to each point of a manifold. These vectors may vary in orientation and magnitude to reflect the local geometry and properties of the medium. Although the global domain may be curved (non-Euclidean), each point can be approximated as locally flat. Moving frames leverage this idea by aligning orthonormal vectors within the tangent plane at each point.

Moving frames encode both the **shape** and the **medium** of the domain. **Shape:** Modeled through variations in the directions of neighboring frames. **Medium:** Represented by the magnitudes and directions of the frame vectors, which align with anisotropy via the eigendecomposition of material tensors.

Comparing vectors at different points requires **parallel transport**, which preserves angles along the transport path. The **connection** quantifies the rate of angular change in moving frames and plays a role analogous to that of Christoffel symbols in differential geometry (Spivak 1999).

The **connection** further characterizes the curvature of wave trajectories in propagation models. Directional derivatives of the connection distinguish between uniformly curved, converging, and diverging trajectories, forming the basis for curvature tensor analysis (e.g., the Riemann tensor). In numerical methods, moving frames are constructed at discrete grid points (cf. Fischer et al. 2005). Although originally formulated in continuous geometry, they remain effective in numerical PDE solvers as long as the induced errors remain small relative to discretization errors. Implementing PDEs on curved surfaces requires the use of **covariant derivatives and integrals**, as classical differential operators are insufficient (see do Carmo 1976; Lee 2018). The moving frame system enables such computations by providing a local Euclidean reference at each point.

In summary, moving frames bridge global geometry and local computation. They provide a powerful framework for analyzing and solving PDEs on complex geometries by offering local orthonormality, anisotropy alignment, and compatibility with numerical discretizations.

1.6 Exercises and Conceptual Applications

1. Suppose a 2D Cartesian coordinate system is rotated counterclockwise by an angle θ. Derive the transformation matrix that maps a point (x, y) in the original system to (x', y') in the rotated system. *Hint: Use trigonometric identities involving* $\cos\theta$ *and* $\sin\theta$.
2. Two observers, A and B, are located 2 meters apart on the x-axis. If A observes an object 3 meters to their right, what is the position of the object relative to B's local coordinate system assuming they share the same axis orientation?
3. Explain how a misunderstanding can arise if two observers in a space use different local referencing systems without a protocol for translation. Provide a concrete example using numerical positions and suggest a method to resolve it.
4. Given a point $P = (5, 4)$ in a global coordinate system, find its coordinates in a local coordinate system whose origin is shifted to (2, 1) but retains the same axis orientation.
5. Compare and contrast global and local referencing systems. List three advantages and three disadvantages of each in the context of modeling physical or biological phenomena.
6. Suppose you are walking on a curved path while carrying a coordinate frame aligned with your forward direction. Describe qualitatively how the orientation of your coordinate system would change compared to a fixed global system. *Hint: Think about how your forward direction defines your local "x"-axis.*

Chapter 2
Constructing Moving Frames for Curved Geometries

"The object of all science, whether natural science or psychology, is to co-ordinate our experiences and to bring them into a logical system."

—Albert Einstein

Moving frames are local orthonormal coordinate systems that provide a reference similar to the Euclidean frame in the neighborhood of a point P. Unlike the classical Cartesian coordinate system, moving frames are constructed at every grid point. The tangent vectors of moving frames may vary in direction and magnitude. In an isotropic medium, their orientations and lengths can appear random; however, they still lie on the tangent plane or tangent volume of the domain to represent the shape of that domain. The direction and magnitude of moving frames can be organized or aligned by a signal, such as a wave or light propagation described by PDEs.

Constructing moving frames on a curved surface involves two main stages. In the first stage, we construct an orthonormal set of tangent vectors and place two of them—($\mathbf{e}_1$ and $\mathbf{e}_2$)—on the tangent plane. Since the surface normal vector is orthogonal to the tangent plane, the third vector ($\mathbf{e}_3$) is aligned with the surface normal vector (see Fig. 2.1a). The orthogonality of the moving frame guarantees that the remaining two vectors span the tangent plane.

In the second stage, the first tangent vector on the plane is readjusted according to the anisotropy of the medium or the direction of wave propagation. Once the first and third vectors are fixed, the second is determined via the right-hand rule to satisfy the following relations:

$$\mathbf{e}_1 = \mathbf{e}_2 \times \mathbf{e}_3, \tag{2.1}$$

$$\mathbf{e}_2 = \mathbf{e}_3 \times \mathbf{e}_1, \tag{2.2}$$

$$\mathbf{e}_3 = \mathbf{e}_1 \times \mathbf{e}_2. \tag{2.3}$$

Constructing moving frames that satisfy these relations is relatively straightforward, both mathematically and computationally. However, the most challenging

S. Chun, *Moving Frames for the Numerical Solution of Partial Differential Equations in Complex Domains*, Springer Asia Pacific Mathematics Series 7,
https://doi.org/10.1007/978-981-95-4495-0_2

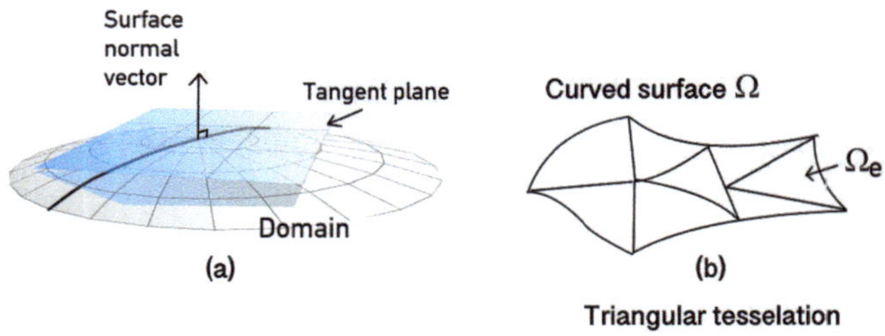

Fig. 2.1 (**a**) Surface normal vector and tangent plane associated with a domain Ω. (**b**) A curved element Ω_e from the tessellation of the curved domain Ω

aspect of moving frames lies in their **differentiability**. This refers to the existence of at least first- or second-order derivatives of the frame vectors. Since the tangent vector $\mathbf{e}_i$ can be interpreted as the first derivative of the domain along a specific direction, the first and second differentiability of the frame corresponds to the second and third differentiability of the domain, respectively. First-order differentiability is typically required for expressing derivatives of vectors expanded in the moving frame, while second-order differentiability is needed for computing curvatures—such as relative acceleration or the Riemann curvature tensor—for geometric analysis in biological or physical systems.

Nevertheless, global differentiability is not necessary. It is sufficient for the moving frames to be differentiable within each tessellated element. As illustrated in Fig. 2.1b, the tessellated element Ω_e is a non-overlapping subdomain of the full domain Ω, defined by:

$$\Omega = \bigcup_e \Omega_e, \quad \Omega_i \cap \Omega_j = \emptyset \ \text{ if } i \neq j. \tag{2.4}$$

Although adjacent elements Ω may share boundaries (lines or points), these intersections are treated as empty sets due to having zero measure. In element-based schemes, the differentiation of moving frames is performed only within each element, which is sufficient for achieving differentiability across the entire domain.

This section explains how moving frames can be constructed at every grid point. We begin with the mathematical formulation originally proposed by Cartan and then discuss several numerical methods for constructing moving frames in discrete domains.

2.1 Mathematical Foundations: Trihedrons and Structure Equations

We consider a three-dimensional domain for general mathematical representation, though this construction easily extends to an arbitrary n-dimensional domain. Let $\mathbf{P} = (x^p, y^p, z^p)$ be an arbitrary point in the three-dimensional space $\mathbb{R}^3$. At point $\mathbf{P}$, we construct a set of orthonormal vectors $\mathbf{e}_i$ satisfying the following properties:

$$\mathbf{e}_i \cdot \mathbf{e}_j = \delta_{ij}, \quad \text{(orthogonality)}, \tag{2.5}$$

$$\|\mathbf{e}_j\| = 1, \quad \text{(in unit length)}, \tag{2.6}$$

where δ_{ij} is the Kronecker delta, equal to one if $i = j$ and zero otherwise. Let the tangent vectors of the Cartesian coordinate axes be denoted as $(\mathbf{x}, \mathbf{y}, \mathbf{z})$. Equations define the orthonormality of the frame vectors. Let α^i, β^i, and γ^i denote the angles between $\mathbf{e}_i$ and the Cartesian axes:

$$\alpha^i = \cos^{-1}(\mathbf{e}_i \cdot \mathbf{x}), \tag{2.7}$$

$$\beta^i = \cos^{-1}(\mathbf{e}_i \cdot \mathbf{y}), \tag{2.8}$$

$$\gamma^i = \cos^{-1}(\mathbf{e}_i \cdot \mathbf{z}), \tag{2.9}$$

Since each vector $\mathbf{e}_i$ has unit length, it can be expressed as a function of both its spatial position and directional angles:

$$\mathbf{e}_i = \mathbf{e}_i(x^p, y^p, z^p, \alpha^i, \beta^i, \gamma^i). \tag{2.10}$$

This implies that $\mathbf{e}_i$ depends on its location and orientation relative to the Cartesian axes.

A moving frame is defined as a trihedron $\mathscr{T}$ located at the origin $\mathbf{P}$, whose edges are represented by orthonormal tangent vectors $\mathbf{e}_i$, as illustrated in Fig. 2.2. Although the directions of the vectors, expressed via $\alpha^i, \beta^i, \gamma^i$, are not predetermined, moving frames may vary from point to point. For simplicity, the dependency on spatial location is omitted in the notation.

Now consider another trihedron $\mathscr{T}'$ located infinitesimally close to $\mathscr{T}$. The position of $\mathbf{P}' = (x'^p, y'^p, z'^p)$ relative to $\mathbf{P}$ is given by:

$$\mathbf{P}' - \mathbf{P} = d\mathbf{P} = (dx, dy, dz), \tag{2.11}$$

where the symbol d represents the difference between the two points as follows:

$$(dx, dy, dz) = (x_p - x'_p, y_p - y'_p, z_p - z'_p). \tag{2.12}$$

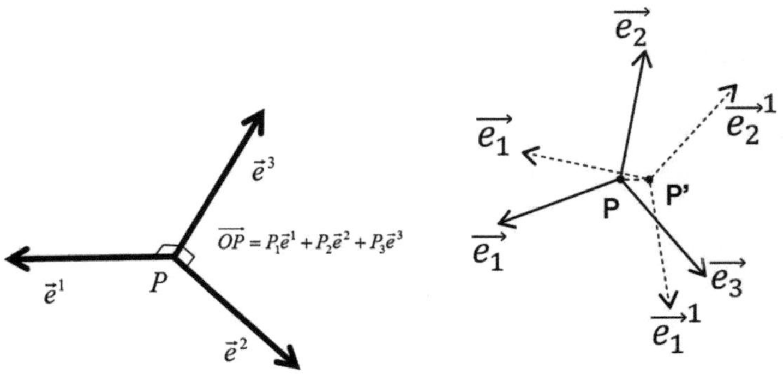

Fig. 2.2 Moving frames $\{\mathbf{e}_1, \mathbf{e}_2, \mathbf{e}_3\}$ at point $\mathbf{P}$ and $\{\mathbf{e}'_1, \mathbf{e}'_2, \mathbf{e}'_3\}$ at an infinitesimally neighboring point $\mathbf{P}'$

The change in the orthonormal vectors from $\mathscr{T}$ to $\mathscr{T}'$ is given by:

$$d\mathbf{e}_i = (\mathbf{e}_i)' - \mathbf{e}_i = (e_i^x, e_i^y, e_i^z). \tag{2.13}$$

This vector difference can alternatively be expressed in terms of angular differences:

$$d\mathbf{e}_i = (d\alpha^i, d\beta^i, d\gamma^i). \tag{2.14}$$

The distance between the points $\mathbf{P}$ and $\mathbf{P}'$ remains unknown unless the geometry of the domain between them is explicitly defined. However, for infinitesimally close points, the Euclidean difference can approximate this distance, as shown in Eq. (2.11). Equations (2.11) and (2.14) describe the positional and directional differences between the trihedrons $\mathscr{T}$ and $\mathscr{T}'$, expressed in Cartesian coordinates.

An interesting geometric perspective arises when $d\mathbf{P}$ and $d\mathbf{e}_i$ are represented using the moving frame vectors at $\mathbf{P}$. Let us expand the infinitesimal displacement of the trihedron $\mathscr{T}'$ relative to $\mathscr{T}$ in terms of the orthonormal tangent vectors $\mathbf{e}_i$ of $\mathscr{T}$. First, the position difference is expressed as:

$$d\mathbf{P} = \omega_1\mathbf{e}_1 + \omega_2\mathbf{e}_2 + \omega_3\mathbf{e}_3, \tag{2.15}$$

where ω_i are the **components of translation**, representing the displacement of the origin. In Eq. (2.15), each ω_i corresponds to a component of the displacement vector from $\mathbf{P}$ to $\mathbf{P}'$. Similarly, the angular difference between $\mathbf{e}_i$ and $\mathbf{e}'_i$ can be written as:

$$d\mathbf{e}_1 = \omega_{11}\mathbf{e}_1 + \omega_{21}\mathbf{e}_2 + \omega_{31}\mathbf{e}_3, \tag{2.16}$$

$$d\mathbf{e}_2 = \omega_{12}\mathbf{e}_1 + \omega_{22}\mathbf{e}_2 + \omega_{32}\mathbf{e}_3, \tag{2.17}$$

$$d\mathbf{e}_3 = \omega_{13}\mathbf{e}_1 + \omega_{23}\mathbf{e}_2 + \omega_{33}\mathbf{e}_3, \tag{2.18}$$

where ω_{ij} are called **components of rotation**, representing the relative angular difference between $\mathbf{e}_i$ and $\mathbf{e}'_i$. These components form a 3×3 matrix. Equations (2.15) through (2.18) are known as the **structure equations of space** (Cartan 2001a; Ivey and Landsberg 2003).

A three-dimensional trihedron $\mathscr{T}$ has six degrees of freedom: three for its position and three for its orientation. Initially, we have twelve parameters—three ω_i and nine ω_{ij}—but some of the ω_{ij} satisfy symmetry relations:

$$d\mathbf{e}_i \cdot \mathbf{e}_i = 0, \tag{2.19}$$

$$d\mathbf{e}_i \cdot \mathbf{e}_j + d\mathbf{e}_j \cdot \mathbf{e}_i = 0. \tag{2.20}$$

Substituting the expressions for $d\mathbf{e}_i$, we get:

$$\omega_{ii} = 0, \tag{2.21}$$

$$\omega_{ij} + \omega_{ji} = 0. \tag{2.22}$$

The reduced structure equations are:

$$d\mathbf{e}_1 = \qquad\qquad \omega_{21}\mathbf{e}_2 + \omega_{31}\mathbf{e}_3, \tag{2.23}$$

$$d\mathbf{e}_2 = -\omega_{21}\mathbf{e}_1 + \qquad\qquad \omega_{32}\mathbf{e}_3, \tag{2.24}$$

$$d\mathbf{e}_3 = -\omega_{31}\mathbf{e}_1 - \omega_{32}\mathbf{e}_2, \tag{2.25}$$

This coordinate system, formed by the origin $\mathbf{P}$ and the orthonormal basis vectors $\mathbf{e}_i$, is called a **moving frame**, a translation of the French **repère mobile** introduced by Élie Cartan. While Cartan also considered oblique trihedrons, we focus exclusively on orthonormal frames. To compute the parameters, we use dot products:

$$\omega_i = d\mathbf{P} \cdot \mathbf{e}_i = e_i^x dx + e_i^y dy + e_i^z dz, \tag{2.26}$$

$$\omega_{ij} = d\mathbf{e}_j \cdot \mathbf{e}_i = e_i^x d\alpha^j + e_i^y d\beta^j + e_i^z d\gamma^j, \tag{2.27}$$

Thus, six parameters fully describe the trihedron $\mathscr{T}'$ in either moving frame or Cartesian coordinates:

$$(\omega_1, \omega_2, \omega_3, \omega_{21}, \omega_{31}, \omega_{32}) \iff (x', y', z', \alpha', \beta', \gamma'). \tag{2.28}$$

The differences in position and orientation between $\mathscr{T}$ and $\mathscr{T}'$ are thus fully characterized by the moving frame at $\mathbf{P}$. Translational components ω_i describe Euclidean distances, while rotational components ω_{ij} capture the local geometric deformation, defining the **shape** of the domain.

2.2 Numerical Construction of Moving Frames

Constructing moving frames at every grid point in a Euclidean domain is computationally simple and can be achieved in many ways. However, special attention is required when the domain is a curved surface or a curved volume, as a restriction of the Euclidean space. Moreover, moving frames for PDEs often require the differentiability of the tangent vectors. The computational goals in constructing moving frames are summarized as follows:

1. The two tangent vectors $\mathbf{e}_1$ and $\mathbf{e}_2$ of the moving frame should lie on the tangent plane of the domain. The third tangent vector $\mathbf{e}_3$ is aligned with the unit surface normal vector $\widehat{\mathbf{k}} = \mathbf{k}/\|\mathbf{k}\|$.
2. The tangent vectors of the moving frame should be at least first differentiable within each tessellated element of the domain. For the computation of curvature, the tangent vectors should be second differentiable.

In this section, we consider a two-dimensional curved surface as the domain. The construction of moving frames in three-dimensional curved volumes can be achieved in a similar manner. Examples of three distinctive construction methods are given below.

2.2.1 *Method 1: Orthogonalization from Surface Normals*

The first method uses the surface normal vector, denoted by $\mathbf{k}$. When $\mathbf{k}$ is easily derived analytically or computationally, two tangent vectors of the moving frame are expanded on the tangent plane T_pM, orthogonal to the surface normal vector, as follows:

$$\mathbf{e}_1 \cdot \mathbf{k} = 0 \quad \text{and} \quad \mathbf{e}_2 \cdot \mathbf{k} = 0. \tag{2.29}$$

Additional alignment of $\mathbf{e}_1$ and $\mathbf{e}_2$ may be required if differentiability of the moving frames is needed within each curved surface element. In general, a vector can be expanded by the three tangent vectors as follows:

$$\mathbf{v} = v^1\mathbf{e}_1 + v^2\mathbf{e}_2 + v^3\mathbf{e}_3. \tag{2.30}$$

The expansion of the vector using only the first two terms, i.e., $\tilde{\mathbf{v}} = v^1\mathbf{e}_1 + v^2\mathbf{e}_2$, corresponds to the projection of the vector onto the tangent plane T_pM, because the moving frames are orthonormal:

$$x = \cos\theta\cos\phi, \tag{2.31}$$

$$y = \cos\theta\sin\phi, \tag{2.32}$$

$$z = \sin\theta, \tag{2.33}$$

where θ and ϕ denote the polar angle and the azimuthal angle, respectively. The surface normal vector is expressed as

$$\mathbf{k} = (\cos\theta\cos\phi, \cos\theta\sin\phi, \sin\theta). \tag{2.34}$$

Suppose that a velocity vector $\mathbf{v}$ lies on the spherical shell. Then, the following orthogonalization constructs a vector orthogonal to $\mathbf{k}$, which is assigned as $\tilde{\mathbf{e}}_1$:

$$\tilde{\mathbf{e}}_1 = \mathbf{v} - \frac{\mathbf{v}\cdot\mathbf{k}}{\|\mathbf{v}\|\,\|\mathbf{k}\|}. \tag{2.35}$$

The normalization of $\tilde{\mathbf{e}}_1$ yields the first tangent vector $\mathbf{e}_1$:

$$\mathbf{e}_1 = \frac{\tilde{\mathbf{e}}_1}{\|\mathbf{e}_1\|}. \tag{2.36}$$

The third tangent vector $\mathbf{e}_3$ is aligned with the unit surface normal vector $\widehat{\mathbf{k}} = \mathbf{k}/\|\mathbf{k}\|$ after normalization:

$$\mathbf{e}_3 = \mathbf{k}. \tag{2.37}$$

Finally, the second moving frame vector is obtained by taking the cross product of $\mathbf{e}_3$ and $\mathbf{e}_1$:

$$\mathbf{e}_2 = \mathbf{e}_3 \times \mathbf{e}_1. \tag{2.38}$$

The orientation defined by $\mathbf{e}_3 = \mathbf{e}_1 \times \mathbf{e}_2$ should be kept consistent throughout the domain. For example, on the spherical shell, the orientation should either consistently point toward the origin or away from it:

$$\mathbf{e}_1 = \frac{\partial}{\partial\theta} \Big/ \left\| \frac{\partial}{\partial\theta} \right\| = (\sin\theta\cos\phi, \sin\theta\sin\phi, -\cos\theta). \tag{2.39}$$

$$\mathbf{e}_2 = \frac{\partial}{\partial\phi} \Big/ \left\| \frac{\partial}{\partial\phi} \right\| = (-\sin\phi, \cos\phi). \tag{2.40}$$

Note that $\mathbf{e}_1$ and $\mathbf{e}_2$ are orthogonal and normalized. However, these moving frames represent a special case of continuous and differentiable moving frames, except at two singular points: the north and south poles where $\theta = 0$ or $\theta = \pi$. Any rotation of $\mathbf{e}_1$ and $\mathbf{e}_2$ around $\mathbf{k}$ within an element, or their independent rotations around $\mathbf{k}$, is permissible as long as differentiability is preserved.

2.2.2 Method 2: Eigenvector Alignment via Diffusion Tensor

The diffusion tensor also provides a convenient way to construct moving frames. In particular, the diffusion tensor derived from image data is a popular data format used to represent the movement of particles, measured along each coordinate axis (Jones 2011). For a local concentration of particles $n(\mathbf{x}, t)$, Fick's law suggests that the flux of molecules, denoted by $\mathbf{J}$, occurs along the gradient of this concentration. This relation can be expressed using the diffusion tensor $\mathscr{D}$ as follows:

$$\mathbf{J} = -\mathscr{D}\nabla n, \tag{2.41}$$

which is known as the multidimensional version of **Fick's first law**. In the time-dependent diffusion equation, the magnitude and direction of diffusion are modified by the diffusion tensor $\mathscr{D}$ as follows:

$$\frac{\partial n}{\partial t} = -\mathscr{D}\nabla^2 n. \tag{2.42}$$

In the above equations, the diffusion tensor $\mathscr{D}$ plays a critical role in enhancing or reducing the magnitude of the flux in Eq. (2.41) and the diffusion in Eq. (2.42). Moreover, the diffusion tensor is a convenient mathematical tool for representing anisotropy in the medium. In matrix form, the diffusion tensor is given as:

$$\mathscr{D} = \begin{bmatrix} d_{xx} & d_{xy} & d_{xz} \\ d_{yx} & d_{yy} & d_{yz} \\ d_{zx} & d_{zy} & d_{zz} \end{bmatrix}. \tag{2.43}$$

Since the diffusion tensor is symmetric, the off-diagonal elements satisfy

$$d_{xy} = d_{yx}, \tag{2.44}$$

$$d_{xz} = d_{zx}, \tag{2.45}$$

$$d_{xz} = d_{zx}. \tag{2.46}$$

The diffusion tensor, being symmetric, has six independent components, which capture the directional variation in diffusion rates. First, three components are required to specify the direction of the first vector, and the remaining two vectors are subsequently determined by orthonormalization. Second, the other three components represent the magnitudes along each direction. These properties are naturally described by the eigendecomposition of the diffusion tensor:

$$\mathscr{D} = Q\Lambda Q^T, \tag{2.47}$$

where the columns of Q are the eigenvectors of $\mathscr{D}$, denoted by $\mathbf{e}_i$. The diagonal entries of Λ are the corresponding eigenvalues, denoted by λ_i, which represent the scaling of diffusion along each eigenvector. The decomposition can be expressed in matrix form as:

$$Q = [\mathbf{e}_1 \;\; \mathbf{e}_2 \;\; \mathbf{e}_3], \quad \Lambda = \begin{bmatrix} \lambda_1 & 0 & 0 \\ 0 & \lambda_2 & 0 \\ 0 & 0 & \lambda_3 \end{bmatrix}, \tag{2.48}$$

where λ_i are the eigenvalues corresponding to orthonormal eigenvectors in Q.

Applying this decomposition to Fick's law or the diffusion equation, we understand that the diffusion tensor provides three operations to **guide** the gradient flow:

1. $Q^T \nabla n$ decomposes the gradient into components along each $\mathbf{e}_i$. Each component can be obtained via the dot product with the tangent vectors.
2. Each gradient component along the i-th anisotropic direction is scaled by the corresponding eigenvalue λ_i. If $\lambda_i > 1$, the gradient component along $\mathbf{e}_i$ is amplified; if $\lambda_i < 1$, it is attenuated proportionally.
3. Multiplying by Q reconstructs the component-wise scaled gradient flow into the original three-dimensional Cartesian coordinate system.

If the eigenvectors $\mathbf{e}_i$ have unit length and all eigenvalues are equal to one, then the diffusion tensor reduces to the identity matrix.

The directions of the tangent vectors of the moving frame can be arbitrary, constrained only by the geometry of the domain. If the diffusion tensor represents a single dominant fiber direction—where water molecules travel faster along one direction than others—then one of the eigenvalues λ_i will be significantly larger than the others. The diffusion tensor must be differentiable within each tessellated element of the domain to ensure the smoothness of the diffusion operator. Consequently, the differentiability of the tangent vectors of the moving frame is also satisfied.

As an example, suppose that the moving frames are aligned with spherical coordinates, and the anisotropy along the ϕ direction is four times greater than in the other directions, i.e.,

$$\mathbf{e}_1 = \frac{\partial}{\partial r}, \qquad \lambda_1 = 1. \tag{2.49}$$

$$\mathbf{e}_2 = \frac{\partial}{\partial \phi} \Bigg/ \left\| \frac{\partial}{\partial \phi} \right\|, \qquad \lambda_2 = 4. \tag{2.50}$$

$$\mathbf{e}_3 = \frac{\partial}{\partial \theta} \Bigg/ \left\| \frac{\partial}{\partial \theta} \right\|, \qquad \lambda_3 = 1. \tag{2.51}$$

Then, the diffusion tensor is expressed using the eigendecomposition in Eq. (2.47):

$$\mathscr{D} = Q \begin{bmatrix} 1 & 0 & 0 \\ 0 & 4 & 0 \\ 0 & 0 & 1 \end{bmatrix} Q^T, \tag{2.52}$$

where the matrix Q is given by

$$Q = \begin{bmatrix} \cos\theta\cos\phi & -\sin\phi & \sin\theta\cos\phi \\ \cos\theta\sin\phi & \cos\phi & \sin\theta\sin\phi \\ \sin\theta & 0 & -\cos\theta \end{bmatrix}. \tag{2.53}$$

Thus, each column of the matrix Q corresponds to the tangent vectors of the moving frames:

$$\mathbf{e}_1 = (\cos\theta\cos\phi, \cos\theta\sin\phi, \sin\theta), \tag{2.54}$$

$$\mathbf{e}_2 = (-\sin\phi, \cos\phi, 0), \tag{2.55}$$

$$\mathbf{e}_3 = (\sin\theta\cos\phi, \sin\theta\sin\phi, -\cos\theta). \tag{2.56}$$

2.2.3 Method 3: Orthonormal Frames via Element Mappings

The previous two methods do not guarantee the differentiability of the tangent vectors of moving frames within each element. Consequently, additional procedures for aligning tangent vectors are required. Here, we introduce a more convenient scheme for constructing moving frames on curved surfaces that ensures the differentiability of moving frames, particularly in the context of the finite element method. If the mapping is sufficiently smooth (e.g., C^k), and numerical differentiation is performed accordingly, the resulting moving frames will inherit that differentiability. This method can easily be adapted to any numerical scheme that uses a tessellation of the domain mapped from a standard element (Chun 2012, 2014a).

Consider a curved domain of a two-dimensional surface Ω. Let Ω_h denote the approximation of Ω, known as the **mesh**. Suppose that the maximum edge length of Ω_h is h. Let the tessellated approximate curved elements of Ω_h be denoted by Ω_e^h, such that

$$\Omega_h = \bigcup_e \Omega_e^h, \quad \Omega_i^h \cap \Omega_i^h = \delta_j^i, \tag{2.57}$$

where δ_j^i is the Kronecker delta, which equals 0 if $i \neq j$ and 1 if $i = j$. Each element Ω_i^h is either a triangular or quadrilateral element, possibly with curvilinear boundaries.

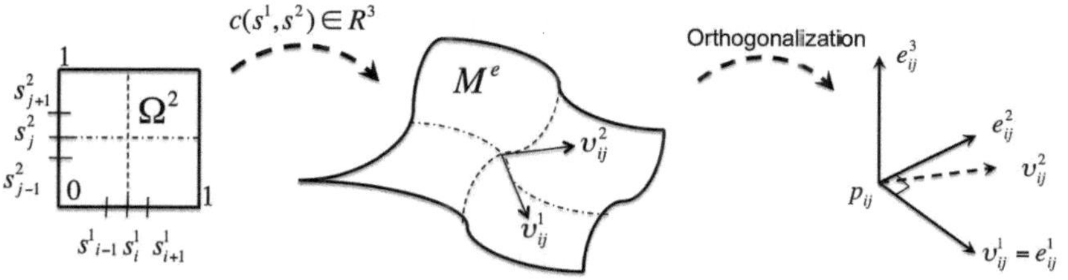

Fig. 2.3 Construction of moving frames using a smooth map **c**. Let Ω^2 be the standard square element with coordinates s^1 and s^2. The map $\mathbf{c}(s^1, s^2)$ sends the standard element Ω^2 to a curved element M^e. Each Euclidean axis s_i is mapped to a curved axis on M^e. The tangent vector of the axis s^1 and s^2 is mapped to tangent vectors (v^1_{ij}, v^2_{ij}) of the curved axis in M^e. These tangent vectors are not necessarily orthogonal nor of unit length. An orthogonalization procedure applied to the tangent vectors produces an orthonormal set of moving frames (Adapted from Chun 2012)

Let $\Omega^2(s^1, s^2)$ denote the square standard element, i.e., the two-dimensional square $[0, 1] \times [0, 1]$, as shown in Fig. 2.3. Denote the two Euclidean axes of Ω^2 by s^1 and s^2, subdivided into N^1 and N^2 subdivisions, respectively.

$$\mathbf{c}(s^1, s^2) : \Omega^2 \to \Omega^h_e = \left\{(c_1, c_2, c_3) \in \mathbf{R}^3,\ c_k = c_k(s^1, s^2) \in C^\infty\right\}, \tag{2.58}$$

where c_k indicates the k-th component in the Cartesian coordinates. The mapping is assumed to be bijective (one-to-one and onto). It is linear if the edges are straight; otherwise, it is polynomial of arbitrary order. The map $\mathbf{c}(s^1, s^2)$ defines the coordinates of grid points in the curved element, matching each grid point in Ω^2 to a unique point in Ω^h_e.

Since the collection of points forms a curve, $\mathbf{c}(s^1, s^2)$ also maps lines: the Euclidean axes s^i in Ω^2 are mapped to curved lines ξ^i in Ω^h_e as follows:

$$\xi^i(j) = \mathbf{c}(s^i(p_j)), \quad \text{for all points } p_j \in s^i. \tag{2.59}$$

Here, the index i denotes the line index, and j denotes the point index along the i-th line.

The function $\mathbf{c}(s^1, s^2)$ also maps the tangent vectors associated with each axis s^i in Ω^2. Let $\partial/\partial s^i$ denote the tangent vector along axis s^i in Ω^2, and $\partial/\partial \xi^i$ the tangent vector along the corresponding curve ξ^i in Ω^h_e. Then, the mapping satisfies:

$$\mathbf{c}\left(\frac{\partial}{\partial s^i}\right) = \frac{\partial}{\partial \xi^i}. \tag{2.60}$$

By the chain rule, the left-hand side of Eq. (2.60) can be expressed as $\partial \mathbf{c}/\partial s^i$, yielding:

$$\frac{\partial \mathbf{c}}{\partial s^i} = \frac{\partial}{\partial \xi^i}. \tag{2.61}$$

In most finite element codes, the differentiation along each axis of the standard element is obtained analytically and stored as a **differentiation matrix**. Therefore, differentiating the map $\mathbf{c}(s^1, s^2)$ is straightforward within a finite element scheme. For triangular elements, deriving the tangent vectors in Ω_e^h proceeds similarly to the quadrilateral case, but with the following relations (Karniadakis and Sherwin 2013):

$$\frac{\partial}{\partial \xi^1} = \left(\frac{2}{1-s^2}\right) \frac{\partial}{\partial s^1}\bigg|_{s^2}, \tag{2.62}$$

$$\frac{\partial}{\partial \xi^2} = \left(\frac{1+s^1}{1-s^2}\right) \frac{\partial}{\partial s^1}\bigg|_{s^2} + \frac{\partial}{\partial s^2}\bigg|_{s^1}. \tag{2.63}$$

For the constructed tangent vectors $\partial/\partial \xi^i$ on Ω_e^h, their differentiability depends solely on the smoothness of the map $\mathbf{c}(s^1, s^2)$. Specifically, any k-th order derivative of the tangent vector corresponds to the $(k+1)$-th order derivative of the map:

$$\left(\frac{\partial}{\partial \xi^i}\right)^k \frac{\partial}{\partial \xi^i} = \frac{\partial^{k+1} \mathbf{c}}{\partial (s^i)^{k+1}}. \tag{2.64}$$

Thus, the differentiability of the tangent vectors of the moving frames is guaranteed if the map $\mathbf{c}(s^1, s^2)$ is sufficiently smooth.

The final step is to orthonormalize the tangent vectors $\partial/\partial \xi^i$, as they are generally neither of unit length nor orthogonal. A common method is the Gram-Schmidt process. A smooth Gram–Schmidt or QR-based orthonormalization is needed to preserve differentiability. Care must be taken when applying this step on meshes to ensure continuity of derivatives. A numerically stable variant (e.g., QR or Householder) is recommended when high regularity is needed across curved meshes.

1. Choose the tangent vector $\partial/\partial \xi^1$ at a point $p_{ij} \in \Omega_e^h$, denoted by $\partial/\partial \xi^1(p_{ij})$. Normalize it to obtain $\mathbf{e}_1(p_{ij})$:

$$\mathbf{e}_1 = \frac{\partial/\partial \xi^1}{\left\|\partial/\partial \xi^1\right\|}. \tag{2.65}$$

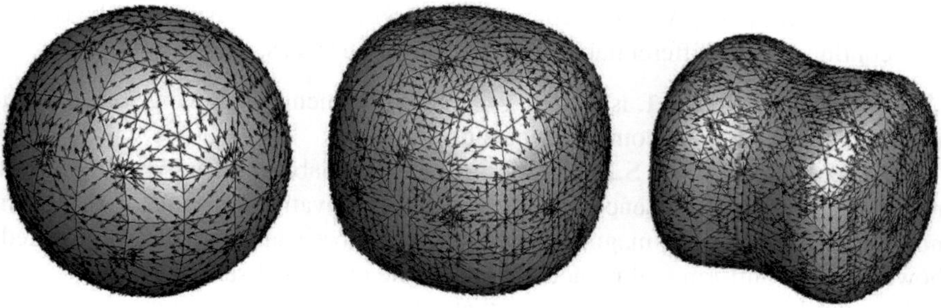

(a) (b) (c)

Fig. 2.4 Construction of moving frames on various curved surfaces: (**a**) a sphere, (**b**) an irregular surface, and (**c**) a nonconvex surface. Each set of moving frames is continuous and differentiable within an element but may not be continuous across element interfaces

2. Obtain the second tangent vector $\mathbf{e}_2(p_{ij})$ by applying the Gram-Schmidt process to $\partial/\partial\xi^2$ with respect to $\mathbf{e}_1(p_{ij})$. Since $\partial/\partial\xi^2$ lies on the tangent plane, orthogonalize it as:

$$\tilde{\mathbf{e}}_2 = \partial/\partial\xi^2 - \left(\frac{\mathbf{e}_1 \cdot \partial/\partial\xi^2}{\|\partial/\partial\xi^2\|}\right)\mathbf{e}_1, \tag{2.66}$$

$$\mathbf{e}_2 = \frac{\tilde{\mathbf{e}}_2}{\|\tilde{\mathbf{e}}_2\|}. \tag{2.67}$$

3. Obtain the third tangent vector $\mathbf{e}_3(p_{ij})$ as the normalized cross-product of $\mathbf{e}_1$ and $\mathbf{e}_2$:

$$\mathbf{e}_3 = \frac{\mathbf{e}_1 \times \mathbf{e}_2}{\|\mathbf{e}_1 \times \mathbf{e}_2\|}. \tag{2.68}$$

Examples of moving frames constructed on various surfaces—a sphere, an irregular surface with varying curvature, and a nonconvex surface with regions of positive and negative curvature—are shown in Fig. 2.4.

2.3 Properties Required for Numerical Fidelity

Moving frames should be constructed using the previously mentioned methods to satisfy the following properties necessary for the numerical solution of PDEs. For each i, where $1 \leq i \leq 3$, we have:

S.1 Moving frames $\mathbf{e}_i$ are orthogonal and of unit length.
S.2 Moving frames $\mathbf{e}_i$ are differentiable within each element Ω_e^h.

S.3 Moving frames $\mathbf{e}_i$, which are piecewise C^k within each element, may be neither continuous nor differentiable across element interfaces, $\partial \Omega_e^h$.

The first condition, **S.1**, is not the general requirement for moving frames but it may make the follow-up computations much easier.

The second condition, **S.2**, requiring the differentiability of moving frames, is necessary due to the presence of first or second derivatives of tangent vectors in numerical schemes. The magnitude of tangent vectors can be uniformly rescaled; however, such uniform scaling does not affect the numerical scheme.

The third condition, **S.3**, specifies that while tangent vectors must be continuously differentiable within each element, continuity and differentiability are not required across element interfaces. It is important to emphasize that variables and vectors can still be differentiable within each element, even if discontinuities occur across elements. In Fig. 2.9a, moving frames are differentiable inside each Ω_i but are not differentiable across elements. In contrast, in Fig. 2.9b, moving frames are continuous and differentiable both within and across elements.

The **non-differentiability** of moving frames across element interfaces implies the use of discontinuous coordinate systems on the domain. Discontinuous moving frames can locally avoid singularities in the field, allowing smooth definitions within each element without violating topological constraints. The Hairy Ball theorem states (Renteln 2013; Eigenberg 1979):

There is always at least one point on the sphere where a continuous tangent vector field vanishes.

In other words, continuous moving frames on a sphere must vanish at least at one point, resulting in a singularity. For instance, spherical coordinates exhibit singularities at the north and south poles. Constructing moving frames that are discontinuous across elements establishes a **discontinuous tangent vector field**, thereby eliminating geometric singularities. These discontinuous frames ensure compatibility with DG or hybridized FEM frameworks.

2.3.1 Benefits of Smooth and Orthonormal Frames in PDE Solvers

The orthonormality and differentiability of moving frames are not strictly required but are highly desirable for numerical schemes. We discuss the reasons below.

(I) The orthonormality of moving frames simplifies the expansion of vectors. Consider a velocity vector $\mathbf{v}$ on a curved mesh Ω_e^h expanded in orthonormal moving frames $\mathbf{e}_i$:

$$\mathbf{v} = v^1\mathbf{e}_1 + v^2\mathbf{e}_2. \tag{2.69}$$

Alternatively, $\mathbf{v}$ can be expanded using tangent vectors $\boldsymbol{\nu}_i$ associated with curved coordinate axes ξ^i:

$$\mathbf{v} = \bar{v}^1 \boldsymbol{\nu}_1 + \bar{v}^2 \boldsymbol{\nu}_2. \tag{2.70}$$

In general, the tangent vectors $\boldsymbol{\nu}_i$ are not orthogonal, i.e., $\boldsymbol{\nu}_1 \cdot \boldsymbol{\nu}_2 \neq 0$. With orthonormal moving frames, the expansion coefficients are simply:

$$v^i = \mathbf{v} \cdot \mathbf{e}_i. \tag{2.71}$$

However, obtaining $\bar{v}^i$ in Eq. (2.70) requires solving a linear system:

$$\begin{bmatrix} \|\boldsymbol{\nu}_1\|^2 & \boldsymbol{\nu}_1 \cdot \boldsymbol{\nu}_2 \\ \boldsymbol{\nu}_1 \cdot \boldsymbol{\nu}_2 & \|\boldsymbol{\nu}_2\|^2 \end{bmatrix} \begin{bmatrix} \bar{v}^1 \\ \bar{v}^2 \end{bmatrix} = \begin{bmatrix} \mathbf{v} \cdot \boldsymbol{\nu}_1 \\ \mathbf{v} \cdot \boldsymbol{\nu}_2 \end{bmatrix}. \tag{2.72}$$

Thus, orthonormal moving frames substantially simplify the computation of vector expansions.

(II) The unit length of tangent vectors transfers the magnitude information to the vector coefficients. Suppose the tangent vectors are orthogonal, $\boldsymbol{\nu}_1 \cdot \boldsymbol{\nu}_2 = 0$. Comparing Eqs. (2.69) and (2.70), we find:

$$v^i = \bar{v}^i \|\boldsymbol{\nu}_i\|, \quad \text{for any integer } i. \tag{2.73}$$

This equation implies that the vector coefficients v^i of the moving frames are obtained by multiplying the coefficients of the vectors in the unit tangent basis ($\bar{v}^i$) by the magnitudes of the corresponding tangent vectors ($\|\boldsymbol{\nu}_i\|$). Using unit-length tangent vectors $\mathbf{e}_i$ does not remove any geometric information about the domain. Instead, it transfers this information from the basis vectors to the vector coefficients. This simple trick is convenient in scientific computation, as computing the magnitude of tangent vectors on general curved surfaces is usually challenging and computationally expensive.

In the vector representation with orthonormal moving frames, as shown in Eq. (2.69), a vector is projected onto the unit tangent vectors of the moving frames. The vector coefficients then inherently incorporate the magnitudes of the tangent vectors and are used directly in the numerical solution of PDEs. In this case, the magnitudes of the tangent vectors do not need to be computed explicitly but are implicitly accounted for in the numerical scheme. If there is any geometric error, it affects the vector representation during the initial projection.

In contrast, for the vector representation using general tangent vectors, as shown in Eq. (2.70), the vector must be projected onto the (possibly non-unit) tangent vectors of the domain. This approach requires the tangent vectors to be computed in advance. Any error in the tangent vectors directly impacts the numerical solution of PDEs, particularly in long-time integrations. Although analytically derived tangent vectors can be used instead of numerically

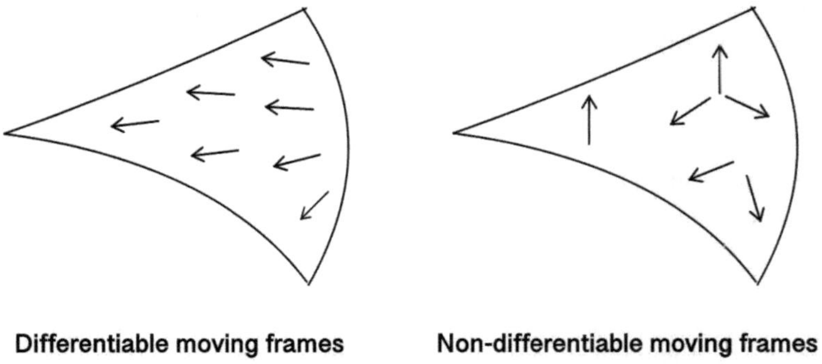

Fig. 2.5 Illustrations on the differentiable and non-differentiable moving frames in an element

computed ones, they may not be consistent with the differential operators. This inconsistency arises because the differential operators must be defined on the approximated domain based on the distribution of grid points where differentiation occurs, which may differ from analytically derived tangent vectors.

(III) The differentiability of moving frames is crucial when computing differential operators (Fig. 2.5). For a vector $\mathbf{v} = (v^x, v^y, v^z)$ in Euclidean space, the divergence operator is defined as:

$$\nabla \cdot \mathbf{v} = \frac{\partial v^x}{\partial x} + \frac{\partial v^y}{\partial y} + \frac{\partial v^z}{\partial z}. \tag{2.74}$$

Expanding $\mathbf{v}$ in moving frames:

$$\mathbf{v} = v^1 \mathbf{e}_1 + v^2 \mathbf{e}_2, \tag{2.75}$$

the divergence turns out to be:

$$\nabla \cdot \mathbf{v} = \left(\nabla v^1 \cdot \mathbf{e}_1 + v^1 \nabla \cdot \mathbf{e}_1 \right) + \left(\nabla v^2 \cdot \mathbf{e}_2 + v^2 \nabla \cdot \mathbf{e}_2 \right). \tag{2.76}$$

Thus, $\nabla \cdot \mathbf{e}_i$ must exist, requiring the differentiability of moving frames within each element. While globally differentiable moving frames are not necessary, differentiability within each tessellated element Ω_e^h suffices.

Although divergence on curved surfaces should be computed covariantly, for sufficiently small elements, the Euclidean divergence provides a good approximation. Roughly speaking, differentiability ensures that tangent vectors vary smoothly within each element, as illustrated in Fig. 2.7.

(IV) **Fourth**: Discontinuity across elements can be advantageous for avoiding geometric singularities.

While continuity across elements is not required, allowing discontinuities is beneficial for handling closed surfaces such as spheres. The spherical coordinate system inherently has singularities at the poles, where tangent

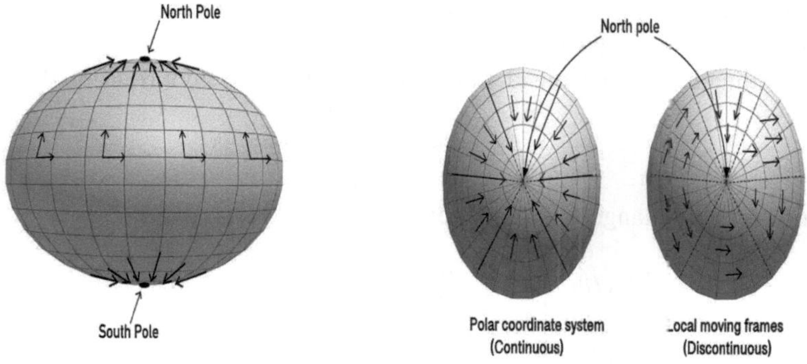

(a) Singularities of the sphere

(b) Avoiding singularity

Fig. 2.6 Singularities in spherical coordinates and their resolution via discontinuous moving frames. (**a**) On a closed sphere, a continuous global reference system produces tangent vectors that are not differentiable at certain points, giving rise to singularities. (**b**) For instance, the polar coordinate system exhibits a singularity at the North Pole. Such singularities can be removed by employing local moving frames, which are continuous within each element but may be discontinuous across element interfaces

vectors cannot be differentiated. According to the Hairy Ball theorem, all closed surfaces exhibit such singularities.

Moving frames, being pointwise-constructed tangent vectors, can naturally be discontinuous across elements. Discontinuous moving frames avoid global singularities, as shown in Fig. 2.6b, by allowing piecewise smoothness and local differentiability.

2.4 Examples of Frame Construction on Common Geometries

2.4.1 Flat-Space Frames: Cartesian and Polar Systems

Consider the Cartesian coordinate system, denoted by (x, y, z). Let its associated tangent vectors be $(\mathbf{x}, \mathbf{y}, \mathbf{z})$. For a two-dimensional plane, the Cartesian coordinate system satisfies the conditions for moving frames because the tangent vectors are orthonormal, i.e.,

$$\mathbf{e}_1 = \mathbf{x}, \tag{2.77}$$

$$\mathbf{e}_2 = \mathbf{y}, \tag{2.78}$$

$$\mathbf{e}_3 = \mathbf{z}. \tag{2.79}$$

The constructed moving frames are continuous and differentiable everywhere, as is the Cartesian coordinate system. Moving frames can also be constructed from the polar coordinate system. Consider the polar coordinate system (r, θ, z), defined as:

$$\mathbf{x} = (r, \theta, z) = (r\cos\theta, r\sin\theta, z). \tag{2.80}$$

The corresponding tangent vectors are:

$$\begin{bmatrix} \partial\mathbf{x}/\partial r \\ \partial\mathbf{x}/\partial\theta \\ \partial\mathbf{x}/\partial z \end{bmatrix} = \begin{bmatrix} \boldsymbol{r} \\ \boldsymbol{\theta} \\ z \end{bmatrix} = \begin{bmatrix} \cos\theta & \sin\theta & 0 \\ -r\sin\theta & r\cos\theta & 0 \\ 0 & 0 & 1 \end{bmatrix}. \tag{2.81}$$

The polar moving frames are consequently constructed by normalizing the tangent vectors:

$$\mathbf{e}_1 = \boldsymbol{r}/\|\boldsymbol{r}\|, \tag{2.82}$$

$$\mathbf{e}_2 = \boldsymbol{\theta}/\|\boldsymbol{\theta}\|, \tag{2.83}$$

$$\mathbf{e}_3 = z/\|z\|, \tag{2.84}$$

or, in matrix form:

$$\begin{bmatrix} \mathbf{e}_1 \\ \mathbf{e}_2 \\ \mathbf{e}_3 \end{bmatrix} = \begin{bmatrix} \cos\theta & \sin\theta & 0 \\ -\sin\theta & \cos\theta & 0 \\ 0 & 0 & 1 \end{bmatrix}. \tag{2.85}$$

The polar moving frames are differentiable everywhere except at the origin (0, 0, 0). However, similar to the Cartesian moving frames, they are continuous and differentiable across element interfaces (Fig. 2.7).

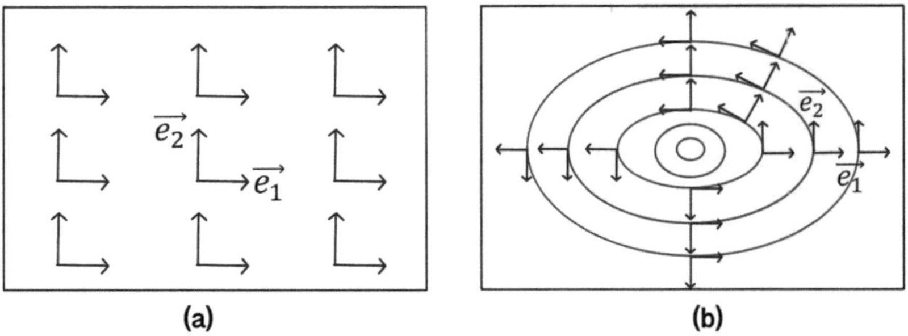

Fig. 2.7 Illustrations of Cartesian and polar moving frames in the plane. The Cartesian moving frames $\{\mathbf{e}_1, \mathbf{e}_2\}$ are constant, whereas the polar moving frames $\{\mathbf{e}'_1, \mathbf{e}'_2\}$ change with position. (**a**) Cartesian moving frames. (**b**) Polar moving frames

2.4.2 Spherical Coordinates and Regularization at the Poles

A well-known construction of moving frames on a sphere is to align them with the spherical coordinate axes. The spherical coordinate system is orthogonal but, like the polar system, its tangent vectors are not of unit length, and thus normalization is required. Consider the spherical coordinate system (r, θ, ϕ), defined by:

$$\mathbf{x} = (x, y, z) = (r\cos\theta\cos\phi, r\cos\theta\sin\phi, r\sin\theta). \tag{2.86}$$

The tangent vectors are:

$$\begin{bmatrix} \partial\mathbf{x}/\partial r \\ \partial\mathbf{x}/\partial\theta \\ \partial\mathbf{x}/\partial\phi \end{bmatrix} = \begin{bmatrix} \boldsymbol{r} \\ \boldsymbol{\theta} \\ \boldsymbol{\phi} \end{bmatrix} = \begin{bmatrix} \cos\theta\cos\phi & \cos\theta\sin\phi & \sin\theta \\ -r\cos\theta\sin\phi & r\cos\theta\cos\phi & 0 \\ r\sin\theta\cos\phi & r\sin\theta\sin\phi & -r\cos\theta \end{bmatrix}. \tag{2.87}$$

The orthonormal spherical moving frames are obtained by normalization:

$$\mathbf{e}_1 = \boldsymbol{\theta}/\|\boldsymbol{\theta}\|, \tag{2.88}$$

$$\mathbf{e}_2 = \boldsymbol{\phi}/\|\boldsymbol{\phi}\|, \tag{2.89}$$

$$\mathbf{e}_3 = \boldsymbol{r}/\|\boldsymbol{r}\|, \tag{2.90}$$

or, in the matrix form, we have

$$\begin{bmatrix} \mathbf{e}_1 \\ \mathbf{e}_2 \\ \mathbf{e}_3 \end{bmatrix} = \begin{bmatrix} \cos\theta\cos\phi & \cos\theta\sin\phi & \sin\theta \\ -\sin\phi & \cos\phi & 0 \\ \sin\theta\cos\phi & \sin\theta\sin\phi & -\cos\theta \end{bmatrix}. \tag{2.91}$$

The spherical moving frames are continuous and differentiable everywhere except at the poles ($\theta = 0$ or $\theta = \pi$), but remain continuous and differentiable across element interfaces, as shown in Fig. 2.9.

2.4.3 LOCAL Frames: Piecewise Smooth Construction per Element

Moving frames constructed by differentiating the mapping from the standard element (Method 3) yield moving frames that are differentiable within each element. No further alignment is required within the element. However, the direction of the first tangent vector $\mathbf{e}_1$ remains arbitrary on the tangent plane. Consequently, the moving frames are generally discontinuous across element interfaces, as shown in Fig. 2.8b. We refer to these as **LOCAL moving frames**. The term 'LOCAL'

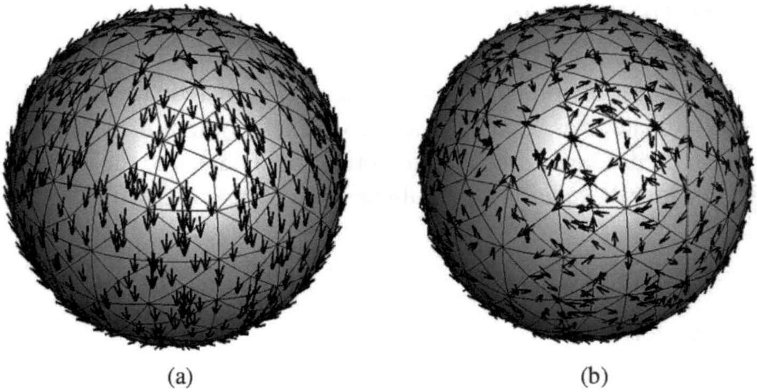

Fig. 2.8 Directional consistency of the first tangent vector $\mathbf{e}_1$ in spherical versus local frame constructions. For spherical moving frames, $\mathbf{e}_1$ remains continuously aligned with the spherical coordinate axis. In contrast, for local moving frames, $\mathbf{e}_1$ is discontinuous across element interfaces. (**a**) Spherical moving frames. (**b**) LOCAL moving frames

indicates that moving frames are **locally** continuous and differentiable within each element.

The accuracy of covariant differentiation using moving frames often depends on how **flat** the moving frames are on each curved element. Mathematically, flatness corresponds to moving frames with zero covariant divergence and zero Christoffel symbols (connection coefficients). Minimizing covariant divergence of the frames reduces geometric distortion, which can improve accuracy in differential operator evaluation.

LOCAL moving frames are not Euclidean frames on a curved surface (Fig. 2.9). Thus, they generally exhibit non-negligible divergence for the first or second tangent vectors:

$$|\nabla \cdot \mathbf{e}_i| \geq \varepsilon > 0, \quad \text{for at least one } i \in \{1, 2\}, \tag{2.92}$$

where ε is a small positive number.

The third tangent vector $\mathbf{e}_3$ is aligned along the surface normal vector. Thus, the covariant divergence $\nabla \cdot \mathbf{e}_3$ represents the curvature of the domain. In differentiation schemes, the divergence of unit tangent vectors governs the deviation between covariant divergence and Euclidean divergence. If $\nabla \cdot \mathbf{e}_i \approx 0$, the covariant and Euclidean divergences become nearly identical. Reducing the divergence in Eq. (2.92) is therefore critical for achieving higher numerical accuracy.

2.4.4 Minimizing Covariant Divergence for Frame Stability

A set of curved axes with zero covariant divergence is referred to as a **normal coordinate system**. Constructing an exact normal coordinate system on an arbitrary

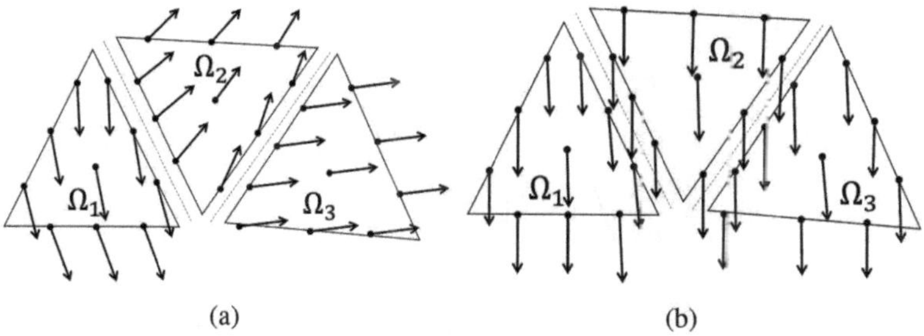

Fig. 2.9 Another illustration comparing between (**a**) LOCAL moving frames and (**b**) spherical moving frames. LOCAL moving frames may be discontinuous across element interfaces

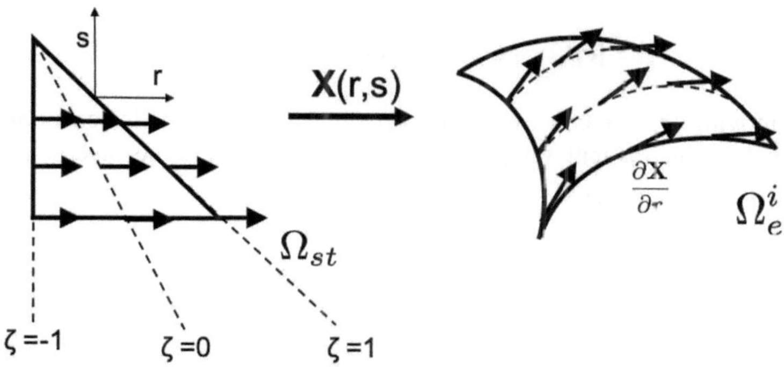

Fig. 2.10 Mapping $\mathbf{X}(r, s)$ from the standard triangular element Ω_{st} to a curved triangular element Ω_e^i. The reference coordinates r and s in Ω_{st} are mapped to curved coordinate lines on Ω_e^i, with tangent vectors $\partial\mathbf{X}/\partial r$ and $\partial\mathbf{X}/\partial s$

surface is challenging, even on simple surfaces like spheres However, a local approximation can be achieved by minimizing the divergence of the tangent vectors.

If the element size is sufficiently small, the divergence in Eq. (2.92) becomes very small, aligning with the intuition that small neighborhoods better approximate Euclidean behavior. Nevertheless, for relatively larger elements, it is desirable to explicitly minimize the divergence of moving frames.

The following simple algorithm is proposed to construct a first tangent vector $\mathbf{e}_1$ with minimized covariant divergence (Fig. 2.10):

1. For a triangular element mapped from standard coordinates (r, s), compute unit tangent vectors along three edge directions: r-axis, s-axis, and ζ-axis ($\zeta = r + s$ direction).
2. For each direction, calculate the covariant divergence of the associated tangent vector.
3. Select the direction with the smallest covariant divergence to define $\mathbf{e}_1$.
4. Construct $\mathbf{e}_2$ orthogonal to $\mathbf{e}_1$ and aligned with the surface tangent plane.

In detail, compute:

$$\mathbf{e}_{E_1} = \frac{d\mathbf{X}}{dr} \Big/ \left\| \frac{d\mathbf{X}}{dr} \right\|, \tag{2.93}$$

$$\mathbf{e}_{E_2} = \frac{d\mathbf{X}}{ds} \Big/ \left\| \frac{d\mathbf{X}}{ds} \right\|, \tag{2.94}$$

$$\mathbf{e}_{E_3} = \frac{1}{2}\left(\frac{d\mathbf{X}}{dr} + \frac{d\mathbf{X}}{ds}\right) \Big/ \left\| \frac{1}{2}\left(\frac{d\mathbf{X}}{dr} + \frac{d\mathbf{X}}{ds}\right) \right\|. \tag{2.95}$$

By comparing the covariant divergence of the above unit tangent vectors, we determine the direction of $\mathbf{e}_1$ as follows:

$$\mathbf{e}_1 = \{\mathbf{e}_{E_k} | \nabla \cdot \mathbf{e}_{E_k} \text{ is the smallest for } 1 \leq k \leq 3\}. \tag{2.96}$$

The moving frames constructed by Eq. (2.96) exhibit smaller covariant divergence in the unit tangent vectors. As shown in Fig. 2.11, the divergence of the

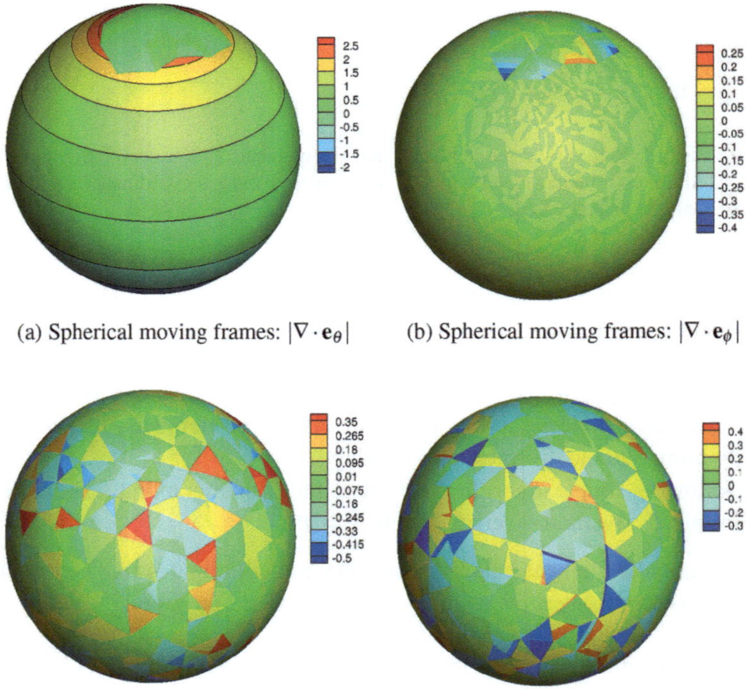

(a) Spherical moving frames: $|\nabla \cdot \mathbf{e}_\theta|$ (b) Spherical moving frames: $|\nabla \cdot \mathbf{e}_\phi|$

(c) LOC moving frames: $|\nabla \cdot \mathbf{e}_1|$ (d) LOC moving frames: $|\nabla \cdot \mathbf{e}_2|$

Fig. 2.11 Covariant divergence of spherical and local moving frames on the sphere. The spherical frames are denoted by $\mathbf{e}_\theta$ and $\mathbf{e}_\phi$, while the local frames are denoted by **e**1 and **e**2. (**a**) Divergence of $\mathbf{e}_\theta$. (**b**) Divergence of $\mathbf{e}_\phi$. (**c**) Divergence of $\mathbf{e}_1$. (**d**) Divergence of $\mathbf{e}_2$

first tangent vector of spherical moving frames, denoted by $\mathbf{e}_\theta$, is relatively large, particularly near the poles. By contrast, the divergence of the second tangent vector, $\mathbf{e}_\phi$, is significantly smaller—almost ten times smaller than that of $\mathbf{e}_\theta$.

In comparison, the divergence $\nabla \cdot \mathbf{e}_i$ for LOCAL moving frames is neither particularly large nor small relative to the spherical moving frames. LOCAL moving frames do not eliminate the divergence of the moving frames entirely. Instead, the magnitude of divergence is approximately distributed equally between the two tangent vectors $\mathbf{e}_1$ and $\mathbf{e}_2$.

The relatively smaller magnitudes of divergence in the tangent vectors suggest that the resulting numerical error from nontrivial divergence is of a lower order than the discretization error. Thus, the associated error can be sufficiently reduced by increasing the spatial resolution.

Consequently, LOCAL moving frames achieve slightly better accuracy in the numerical solution of PDEs. However, the improvement in accuracy remains relatively modest. The above algorithm only considers the average of two axes, r and s, for constructing the third direction. A more rigorous optimization procedure involving arbitrary combinations of axes could potentially yield flatter moving frames with even smaller covariant divergence, bringing them closer to an ideal normal coordinate system.

2.4.5 Combining LOCAL and Spherical Frames: LOCSPH

Another important construction is the combination of spherical moving frames and LOCAL moving frames, referred to as **LOCSPH moving frames** (short for LOCAL Sphere).

Let Ω_e^0 denote the original domain and Ω_e^h the approximated domain obtained by tessellation. In numerical schemes, grid points lie on the approximated domain Ω_e^h and may deviate slightly from the original surface Ω_e^0. Consequently, the surface normal vector on Ω_e^h, denoted by $\mathbf{k}^h$, is not necessarily consistent with the exact surface normal vector $\mathbf{k}^0$ of Ω_e^0.

LOCAL moving frames accurately represent the approximated domain Ω_e^h, while spherical moving frames represent the exact geometry, independent of the discretization error. Thus, spherical moving frames are constructed analytically and do not depend on the approximated grid unless projection is explicitly enforced.

To construct LOCSPH moving frames, the only adjustment needed from LOCAL moving frames is to realign the third unit tangent vector $\mathbf{e}_3$ with $\mathbf{k}^0$, the surface normal vector of the original domain. Re-orthonormalization must be performed afterward to maintain orthogonality.

The LOCSPH moving frames inherit properties from both spherical and LOCAL frames:

- Like spherical frames, $\mathbf{e}_3$ aligns with the original surface normal $\mathbf{k}^0$.

Table 2.1 Comparison of spherical, LOCSPH, and LOCAL moving frames on a sphere

	Spherical	LOCSPH	LOCAL
Surface normal vector	$\mathbf{r}$[a]	$\mathbf{r}$[a]	Analytically unknown
Continuity across elements[b]	Yes	No	No

[a] $\mathbf{r}$ denotes the radial vector of the spherical coordinate system
[b] Refers to continuity of frames across element interfaces

- Like LOCAL frames, LOCSPH moving frames are generally discontinuous across element interfaces.

See Table 2.1 for a comparison among spherical, LOCAL, and LOCSPH moving frames.

A practical construction scheme is as follows:

1. Replace the third tangent vector $\mathbf{e}_3$ with $\mathbf{k}^0$, denoted as $\mathbf{e}_3^{\text{LOCSPH}}$.
2. Project $\mathbf{e}_1$ onto the plane orthogonal to $\mathbf{e}_3^{\text{LOCSPH}}$ and normalize, denoting the result as $\mathbf{e}_1^{\text{LOCSPH}}$.
3. Define $\mathbf{e}_2^{\text{LOCSPH}}$ by the cross product: $\mathbf{e}_2^{\text{LOCSPH}} = \mathbf{e}_3^{\text{LOCSPH}} \times \mathbf{e}_1^{\text{LOCSPH}}$.

The LOCSPH moving frames thus represent the original domain Ω_e^0, even though the numerical computation occurs on Ω_e^h. Representing the geometry of the original domain is crucial because all surface derivatives are directional derivatives along the surface:

$$\text{Directional derivative of } u \text{ along } \mathbf{v} = \nabla u \cdot \mathbf{v}, \tag{2.97}$$

where u is a scalar field and $\mathbf{v}$ is a tangent vector to the surface.

Even when grid points are slightly misallocated due to approximation, derivatives must be taken along the original geometric directions. Otherwise, errors accumulate, particularly for long-time integration or higher-order PDEs. In later chapters, we will explore how preserving the original path of differentiation, called the **connection** (Kobayashi and Nomizu 1969), improves numerical accuracy.

2.5 [MATLAB] Grid-Based Moving Frame Generation

The following MATLAB scripts generate a set of orthonormal moving frames on both a spherical and a pseudospherical shell of unit radius ($r = 1$), defined using parametric equations in spherical and curved coordinates (Fig. 2.12).

For the **sphere**, the surface is parameterized by:

$$x = r \sin\theta \cos\phi, \tag{2.98}$$

$$y = r \sin\theta \sin\phi, \tag{2.99}$$

$$z = r \cos\theta, \tag{2.100}$$

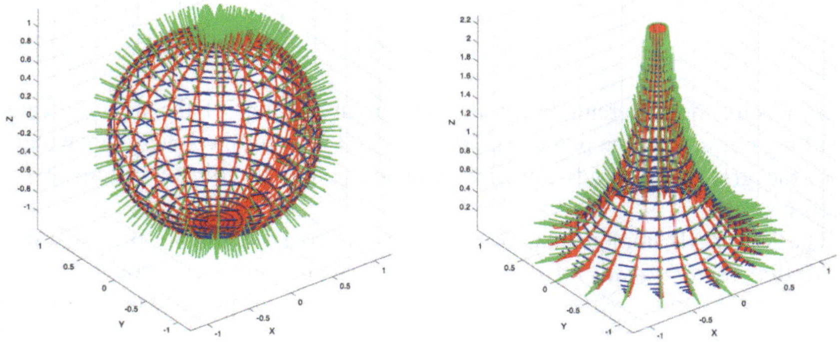

(a) Moving frames on a unit sphere (b) Moving frames on a pseudosphere

Fig. 2.12 Grid points and orthonormal moving frames constructed on (**a**) the sphere and (**b**) the pseudosphere. A small region near the poles is omitted to preserve smoothness and prevent coordinate singularities

where $\theta \in (0, \pi)$ is the polar angle and $\phi \in [0, 2\pi)$ is the azimuthal angle.

For the **pseudosphere**, the surface is defined using:

$$x = r\,\mathrm{sech}\,\theta \cos\phi, \tag{2.101}$$

$$y = r\,\mathrm{sech}\,\theta \sin\phi, \tag{2.102}$$

$$z = r(\theta - \tanh\theta), \tag{2.103}$$

where $\theta > 0$ and $\phi \in [0, 2\pi)$. This surface has constant negative Gaussian curvature and is commonly used as a model of hyperbolic geometry.

To simplify computations, the moving frames are initially aligned with the coordinate directions. That is, the first and second tangent vectors, $\mathbf{e}_1$ and $\mathbf{e}_2$, are constructed based on derivatives of the surface parametrization with respect to ϕ and θ, respectively.

For the **sphere**, the tangent vectors aligned with the (ϕ, θ) directions are:

$$\mathbf{e}_1 = [-\sin\phi, \cos\phi, 0], \tag{2.104}$$

$$\mathbf{e}_2 = [\cos\theta\cos\phi, \cos\theta\sin\phi, -\sin\theta]. \tag{2.105}$$

For the **pseudosphere**, the corresponding tangent directions are:

$$\mathbf{e}_1 = [-\sin\phi, \cos\phi, 0], \tag{2.106}$$

$$\mathbf{e}_2 = [-\,\mathrm{sech}\,\theta\cos\phi, -\,\mathrm{sech}\,\theta\sin\phi, \tanh\theta]. \tag{2.107}$$

The third vector $\mathbf{e}_3$ is then obtained by the cross product:

$$\mathbf{e}_3 = \mathbf{e}_1 \times \mathbf{e}_2. \tag{2.108}$$

To ensure the differentiability and numerical stability of the moving frames, small regions around the poles (e.g., $\theta = 0$ and $\theta = \pi$ for the sphere) are excluded from the grid. This avoids singular behavior of the coordinate system at these locations.

The MATLAB implementations used for this construction are provided in the Appendix: *SetupParameter.m* defines the physical and numerical parameters, while *Generategrid.m* and *CreateGridTangentVec.m* generate the grid and tangent vectors.

```
function ch2MATLAB(Nphi, Ntheta, shape)

[grids,movingframes] = CreateGridTangentVec(Nphi, Ntheta, 1, shape);

% Visualize the grids and movingframes
Displaymovingframes(grids,movingframes);

end

function Displaymovingframes(grids,movingframes)

[Nphip, Nthetap, ~] = size(movingframes);

%% Plot grid points on the sphere (optional visualization)
x_grid = grids(:,:,1);
y_grid = grids(:,:,2);
z_grid = grids(:,:,3);

figure;
plot3(x_grid(:), y_grid(:), z_grid(:), 'ro');  % Red dots represent grid
     points
hold on;
axis equal;
grid on;
xlabel('X'); ylabel('Y'); zlabel('Z');

for i = 1:Nphip
  for j = 1:Nthetap

        % Cartesian coordinates at the current grid point
        xp = grids(i,j,1);
        yp = grids(i,j,2);
        zp = grids(i,j,3);

        e_phi = movingframes(i, j, 1, :);
        e_theta = movingframes(i, j, 2, :);
        e_r = movingframes(i, j, 3, :);

        % Plot moving frame vectors at the current grid point
        quiver3(xp, yp, zp, e_phi(1), e_phi(2), e_phi(3), 0.2, 'b', '
             LineWidth', 2); % Azimuthal (phi) direction
        quiver3(xp, yp, zp, e_theta(1), e_theta(2), e_theta(3), 0.2, 'r', '
             LineWidth', 2); % Polar (theta) direction
        quiver3(xp, yp, zp, e_r(1), e_r(2), e_r(3), 0.2, 'g', 'LineWidth',
             2); % Radial direction

  end
end

%% Add legend to the plot (if displayed)
```

```
legend({'Grid Points', ...
    'X-axis', 'Y-axis', 'Z-axis', ...
    'Azimuthal Direction ($e_{\phi}$)', ...
    'Polar Direction ($e_{\theta}$)', ...
    'Radial Direction ($e_{r}$)'}, ...
    'Location', 'northeastoutside','FontSize',20,'Interpreter','latex');

end
```

Listing 2.1 Main script for generating grids and moving frames on sphere and pseudosphere

```
function [grids,movingframes] = CreateGridTangentVec(Nphi, Ntheta, normalized, shape)
% ch2MATLAB.m
% This function visualizes a spherical shell with discretized grid points
% and modified tangent vectors, known as moving frames.
%
% Inputs:
%   Nphi       - Number of discretization points along the azimuthal angle (phi)
%   Ntheta     - Number of discretization points along the polar angle (theta)
%   normalized- Flag to normalize frame vectors (default: 1, normalized)
%   alpha      - Interpolation factor controlling frame deformation (default: 0.0)
%   display    - Flag to enable plot display (default: 0, no plot)
%
% Output:
%   movingframes - A 4-dimensional array storing moving frame vectors
%
% Usage example:
%   movingframes = ch2MATLAB(20, 20, 1, 0.1, 1);

% Set default values for optional arguments if not provided
if nargin < 3
    normalized = 1; % Default normalization flag
end

if nargin < 4
    shape = "sphere"; % Default shape option
end

if nargin < 5
    alpha = 0.0; % Default deviation factor
end

%% --- Grid and Parameter Setup ---
SetupParameter; % Set global parameters (e.g., radius R, grid dimensions)

% --- Generate theta and phi grids using Spectral Element Method (SEM) nodes ---
[phigrid, thetagrid, ~, ~] = Generategrid(Nphi, Ntheta);

% --- Create 2D meshgrid of spherical coordinates ---
[PHI, TH] = meshgrid(phigrid, thetagrid);

grids = zeros(Nphip, Nthetap, spacedim);
% Convert spherical coordinates (phi, theta) to Cartesian coordinates (x, y, z)
if(shape=="sphere")
    grids(:,:,1) = R .* sin(TH') .* cos(PHI');
    grids(:,:,2) = R .* sin(TH') .* sin(PHI');
    grids(:,:,3) = R .* cos(TH');

elseif(shape=="deformedsphere")
    grids(:,:,1) = R .* sin(TH') .* cos(PHI');
```

```
    grids(:,:,2) = 2.* R .* sin(TH') .* sin(PHI');
    grids(:,:,3) = R .* cos(TH');

elseif(shape=="pseudosphere")
    grids(:,:,1) = R .* sech(TH') .* cos(PHI');
    grids(:,:,2) = R .* sech(TH') .* sin(PHI');
    grids(:,:,3) = R .* (TH'-tanh(TH'));

elseif(shape=="deformedpseudosphere")
    grids(:,:,1) = R .* sech(TH') .* cos(PHI');
    grids(:,:,2) = 2.* R .* sech(TH') .* sin(PHI');
    grids(:,:,3) = R .* (TH'-tanh(TH'));
end

% Initialize arrays to store moving frame vectors
movingframes = zeros(Nphip, Nthetap, mfdim, spacedim); % Dimensions: phi x
     theta x frame index x space dimension

%% Compute and plot tangent vectors (moving frames) at each grid point
for i = 1:Nphip
  for j = 1:Nthetap

        % Current spherical coordinates (phi, theta)
        phi = phigrid(i);
        theta = thetagrid(j);

        % Define standard spherical basis vectors:
        % e_phi (azimuthal), e_theta (polar), and e_r (radial)

        % Sphere
        if(shape=="sphere")
            e_phi   = [-sin(theta)*sin(phi), sin(theta)*cos(phi), 0];
            e_theta = [cos(theta)*cos(phi), cos(theta)*sin(phi), -sin(theta)
                 ];
            e_r     = [sin(theta)*cos(phi), sin(theta)*sin(phi), cos(theta)
                 ];

        % deformed sphere
        elseif(shape=="deformedsphere")
            e_phi   = [-sin(theta)*sin(phi), 2.*sin(theta)*cos(phi), 0];
            e_theta = [cos(theta)*cos(phi), 2.*cos(theta)*sin(phi), -sin(
                 theta)];
            e_r     = [sin(theta)*cos(phi), 2.*sin(theta)*sin(phi), cos(
                 theta)];

        % Pseudosphere
        elseif(shape=="pseudosphere")
            e_phi   = [-sech(theta).*sin(phi), sech(theta).*cos(phi), 0.0];
            e_theta = [-sech(theta).*tanh(theta).*cos(phi), -sech(theta).*
                 tanh(theta).*sin(phi), tanh(theta).*tanh(theta)];
            e_r     = [sech(theta)*cos(phi), sech(theta)*sin(phi), theta -
                 tanh(theta)];

        % deformed pseudosphere
        elseif(shape=="deformedpseudosphere")
            e_phi   = [-sech(theta).*sin(phi), 2.*sech(theta).*cos(phi),
                 0.0];
            e_theta = [-sech(theta).*tanh(theta).*cos(phi), -2.*sech(theta)
                 .*tanh(theta).*sin(phi), tanh(theta).*tanh(theta)];
            e_r     = [sech(theta)*cos(phi), 2.*sech(theta)*sin(phi), theta
                 - tanh(theta)];
        end

        % Normalize vectors if required
        if(normalized==1)
            e_phi   = e_phi ./ norm(e_phi);
            e_theta = e_theta ./ norm(e_theta);
```

```
            e_r      = e_r ./ norm(e_r);
        end

        % Store the computed frame vectors in the array
        movingframes(i, j, 1, :) = e_phi;
        movingframes(i, j, 2, :) = e_theta;
        movingframes(i, j, 3, :) = e_r;
    end
end

end
```

Listing 2.2 Subroutine to create grid points and tangent vectors

2.6 Summary

This chapter introduced the construction of **moving frames**—local orthonormal coordinate systems—on curved surfaces and their foundational role in formulating differential operators and numerically solving partial differential equations (PDEs). The main insights are summarized as follows:

Moving frames $\{\mathbf{e}_1, \mathbf{e}_2, \mathbf{e}_3\}$ are **orthonormal bases** constructed at each grid point. The tangent vectors $\mathbf{e}_1$ and $\mathbf{e}_2$ span the local tangent plane, while $\mathbf{e}_3$ is aligned with the surface normal vector. The relations $\mathbf{e}_1 = \mathbf{e}_2 \times \mathbf{e}_3$, $\mathbf{e}_2 = \mathbf{e}_3 \times \mathbf{e}_1$, and $\mathbf{e}_3 = \mathbf{e}_1 \times \mathbf{e}_2$ ensure orthogonality and orientation. The variation of moving frames between adjacent points captures both positional and angular deformations of the domain. **Structure equations** involving translation (ω_i) and rotation (ω_{ij}) components characterize these variations across neighboring elements.

The following methods can be employed to construct moving frames on general curved surfaces:

- *Method 1: Surface Normal Vector*—constructs frames using the known surface normal vector **k**.
- *Method 2: Diffusion Tensor*—aligns frames with the eigenvectors of an anisotropic diffusion tensor.
- *Method 3: Mapping from Standard Elements*—ensures differentiability by computing tangent vectors via analytic mappings and orthonormalizing them.

For PDE solvers, moving frames must be **differentiable within each element** but need **not be continuous across element boundaries. Discontinuous frames** help avoid singularities—especially on closed surfaces such as spheres—thus satisfying constraints like the Hairy Ball theorem. Discontinuous frames across elements offer several benefits:

- Simplify vector expansions: coefficients become inner products $v_i = \mathbf{v} \cdot \mathbf{e}_i$.
- Transfer geometric complexity to the vector coefficients, avoiding explicit computation of metric tensor magnitudes.
- Facilitate accurate computation of differential operators (e.g., divergence).

Several types of moving frames are commonly used in practice:

- *Spherical Frames:* Globally smooth but singular at poles.
- *LOCAL Frames:* Differentiable within each element, discontinuous across interfaces.
- *LOCSPH Frames:* A hybrid approach combining LOCAL and spherical frames; aligns the normal vector with the exact surface geometry while retaining local frame construction.

A practical algorithm selects the frame direction with the **lowest divergence** among edge-aligned axes in triangular elements, improving numerical stability and convergence. Code examples demonstrated how to generate and visualize moving frames on a spherical shell using spectral methods and parametric interpolation.

Overall, this chapter established the principles and practical techniques for constructing moving frames on curved geometries, enabling **high-order numerical schemes** that combine geometric fidelity with computational efficiency.

2.7 Exercises and Conceptual Applications

1. Given a point on a curved surface with known surface normal vector $\mathbf{n}$ and a preferred propagation direction $\mathbf{v}$ lying in the tangent plane, construct an orthonormal basis $\{\mathbf{e}_1, \mathbf{e}_2, \mathbf{e}_3\}$ using the cross product rules introduced in this chapter.
2. Let $\mathbf{e}_3$ be a surface normal vector at point P, and let $\mathbf{e}_1$ be a tangent vector. Prove that the vector $\mathbf{e}_2 = \mathbf{e}_3 \times \mathbf{e}_1$ lies in the tangent plane and is orthogonal to both $\mathbf{e}_1$ and $\mathbf{e}_3$.
3. Explain the connection between the differentiability of moving frames and the differentiability of the underlying curved domain. Why is first-order differentiability sufficient for directional derivatives, while second-order is required for curvature computations?
4. Describe a method for ensuring that the numerically constructed moving frames are smooth across a tessellated curved surface. What challenges arise when neighboring elements are not smoothly connected?
5. Suppose you are given an anisotropic tensor field on a surface. How would you align the first tangent vector $\mathbf{e}_1$ with the principal direction of anisotropy? Discuss the role of eigendecomposition in this process.
6. Write a pseudocode for constructing moving frames on a curved triangular mesh. Your algorithm should include: (1) computing the normal vector, (2) determining a primary tangent direction (e.g., via gradient or propagation direction), and (3) constructing the remaining orthonormal vectors.

Chapter 3
Covariant Derivatives on Curved Surfaces via Moving Frames

> *"All beginnings are obscure. Inasmuch as the mathematician operates with his conceptions along strict and formal lines, he, above all, must be reminded from time to time that the origins of things lie in greater depths than those to which his methods enable him to descend. Beyond the knowledge gained from the individual sciences, there remains the task of comprehending. In spite of the fact that the views of philosophy sway from one system to another, we cannot dispense with it unless we are to convert knowledge into a meaningless chaos."*
>
> *—Herman Weyl*

In this chapter, we investigate covariant derivatives on curved surfaces using orthonormal moving frames. We compare the traditional formulation involving Christoffel symbols and the metric tensor with a novel approach based on connection 1-forms. The methods are validated on the sphere.

Covariant derivatives generalize the concept of directional derivatives to curved spaces, allowing for consistent differentiation of vectors across a manifold. Classical approaches involve computing the metric tensor and Christoffel symbols, but these computations become difficult for complex geometries.

An alternative is to use orthonormal moving frames, where local bases at each point are constructed to simplify differentiation without explicitly requiring the metric tensor. However, several important issues must be addressed, including holonomicity and local Euclidean approximations.

Suppose that randomly oriented moving frames are placed throughout a domain. The directions of the first tangent vectors are not necessarily smooth or differentiable. We may ask about the benefits of such a *local* and *broken* coordinate system. However, one fundamental concept of calculus must be redefined: **differentiation**.

Normally, differentiation requires a specific direction because it measures how scalar or vector fields change along a given direction. In the Cartesian coordinate system, differentiation with respect to the x, y, and z coordinate axes is straightforward. The Cartesian coordinate axes have the same direction everywhere in the domain, enabling measurements of changes along fixed directions. However, in

S. Chun, *Moving Frames for the Numerical Solution of Partial Differential Equations in Complex Domains*, Springer Asia Pacific Mathematics Series 7,
https://doi.org/10.1007/978-981-95-4495-0_3

local reference systems like moving frames, this becomes problematic because there is no globally universal direction.

3.1 Differentiating Scalar Fields in Curved Frames

3.1.1 Classical Differentiation in Flat Space

Consider Fig. 3.1. For the Cartesian coordinate system, differentiation along the x- or y-axis has a definite meaning. Consider the differentiation of a scalar field $u = u(x, y)$ along the x-axis, denoted by $\partial u/\partial x$. Let the distance between points P_1 and P_2 be h_{12}. Then, the rate of change of u along the x-axis at P_1 is approximated as:

$$\frac{\partial u}{\partial x}(P_1) \approx \frac{u(P_1) - u(P_2)}{h_{12}}. \tag{3.1}$$

The exact value is obtained by letting P_2 approach P_1:

$$\frac{\partial u}{\partial x}(P_1) = \lim_{P_2 \to P_1} \frac{u(P_1) - u(P_2)}{h_{12}}. \tag{3.2}$$

In the setting of moving frames at P_1 (see Fig. 3.1), the partial derivative $\partial u/\partial x$ cannot be expressed in the same manner as Eq. (3.1) because the x-axis is not globally defined. Instead, we must use the directional derivative. The directional derivative of u along a vector $\mathbf{v}$ is denoted by $\nabla_{\mathbf{v}} u$ and defined as:

$$\nabla_{\mathbf{v}} u = \mathbf{v} \cdot \nabla u, \tag{3.3}$$

where ∇u is the gradient of u. In three-dimensional Euclidean space, it expands as:

$$\nabla u = \frac{\partial u}{\partial x}\mathbf{x} + \frac{\partial u}{\partial y}\mathbf{y} + \frac{\partial u}{\partial z}\mathbf{z}. \tag{3.4}$$

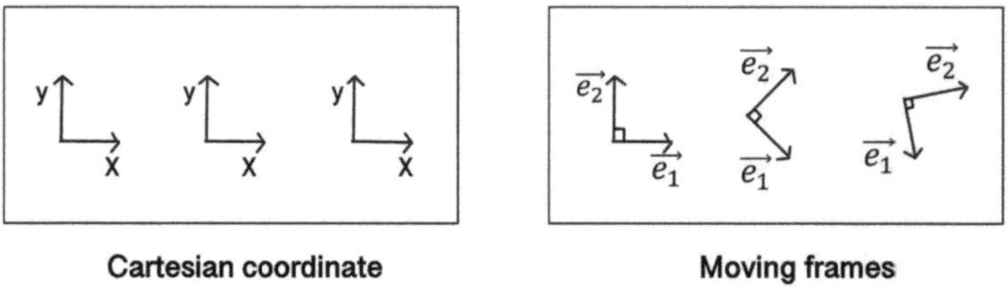

Fig. 3.1 Scalar differentiation in the Cartesian coordinate system versus moving frames. The Cartesian system employs identical reference frames at all points, so differentiation along each axis always corresponds to a fixed direction. Moving frames, however, can vary in orientation or definition depending on location. Thus, differentiation along moving frames measures how the scalar variable changes along each locally defined direction

3.1.2 Directional Derivatives in Local Frames

In moving frames, the gradient at P_1 can be approximated by:

$$\nabla u(P_1) \approx \left(\frac{u(P_1) - u(P_2)}{h_{12}} \mathbf{e}_1 + \frac{u(P_1) - u(P_4)}{h_{14}} \mathbf{e}_2 \right), \tag{3.5}$$

where P_2 and P_4 lie along $\mathbf{e}_1$ and $\mathbf{e}_2$, respectively, and h_{12} and h_{14} denote the corresponding distances. Using this gradient, the partial derivative $\partial u/\partial x$ is approximated as:

$$\frac{\partial u}{\partial x}(P_1) = \mathbf{x} \cdot \nabla u \approx \mathbf{x} \cdot \left(\frac{u(P_1) - u(P_2)}{h_{12}} \mathbf{e}_1 + \frac{u(P_1) - u(P_4)}{h_{14}} \mathbf{e}_2 \right). \tag{3.6}$$

If $\mathbf{e}_1$ and $\mathbf{e}_2$ align with the x- and y-axes, respectively, Eq. (3.6) reduces to Eq. (3.1).

In discontinuously aligned moving frames, difficulties arise particularly in evaluating directional derivatives along the first tangent vector $\mathbf{e}_1$. In Fig. 3.1, moving frames are constructed at points from P_1 to P_4. Since $\mathbf{e}_1$ varies between points, the directional derivative along $\mathbf{e}_1$ differs at each location.

At point P_i, the directional derivative of u along $\mathbf{e}_1$ is approximated by:

$$\mathbf{e}_1(P_i) \cdot \nabla u(P_i) \approx \mathbf{e}_1(P_i) \cdot \left(\frac{u(P_i) - u(P_j)}{h_{ij}} \mathbf{e}_1(P_i) + \frac{u(P_i) - u(P_k)}{h_{ik}} \mathbf{e}_2(P_i) \right), \tag{3.7}$$

where P_j and P_k are the neighborhood points along the direction of $\mathbf{e}_1$ and $\mathbf{e}_2$ at point P_i, respectively. h_{ij} and h_{ik} correspond to the distance between P_i and P_j and between P_i and P_k, respectively. By performing the dot product in Eq. (3.7), we obtain

$$\mathbf{e}_1(P_i) \cdot \nabla u(P_i) \approx \frac{u(P_i) - u(P_j)}{h_{ij}}, \tag{3.8}$$

where P_j lies along $\mathbf{e}_1$ at P_i. In contrast, directional derivatives along a fixed vector $\mathbf{v}$ are consistent. Expanding $\mathbf{v}$ at P_i:

$$\mathbf{v}(P_i) = v^1(P_i)\mathbf{e}_1(P_i) + v^2(P_i)\mathbf{e}_2(P_i), \tag{3.9}$$

the directional derivative becomes:

$$\mathbf{v} \cdot \nabla u \approx \left(v^1 \mathbf{e}_1 + v^2 \mathbf{e}_2 \right) \cdot \left(\frac{u(P_i) - u(P_j)}{h_{ij}} \mathbf{e}_1 + \frac{u(P_i) - u(P_k)}{h_{ik}} \mathbf{e}_2 \right). \tag{3.10}$$

Thus, we obtain

$$\mathbf{v} \cdot \nabla u \approx v^1 \left(\frac{u(P_i) - u(P_j)}{h_{ij}} \right) + v^2 \left(\frac{u(P_i) - u(P_k)}{h_{ik}} \right). \tag{3.11}$$

Thus, directional derivatives based on $\mathbf{v}$ are invariant under rotations of the moving frame. For an orthonormal frame $\{\mathbf{e}_1, \mathbf{e}_2\}$ at P_i, we expand:

$$\mathbf{v} \cdot \nabla u = ((\mathbf{v} \cdot \mathbf{e}_1)\mathbf{e}_1 + (\mathbf{v} \cdot \mathbf{e}_2)\mathbf{e}_2) \cdot ((\nabla u \cdot \mathbf{e}_1)\mathbf{e}_1 + (\nabla u \cdot \mathbf{e}_2)\mathbf{e}_2). \tag{3.12}$$

By solving the parentheses, we obtain

$$\mathbf{v} \cdot \nabla u = (\mathbf{v} \cdot \mathbf{e}_1)(\nabla u \cdot \mathbf{e}_1) + (\mathbf{v} \cdot \mathbf{e}_2)(\nabla u \cdot \mathbf{e}_2). \tag{3.13}$$

Considering a rotated frame $\{\mathbf{e}'_1, \mathbf{e}'_2\}$:

$$\begin{bmatrix} \mathbf{e}'_1 \\ \mathbf{e}'_2 \end{bmatrix} = \begin{bmatrix} \cos\theta & \sin\theta \\ -\sin\theta & \cos\theta \end{bmatrix} \begin{bmatrix} \mathbf{e}_1 \\ \mathbf{e}_2 \end{bmatrix}, \tag{3.14}$$

it can be shown that $\mathbf{v} \cdot \nabla u$ remains unchanged, confirming independence from the choice of frame:

$$\mathbf{v} \cdot \nabla u = (\mathbf{v} \cdot \mathbf{e}'_1)(\nabla u \cdot \mathbf{e}'_1) + (\mathbf{v} \cdot \mathbf{e}'_2)(\nabla u \cdot \mathbf{e}'_2). \tag{3.15}$$

3.2 Vector Differentiation: Beyond Scalars

Greater subtleties arise in differentiating vectors in local reference systems. Consider a constant vector:

$$\mathbf{v} = a\mathbf{x}, \quad a \in \mathbf{R}. \tag{3.16}$$

In Cartesian coordinates:

$$\frac{\partial \mathbf{v}}{\partial x} = \frac{\partial a}{\partial x}\mathbf{x} = \mathbf{0}. \tag{3.17}$$

However, in moving frames:

$$\mathbf{v} = v^1\mathbf{e}_1 + v^2\mathbf{e}_2, \tag{3.18}$$

and differentiating with respect to x yields:

$$\frac{\partial \mathbf{v}}{\partial x} = \frac{\partial v_i}{\partial x}\mathbf{e}_i + v_i\frac{\partial \mathbf{e}_i}{\partial x}. \tag{3.19}$$

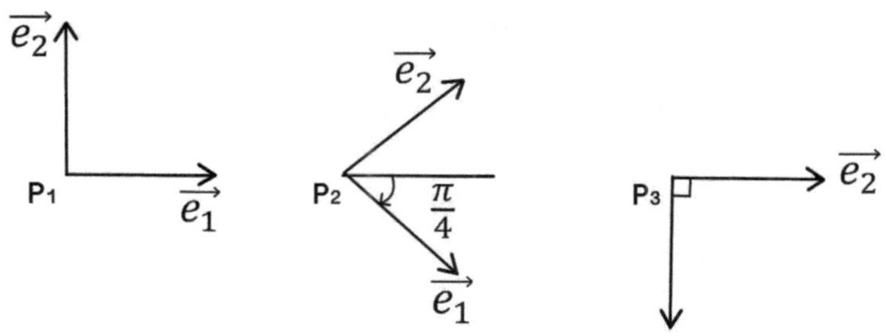

Fig. 3.2 Vector differentiation in the Cartesian coordinate system versus moving frames. Similar to scalar differentiation, the Cartesian system provides fixed directions and identical vector representations throughout the domain, so differentiation can be carried out component-wise. In moving frames, by contrast, the basis vectors themselves may vary from point to point, and simply differentiating the components does not capture the full vector differentiation, even in a Euclidean domain

Unlike the Cartesian case, the second term does not generally vanish. Even when the components v^i are constant, changes in basis vectors must be considered.

Suppose (Fig. 3.2):

$$\mathbf{e}_1(P_1) = (1, 0), \tag{3.20}$$

$$\mathbf{e}_1(P_2) = \left(\cos\frac{\pi}{4}, \sin\frac{\pi}{4}\right), \tag{3.21}$$

$$\mathbf{e}_1(P_3) = (0, -1). \tag{3.22}$$

Assuming $\mathbf{v} = a\mathbf{e}_1$, the differentiation becomes:

$$\frac{\partial \mathbf{v}}{\partial x} = a\frac{\partial \mathbf{e}_1}{\partial x}. \tag{3.23}$$

3.2.1 Why Basis Changes Require a Connection

Thus, even if the vector coefficients are constant, the changing frame must be accounted for. This leads to the notion of **connection**, and the appropriate differentiation is known as the **covariant derivative**.

3.3 Curvature and the Need for Covariant Derivatives

3.3.1 *From Straight Lines to Curved Geometry*

One key difference between a curved domain and a Euclidean domain lies in the behavior of the covariant derivative. This difference is directly related to whether the coordinate axes remain **fixed** or **vary** from point to point within the domain. In a curved domain, the tangent vectors of the coordinate axes change their direction and possibly their magnitude, in contrast to the constant tangent vectors of a Cartesian coordinate system.

3.3.2 *Vector Transport and the Concept of Sameness*

Figure 3.3 compares (a) a curved line, (b) a Euclidean line, (c) a curved surface, and (d) a Euclidean plane. In a curved domain, the direction or magnitude of the tangent vectors associated with the coordinate axes may vary from point to point. Along a straight line, the tangent vector maintains a constant magnitude and direction. In the Euclidean plane, two orthonormal tangent vectors can be constructed at every point, and their magnitudes remain constant. In contrast to the Cartesian coordinate system, the varying axes and accordingly changing tangent vectors in a curved domain require new methodologies for comparing scalars and vectors between neighboring points.

Figure 3.4 illustrates the transport of a vector along (a) a straight line and (b) a circular arc. Consider a vector $\mathbf{v}$ initially placed at location $\mathbf{x}_0$, denoted by $\mathbf{v}(0)$. This vector can be transported from $\mathbf{x}_0$ to a neighboring point $\mathbf{x}_1$ along infinitely many possible paths. If the vector is transported along a straight line between $\mathbf{x}_0$ and $\mathbf{x}_1$, the resulting vector $\mathbf{v}(1)$ at $\mathbf{x}_1$ is identical to $\mathbf{v}(0)$, as shown in Fig. 3.4a. As is well established in vector calculus, there is no ambiguity: $\mathbf{v}(0)$ and $\mathbf{v}(1)$ are indeed the same.

Alternatively, we can transport $\mathbf{v}(0)$ along a circular arc from $\mathbf{x}_0$ to $\mathbf{x}_1$. Suppose the points are expressed in polar coordinates such that

$$\mathbf{x}_0 = (\cos\theta, \sin\theta), \tag{3.24}$$

where $\theta = 0$ for $\mathbf{x}_0$ and $\theta = \pi/2$ for $\mathbf{x}_1$. The vector $\mathbf{v}(0)$ is aligned along the tangent vector of the circular arc at $\mathbf{x}_0$. As we transport the vector along the circular arc to $\mathbf{x}_1$, we observe that the vector undergoes a 90° counterclockwise rotation.

Intuitively, $\mathbf{v}(1)$ should still represent the "same" vector as $\mathbf{v}(0)$ because there is no unnatural change in the vector's magnitude or direction relative to its local tangent frame during transport. Mathematically, they are indeed the same. However, a visual comparison reveals that $\mathbf{v}(1)$ forms a 90° angle with $\mathbf{v}(0)$. This discrepancy highlights the need for a **new perspective** in vector comparison: on a straight line, the visual and mathematical concepts of vector sameness coincide, but on a curved path, the **path** must be taken into account. This path defines the domain on which vector comparison should be based.

The key factor for vector comparison in a general domain is that every vector must be interpreted **relative to the local tangent basis vectors** at each point. Consider a straight line between $\mathbf{x}(0)$ and $\mathbf{x}(1)$, as shown in Fig. 3.3a. Here, the tangent vector is constant along the line, enabling $\mathbf{v}(0)$ and $\mathbf{v}(1)$ to be represented relative to the same tangent basis. Consequently, their magnitudes and directions relative to the tangent vector are identical.

In contrast, when transporting $\mathbf{v}(0)$ along a curved arc to $\mathbf{x}(1)$, $\mathbf{v}(1)$ has a different absolute direction compared to $\mathbf{v}(0)$. However, what remains unchanged is the relative angle between the vector and the local tangent vector at each point: the angle between $\mathbf{v}(0)$ and the tangent vector at $\mathbf{x}(0)$ is the same as the angle between $\mathbf{v}(1)$ and the tangent vector at $\mathbf{x}(1)$. In the example considered, this angle is zero throughout the transport, meaning that $\mathbf{v}(0)$ and $\mathbf{v}(1)$ are unchanged relative to the local tangent basis.

Differentiation in a curved domain must account for changes in the local tangent basis vectors. The covariant derivative captures this concept: it defines a differentiation of vectors that appropriately adjusts for the variation in the tangent basis across the domain.

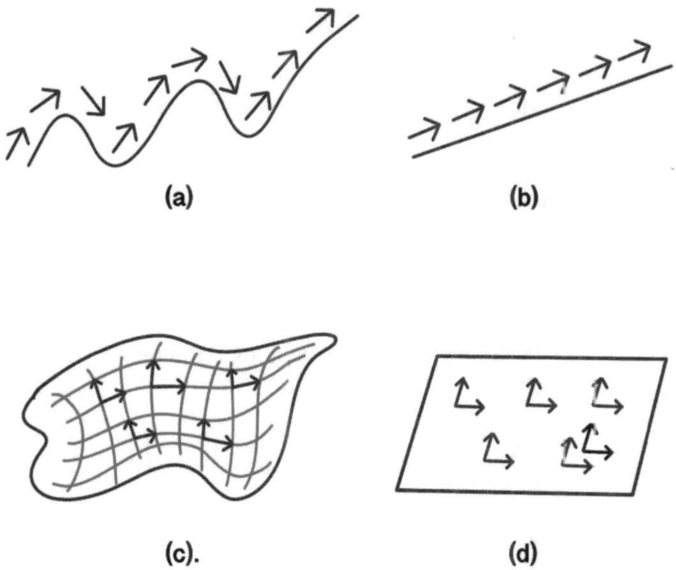

Fig. 3.3 Curved versus Euclidean domains. (**a**) One-dimensional curved domain, where tangent vectors vary along the curve. (**b**) One-dimensional Euclidean domain, where the tangent vector remains aligned in a constant direction. (**c**) Two-dimensional curved domain, where tangent vectors cannot be aligned along constant directions. (**d**) Two-dimensional Euclidean domain, where tangent vectors can be aligned along constant directions

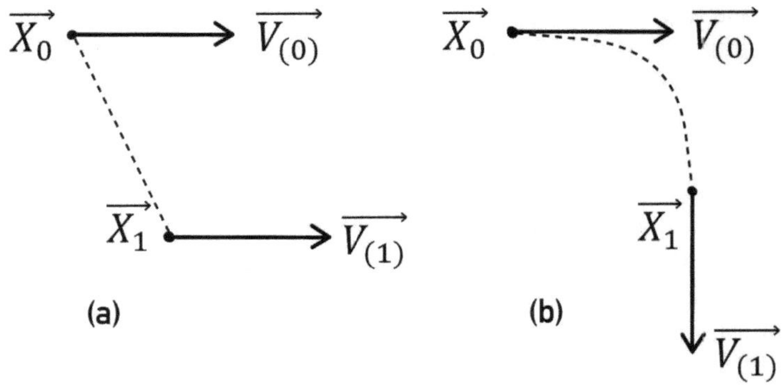

Fig. 3.4 Differentiation between two points depending on the connecting path. (**a**) When $\mathbf{X}_0$ and $\mathbf{X}_1$ are connected by a straight (Euclidean) line, the differentiation of vectors can be obtained by direct comparison. (**b**) When they are connected only by a curved path, the comparison between $\mathbf{V}_0$ and $\mathbf{V}_1$ must be made with respect to the tangent vectors obtained via parallel transport

3.4 Metric-Based Covariant Derivatives

Consider a vector $\mathbf{u}$ in a smooth curved domain Ω. Suppose there exists an orthogonal coordinate axis system x^i within the domain Ω. Then, the vector $\mathbf{u}$ can be expanded in terms of the tangent vectors of the axes as follows:

$$\mathbf{u} = \sum_{i=1}^{2} \widetilde{u}^i \frac{\partial}{\partial x^i} = \widetilde{u}^1 \frac{\partial}{\partial x^1} + \widetilde{u}^2 \frac{\partial}{\partial x^2}. \tag{3.25}$$

Let us define the differentiation of $\mathbf{u}$ with respect to x^j as:

$$\frac{\partial \mathbf{u}}{\partial x^j} \equiv \nabla_j \mathbf{u}. \tag{3.26}$$

Differentiating Eq. (3.25) with respect to x^j and applying the product rule yields

$$\nabla_j \mathbf{u} = \sum_{i=1}^{2} \left[\frac{\partial \widetilde{u}^i}{\partial x^j} \frac{\partial}{\partial x^i} + \widetilde{u}^i \frac{\partial}{\partial x^j} \left(\frac{\partial}{\partial x^i} \right) \right]. \tag{3.27}$$

The first term in Eq. (3.27) is aligned along the tangent vector $\partial/\partial x^i$ and thus lies within the tangent plane. The second term involves the differentiation of the tangent vector $\partial/\partial x^i$ along the x^j direction. The indices i and j may coincide. The variation of the tangent vector along x^j can be expressed as another vector; however, this resulting vector may not remain within the tangent plane. It can have a component orthogonal to the tangent plane, aligned along the surface normal vector, denoted by $\mathbf{k}$. Thus, the second term in Eq. (3.27) can be decomposed as

$$\frac{\partial^2}{\partial x^i \partial x^j} = \sum_{k=1}^{2} \Gamma_{ij}^{k} \frac{\partial}{\partial x^k} + \mathbf{k}, \tag{3.28}$$

where we have introduced a new symbol, Γ_{ij}^{k}, known as the **Christoffel symbol of the second kind**. The Christoffel symbol Γ_{ij}^{k} quantifies how the tangent vector $\partial/\partial x^i$ changes as it moves along the x^j direction. Eliminating the orthogonal component $\mathbf{k}$ in Eq. (3.28) corresponds to projecting the second term onto the tangent plane. Projecting Eq. (3.28) onto the surface domain yields the covariant derivative (Moore 2013):

$$\nabla_j \mathbf{u} = \sum_{i=1}^{2} \left[\frac{\partial \widetilde{u}^i}{\partial x^j} + \sum_{k=1}^{2} \Gamma_{kj}^{i} \widetilde{u}^k \right] \frac{\partial}{\partial x^i}, \tag{3.29}$$

For convenience, we have exchanged the indices i and k in the second term of Eq. (3.29), based on the following identity:

$$\sum_{i=1}^{2} \sum_{k=1}^{2} \Gamma_{ij}^{k} \widetilde{u}^i \frac{\partial}{\partial x^k} = \sum_{i=1}^{2} \sum_{k=1}^{2} \Gamma_{kj}^{i} \widetilde{u}^k \frac{\partial}{\partial x^i}. \tag{3.30}$$

Thus, computing the covariant derivative in Eq. (3.28) requires the metric tensor g_{ij} and its derivative, the Christoffel symbols Γ_{jk}^{i}. The metric tensor, denoted by g_{ij}, describes the **distance** between two points via

$$ds^2 = g_{ij} dx_i dx_j. \tag{3.31}$$

For grid point intervals dx_i and dx_j, the metric tensor g_{ij} provides the actual **length** between grid points. Without g_{ij}, the distance ds^2 reduces to the Euclidean distance formula:

$$ds^2 = \sqrt{dx^2 + dy^2 + dz^2}. \tag{3.32}$$

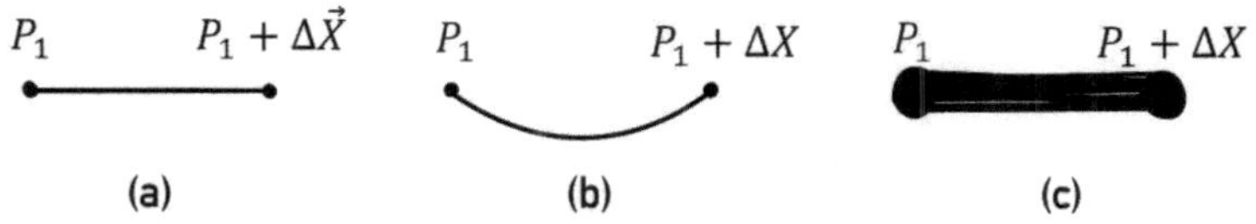

Fig. 3.5 Various paths and arrival times between two points. (**a**) Two points connected by the shortest straight line. (**b**) Two points connected by a curved path. (**c**) Two points connected by straight lines with increased conduction velocity

Suppose that a signal propagates between two grid points at a constant speed. Then, ds is proportional to the travel time between these points. Equation (3.32) applies to isotropic media in Euclidean space without any curvature. However, more complex geometries require different considerations. Figure 3.5 shows two cases: (1) a curved path representing the shortest distance between points in a curved 3D domain (Fig. 3.5b), and (2) a straight but "strengthened" path where the medium properties cause faster signal propagation (Fig. 3.5c). In both cases, Eq. (3.32) is insufficient, and the metric tensor g_{ij} must be used.

The metric tensor is defined as the dot product of the tangent vectors:

$$g_{ij} = \frac{\partial}{\partial x^i} \cdot \frac{\partial}{\partial x^j}, \tag{3.33}$$

The diagonal elements g_{ii} represent the square of the magnitude of the tangent vector along the x^i axis:

$$g_{ii} = \left\| \frac{\partial}{\partial x^i} \right\|^2, \quad 1 \leq i \leq n \tag{3.34}$$

For a two-dimensional domain, the metric tensor g_{ij} is

$$g_{ij} = \begin{bmatrix} g_{11} & g_{12} \\ g_{21} & g_{22} \end{bmatrix}. \tag{3.35}$$

The metric tensor is typically symmetric, satisfying $g_{12} = g_{21}$. For an orthogonal coordinate system, the off-diagonal terms vanish, resulting in

$$g_{ij} = \begin{bmatrix} g_{11} & 0 \\ 0 & g_{22} \end{bmatrix}. \tag{3.36}$$

The Christoffel symbols can be computed by differentiating the metric tensor:

$$\Gamma^k_{ij} = \sum_{m=1}^{2} \frac{g^{km}}{2} \left[\frac{\partial g_{mi}}{\partial x^j} + \frac{\partial g_{mj}}{\partial x^i} - \frac{\partial g_{ij}}{\partial x^m} \right], \tag{3.37}$$

where g^{ij} is the inverse of g_{ij}, given by

$$g^{ij} = \frac{1}{g_{11}g_{22} - g_{12}g_{21}} \begin{bmatrix} g_{22} & -g_{12} \\ -g_{21} & g_{11} \end{bmatrix}. \tag{3.38}$$

For an orthogonal coordinate system, we have

$$g^{ij} = \begin{bmatrix} \frac{1}{g_{11}} & 0 \\ 0 & \frac{1}{g_{22}} \end{bmatrix}. \tag{3.39}$$

For an orthonormal coordinate system, Eq. (3.37) simplifies to

$$\Gamma_{ij}^{k} = \frac{1}{2g_{kk}} \left[\frac{\partial g_{ki}}{\partial x^{j}} + \frac{\partial g_{kj}}{\partial x^{i}} - \frac{\partial g_{ij}}{\partial x^{k}} \right], \tag{3.40}$$

Swapping the indices i and j in Eq. (3.40) does not change the Christoffel symbol. If two indices are the same and the third is different, we obtain

$$\Gamma_{ii}^{k} = -\frac{1}{2g_{kk}} \frac{\partial g_{ii}}{\partial x_{k}}, \tag{3.41}$$

$$\Gamma_{ij}^{i} = \Gamma_{ji}^{i} = \frac{1}{2g_{ii}} \frac{\partial g_{ii}}{\partial x_{j}}. \tag{3.42}$$

Equations (3.41) and (3.42) confirm that the Christoffel symbols describe how the magnitude of the tangent vectors changes along different axes. When all indices are the same, some Christoffel symbol Γ_{ii}^{i} may vanishes due to the opposite signs in Eqs. (3.41) and (3.42), but this is not true in general.

3.5 [MATLAB] Computing Metric and Christoffel Symbols

The following file, *ch3MATLABa.m*, computes the metric tensor g_{ij} and Christoffel symbols Γ_{jk}^{i} on a spherical shell of radius 1.Spectral element methods using Gauss–Lobatto–Legendre (GLL) nodes are employed for the differentiation of tangent vectors along each coordinate axis (Fischer 2023). The m-files, *semhat.m* and *zwgll.m*, are provided in the Appendix.

```
function ch3MATLABa(Nphi, Ntheta, shape)
% ch3MATLABa.m
%
% This function computes and validates the numerical metric tensor and Christoffel
% symbols on a spherical coordinate grid using spectral element methods (SEM).
%
% Inputs:
%   Nphi   - Number of collocation points in the azimuthal (longitude) direction
%   Ntheta - Number of collocation points in the polar (latitude) direction
%
% Outputs:
%   Visualizations and numerical comparisons of metric tensor and Christoffel symbols
%
% Usage example:
%   ch3MATLABa(20, 20);

%% --- Initialization ---
close all; clc;

%% --- Parameter and Grid Setup ---
SetupParameter;  % Load parameters such as radius R, grid extents
```

```
% Generate spherical grid points (phi, theta) using SEM nodes
[phigrid, thetagrid, ~, ~] = Generategrid(Nphi, Ntheta);

% Create 2D meshgrid of spherical coordinates
[TH, ~] = meshgrid(thetagrid, phigrid); % Dimensions: Nphi x Ntheta

% Create grid points and tangent vectors
[grids,curvedcoord] = CreateGridTangentVec(Nphi, Ntheta, 0, shape);

%% --- Compute Numerical Metric Tensor and Christoffel Symbols ---
[gij, Gamma] = Computemetric(Nphi, Ntheta, curvedcoord);
% gij(i,j,a,b): metric tensor component g_ab at (phi_i, theta_j)
% Gamma(i,j,m,a,b): Christoffel symbol Gamma^m_{ab} at (phi_i, theta_j)

%% --- Compare Numerical Results with Analytical Solutions ---

% Analytical metric tensor components
if(shape=="sphere") % sphere
    metric_phi_phi_exact   = sin(TH).^2; % g_phi_phi = sin^2(theta)
    metric_theta_theta_exact = ones(size(TH)); % g_theta_theta = 1

elseif(shape=="pseudosphere")
    metric_phi_phi_exact   = sech(TH).^2; % g_phi_phi = sin^2(theta)
    metric_theta_theta_exact = tanh(TH).^2; % g_theta_theta = 1

end

% Compute absolute differences between numerical and analytical metrics
metric_diff_phi_phi    = abs(gij(:, :, 1, 1) - metric_phi_phi_exact);
metric_diff_theta_theta = abs(gij(:, :, 2, 2) - metric_theta_theta_exact);

% Compute L2 norms of errors
error_metric_phi_phi    = norm(metric_diff_phi_phi(:));
error_metric_theta_theta = norm(metric_diff_theta_theta(:));

% Display metric errors
fprintf('L2␣Error␣in␣g_{\\phi␣\\phi}:␣%e\n', error_metric_phi_phi);
fprintf('L2␣Error␣in␣g_{\\theta␣\\theta}:␣%e\n', error_metric_theta_theta);


%% --- Compare Numerical Christoffel Symbols with Analytical Expressions ---

% Analytical Christoffel symbols
if(shape=="sphere") % sphere
    Gamma_theta_thetatheta_exact = zeros(Nphip,Nthetap);  % \Gamma^t\heta_{\
        theta \theta}
    Gamma_theta_phiphi_exact = -sin(TH) .* cos(TH);  % \Gamma^\theta_{\phi \
        phi}
    Gamma_phi_thetaphi_exact = cot(TH);              % \Gamma^\phi_{\theta \
        phi}

elseif(shape=="pseudosphere")
    Gamma_theta_thetatheta_exact = sech(TH).*sech(TH)./tanh(TH);  % \Gamma^\
        theta_{\theta \theta}
    Gamma_theta_phiphi_exact = sech(TH).*sech(TH)./tanh(TH);  % \Gamma^\
        theta_{\phi \phi}
    Gamma_phi_thetaphi_exact = -tanh(TH);             % \Gamma^phi_{\theta
        \phi}

end

% Compute absolute differences
Gamma_diff_theta_thetatheta = abs(Gamma(:, :, 2, 2, 2) -
     Gamma_theta_thetatheta_exact);
Gamma_diff_theta_phiphi = abs(Gamma(:, :, 2, 1, 1) -
     Gamma_theta_phiphi_exact);
```

```
Gamma_diff_phi_thetaphi = abs(Gamma(:, :, 1, 2, 1) - Gamma_phi_thetaphi_exact);

% Compute normalized L2 errors
error_Gamma_theta_thetatheta = norm(Gamma_diff_theta_thetatheta(:)) / sqrt(numel(Gamma_diff_theta_thetatheta(:)));
error_Gamma_theta_phiphi = norm(Gamma_diff_theta_phiphi(:)) / sqrt(numel(Gamma_theta_phiphi_exact(:)));
error_Gamma_phi_thetaphi = norm(Gamma_diff_phi_thetaphi(:)) / sqrt(numel(Gamma_phi_thetaphi_exact(:)));

% Display Christoffel symbol errors
fprintf('Normalized␣L2␣Error␣in␣\\Gamma^{\\theta}_{\\theta␣\\theta}:␣%e\n', error_Gamma_theta_phiphi);
fprintf('Normalized␣L2␣Error␣in␣\\Gamma^{\\theta}_{\\phi␣\\phi}:␣%e\n', error_Gamma_theta_phiphi);
fprintf('Normalized␣L2␣Error␣in␣\\Gamma^{\\phi}_{\\theta␣\\phi}:␣%e\n', error_Gamma_phi_thetaphi);

%% --- Visualization of Christoffel Symbols ---

% Christoffel symbols for visualization
Gamma_phi_theta_phi   = Gamma(:, :, 1, 2, 1)'; % Gamma^phi_{theta phi}
Gamma_theta_phi_phi   = Gamma(:, :, 2, 1, 1)'; % Gamma^theta_{phi phi}

% Visualize Gamma^phi_{theta phi}
figure;
surf(grids(:,:,1), grids(:,:,2), grids(:,:,3), Gamma_phi_theta_phi, 'EdgeColor', 'none');
title('Christoffel␣Symbol:␣$\Gamma^{\phi}_{\theta␣\phi}$', 'Interpreter', 'latex');
xlabel('x'); ylabel('y'); zlabel('z');
axis equal; view(45,45);
lighting gouraud; camlight;
colormap(parula(256));
colorbar;

% Visualize Gamma^theta_{phi phi}
figure;
surf(grids(:,:,1), grids(:,:,2), grids(:,:,3), Gamma_theta_phi_phi, 'EdgeColor', 'none');
title('Christoffel␣Symbol:␣$\Gamma^{\theta}_{\phi␣\phi}$', 'Interpreter', 'latex');
xlabel('x'); ylabel('y'); zlabel('z');
axis equal; view(45,45);
lighting gouraud; camlight;
colormap(parula(256));
colorbar;

end
```

Listing 3.1 Setting up spherical coordinate system to compute metric tensor and Christoffel symbol

The following file, *Computemetric.m*, derives the metric tensor by computing inner products of tangent vectors. The Christoffel symbols are then obtained by differentiating the metric tensor along each coordinate direction using spectral element methods (Figs. 3.6 and 3.7).

```
function [gij, Gamma] = Computemetric(Nphi, Ntheta, curvedcoord)
% compute_metric.m
%
% This function computes the numerical metric tensor and selected Christoffel symbols
% on a spherical grid using spectral element methods (SEM).
```

```
%
% Inputs:
%   Nphi   - Number of discretization points along the azimuthal (longitude) direction
%   Ntheta - Number of discretization points along the polar (latitude) direction
%
% Outputs:
%   g_grid - Numerical metric tensor at each grid point
%            (dimensions: Nphi x Ntheta x 2 x 2)
%   Gamma  - Selected Christoffel symbols at each grid point
%            (dimensions: Nphi x Ntheta x 2 x 2 x 2)
%
% Usage example:
%   [g_grid, Gamma] = Computemetric(20, 20);

%% --- Initialization and Spectral Setup ---
SetupParameter;  % Load parameters and set grid dimensions

% Generate spectral differentiation matrices for spherical coordinates
[~, ~, Dbphi, Dbtheta] = Generategrid(Nphi, Ntheta);

%% --- Spectral Derivatives of Tangent Vectors ---
% Initialize arrays for spectral derivatives
de_phi_dphi     = zeros(Nphip, Nthetap, mfdim);
de_phi_dtheta   = zeros(Nphip, Nthetap, mfdim);
de_theta_dphi   = zeros(Nphip, Nthetap, mfdim);
de_theta_dtheta = zeros(Nphip, Nthetap, mfdim);

% Compute spectral derivatives along phi and theta directions
for k = 1:mfdim
    de_phi_dphi(:,:,k)     = Dbphi * squeeze(curvedcoord(:,:,1,k));
    de_theta_dphi(:,:,k)   = Dbphi * squeeze(curvedcoord(:,:,2,k));
    de_phi_dtheta(:,:,k)   = squeeze(curvedcoord(:,:,1,k)) * Dbtheta';
    de_theta_dtheta(:,:,k) = squeeze(curvedcoord(:,:,2,k)) * Dbtheta';
end

%% --- Compute Metric Tensor and Christoffel Symbols ---
% Initialize storage for metric tensor and Christoffel symbols
gij = zeros(Nphip, Nthetap, expdim, expdim);
Gamma  = zeros(Nphip, Nthetap, expdim, expdim, expdim);

for i = 1:Nphip
    for j = 1:Nthetap
        % Tangent vectors at current grid point
        e_phi   = squeeze(curvedcoord(i,j,1,:));   % e_phi vector
        e_theta = squeeze(curvedcoord(i,j,2,:));   % e_theta vector

        % Derivatives of tangent vectors at current grid point
        d_theta_e_theta = squeeze(de_theta_dtheta(i,j,:));
        d_theta_e_phi   = squeeze(de_phi_dtheta(i,j,:));
        d_phi_e_phi     = squeeze(de_phi_dphi(i,j,:));
        d_phi_e_theta   = squeeze(de_theta_dphi(i,j,:));

        %% --- Compute Metric Tensor Components ---
        g_phi_phi     = dot(e_phi, e_phi);
        g_phi_theta   = dot(e_phi, e_theta);
        g_theta_theta = dot(e_theta, e_theta);
        g = [g_phi_phi, g_phi_theta; g_phi_theta, g_theta_theta];

        gij(i,j,:,:) = g;

        %% --- Compute Derivatives of Metric Components ---
        d_g_phi_phi_dtheta     = 2 * dot(d_theta_e_phi, e_phi);
        d_g_phi_phi_dphi       = 2 * dot(d_phi_e_phi, e_phi);
        d_g_phi_theta_dtheta   = dot(d_theta_e_theta, e_phi) + dot(e_theta, d_theta_e_phi);
```

```
            d_g_phi_theta_dphi       = dot(d_phi_e_theta, e_phi) + dot(e_theta,
                d_phi_e_phi);
            d_g_theta_theta_dtheta = 2 * dot(d_theta_e_theta, e_theta);
            d_g_theta_theta_dphi     = 2 * dot(d_phi_e_theta, e_theta);

            %% --- Compute Christoffel Symbols using Koszul Formula ---
            % Christoffel symbol Gamma^theta_{phi phi}
            Gamma(i,j,2,1,1) = 0.5 / g_theta_theta * ...
                (2 * d_g_phi_theta_dtheta - d_g_phi_phi_dtheta);

            % Christoffel symbol Gamma^phi_{theta phi}
            Gamma(i,j,1,2,1) = 0.5 / g_phi_phi * ...
                (d_g_phi_phi_dtheta);
        end
    end

end
```

Listing 3.2 Differentiating the tangent vectors by spectral element methods

3.6 Example: Covariant Derivatives on the Sphere

A popular choice of coordinate axes for a spherical shell is the spherical coordinate system, with the azimuthal θ-axis and the polar ϕ-axis. The metric tensor and Christoffel symbols for the spherical coordinate system can be mathematically derived for the sphere. The covariant derivative of a vector $\mathbf{u}$ along the direction of a vector $\mathbf{a}$ is given as follows:

$$\nabla_{\mathbf{a}}\mathbf{u} = (\mathbf{a} \cdot \widehat{\boldsymbol{\theta}})\nabla_{\theta}\mathbf{u} + (\mathbf{a} \cdot \widehat{\boldsymbol{\phi}})\nabla_{\phi}\mathbf{u}, \tag{3.43}$$

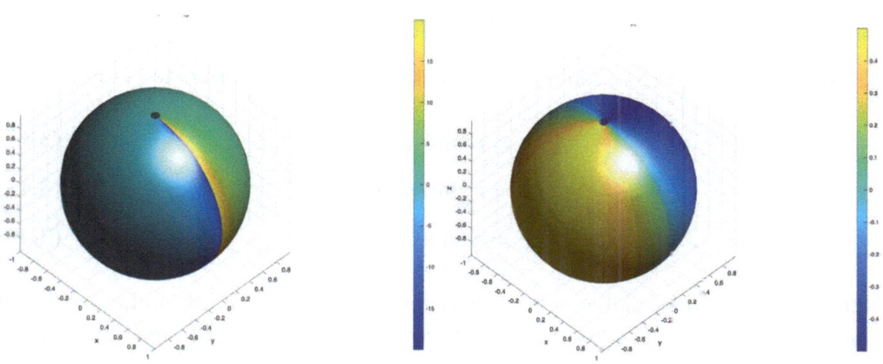

(a) Component of $\Gamma^{\phi}_{\theta\phi}$ (b) Component of $\Gamma^{\theta}_{\phi\phi}$

Fig. 3.6 Christoffel symbols associated with tangent vectors on the unit sphere, computed using spectral element methods. Contour maps of the components (**a**) $\Gamma^{\phi}_{\theta\phi}$ and (**b**) $\Gamma^{\theta}_{\phi\phi}$. The north pole is marked by the dark point

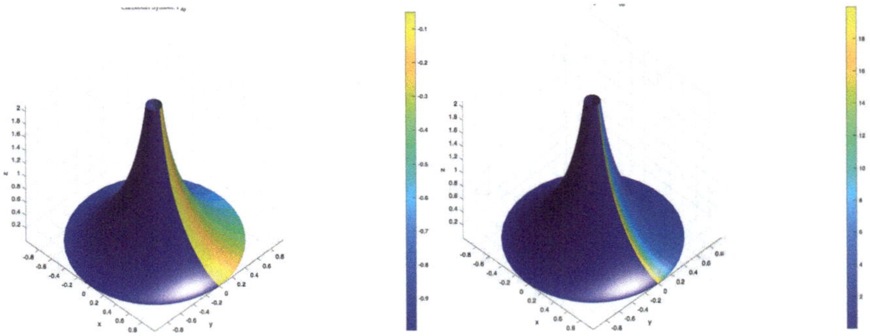

(a) Component of $\Gamma^{\phi}_{\theta\phi}$ (b) Component of $\Gamma^{\theta}_{\phi\phi}$

Fig. 3.7 Christoffel symbols associated with tangent vectors of the unit pseudosphere computed using spectral element methods. Contour maps of the components (**a**) $\Gamma^{\phi}_{\theta\phi}$ and (**b**) $\Gamma^{\theta}_{\phi\phi}$

where $\widehat{\boldsymbol{\theta}}$ and $\widehat{\boldsymbol{\phi}}$ are the tangent vectors of the θ and ϕ axes, respectively, defined as

$$\widehat{\boldsymbol{\theta}} = \frac{\partial}{\partial \theta} \bigg/ \left| \frac{\partial}{\partial \theta} \right|^2, \tag{3.44}$$

$$\widehat{\boldsymbol{\phi}} = \frac{\partial}{\partial \phi} \bigg/ \left| \frac{\partial}{\partial \phi} \right|^2. \tag{3.45}$$

The covariant derivatives $\nabla_\theta \mathbf{u}$ and $\nabla_\phi \mathbf{u}$ are computed by using Eq. (3.29) and the linearity of covariant derivatives:

$$\nabla_{a\mathbf{v}+b\mathbf{w}} \nabla \mathbf{u} = a \nabla_{\mathbf{v}} \mathbf{u} + b \nabla_{\mathbf{w}} \mathbf{u}. \tag{3.46}$$

For a spherical shell of radius r, consider the spherical coordinate system:

$$x = r \sin\theta \cos\phi, \tag{3.47}$$

$$y = r \sin\theta \sin\phi, \tag{3.48}$$

$$z = r \cos\theta. \tag{3.49}$$

Here, r is the radius of the sphere, θ is the polar angle, and ϕ is the azimuthal angle. The locations $\theta = 0$ and $\theta = \pi$ correspond to the north and south poles, respectively, while $\theta = \pi/2$ corresponds to the equator. Differentiating Eqs. (3.47)–(3.49) with respect to θ and ϕ yields the tangent vectors $\partial \mathbf{r}/\partial\theta$ and $\partial \mathbf{r}/\partial\phi$:

$$\frac{\partial \mathbf{r}}{\partial \theta} = \left(\frac{\partial x}{\partial \theta}, \frac{\partial y}{\partial \theta}, \frac{\partial z}{\partial \theta} \right) = (r\cos\theta\cos\phi, r\cos\theta\sin\phi, -r\sin\theta), \tag{3.50}$$

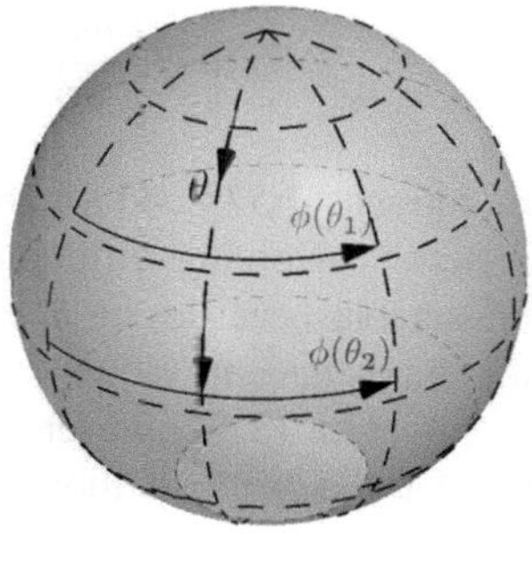

Fig. 3.8 Intuitive explanation for Christoffel symbols. The magnitude of the tangent vectors $\boldsymbol{\phi}$ changes along the θ-axis

$$\frac{\partial \mathbf{r}}{\partial \phi} = \left(\frac{\partial x}{\partial \phi}, \frac{\partial y}{\partial \phi}, \frac{\partial z}{\partial \phi}\right) = r \sin\theta(-\sin\phi, \cos\phi, 0). \tag{3.51}$$

The dot product of the tangent vectors yields the metric tensor g_{ij}:

$$g_{\theta\theta} = \frac{\partial \mathbf{r}}{\partial \theta} \cdot \frac{\partial \mathbf{r}}{\partial \theta} = r^2, \tag{3.52}$$

$$g_{\phi\phi} = \frac{\partial \mathbf{r}}{\partial \phi} \cdot \frac{\partial \mathbf{r}}{\partial \phi} = r^2 \sin^2\theta, \tag{3.53}$$

$$g_{\theta\phi} = g_{\phi\theta} = \frac{\partial \mathbf{r}}{\partial \theta} \cdot \frac{\partial \mathbf{r}}{\partial \phi} = 0. \tag{3.54}$$

The fact that $g_{\theta\phi} = g_{\phi\theta} = 0$ confirms that the spherical coordinate axes are orthogonal. Consequently, using Eqs. (3.41) and (3.42), the only nontrivial Christoffel symbols for constant r are derived as:

$$\Gamma^{\theta}_{\phi\phi} = -\frac{1}{2r^2}\frac{\partial(r^2\sin^2\theta)}{\partial\theta} = -\sin\theta\cos\theta. \tag{3.55}$$

$$\Gamma^{\phi}_{\phi\theta} = \frac{1}{2r^2\sin^2\theta}\frac{\partial(r^2\sin^2\theta)}{\partial\theta} = \frac{\cos\theta}{\sin\theta}. \tag{3.56}$$

The nontrivial Christoffel symbol $\Gamma^{\theta}_{\phi\phi}$ indicates that the θ-axis tangent vector changes with respect to motion along the ϕ-axis. Similarly, $\Gamma^{\phi}_{\phi\theta}$ indicates that the ϕ-axis tangent vector changes along the θ-axis (Fig. 3.8). Other Christoffel symbols vanish because the θ-axis tangent vector is constant along the θ-axis, and the ϕ-axis tangent vector is constant along the ϕ-axis.

The covariant derivatives along the θ- and ϕ-axis are given by:

$$\nabla_\theta \mathbf{u} = \frac{\partial u^\theta}{\partial\theta}\widehat{\boldsymbol{\theta}} + \left[\frac{\partial u^\phi}{\partial\theta} + \Gamma^{\phi}_{\theta\phi}u^\phi\right]\widehat{\boldsymbol{\phi}}, \tag{3.57}$$

$$\nabla_\phi \mathbf{u} = \left[\frac{\partial u^\theta}{\partial \phi} + \Gamma^\theta_{\phi\phi} u^\phi \right] \widehat{\boldsymbol{\theta}} + \left[\frac{\partial u^\phi}{\partial \phi} + \Gamma^\phi_{\theta\phi} u^\theta \right] \widehat{\boldsymbol{\phi}}. \tag{3.58}$$

Equations (3.57) and (3.58) show that the Christoffel symbols mainly appear in terms involving differentiation along or associated with the ϕ-axis, which is reasonable because the ϕ-axis tangent vector has a variable magnitude, whereas the θ-axis tangent vector is constant.

Equation (3.57) presents that the differentiation of u^θ along the θ-axis is the same as the differentiation in the Euclidean space. This happens because the θ-axis' tangent vector has a constant magnitude everywhere. Consider the vector $\mathbf{u}^\theta$ that is aligned along with the θ-axis as shown in Fig. 3.9a. The covariant differentiation of the vector $\mathbf{u}^\theta$ along the θ-axis is obtained without the Christoffel symbols as follows:

$$\nabla_\theta \mathbf{u}^\theta = \frac{\partial u^\theta}{\partial \theta} \widehat{\boldsymbol{\theta}}. \tag{3.59}$$

Equation (3.57) reveals that the differentiation of u^θ along the θ-axis mirrors differentiation in Euclidean space, because the θ-axis tangent vector has constant magnitude. For a vector $\mathbf{u}^\theta$ aligned along the θ-axis (Fig. 3.9a), the covariant derivative along θ-axis is simply:

$$\frac{\partial \mathbf{u}^\theta}{\partial \theta} = 0. \tag{3.60}$$

Similarly, consider a vector $\mathbf{u}^\phi$ aligned along the ϕ-axis, as shown in Fig. 3.10. The magnitude of the ϕ-axis tangent vector changes as it moves along the θ-axis.

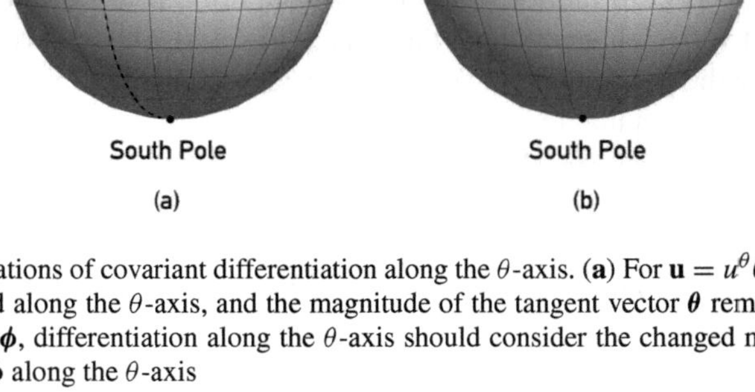

Fig. 3.9 Illustrations of covariant differentiation along the θ-axis. (**a**) For $\mathbf{u} = u^\theta \boldsymbol{\theta}$, the vector can be differentiated along the θ-axis, and the magnitude of the tangent vector $\boldsymbol{\theta}$ remains unchanged. (**b**) For $\mathbf{u} = u^\phi \boldsymbol{\phi}$, differentiation along the θ-axis should consider the changed magnitude of the tangent vector $\boldsymbol{\phi}$ along the θ-axis

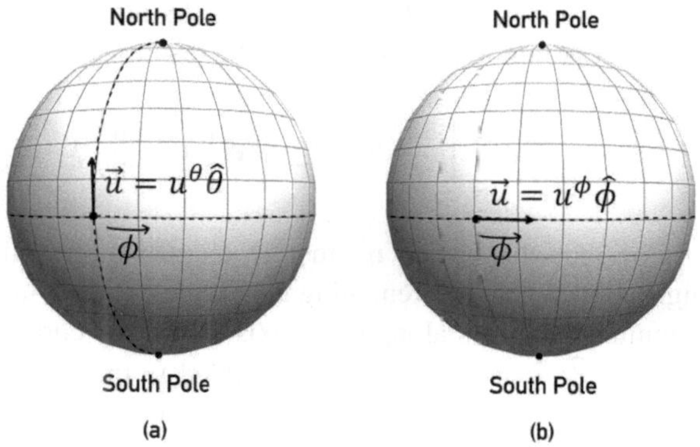

Fig. 3.10 Illustrations of covariant differentiation along the ϕ-axis. (**a**) For $\mathbf{u} = u^\theta \boldsymbol{\theta}$, the vector can be differentiated along the ϕ-axis, and the magnitude of the tangent vector $\boldsymbol{\theta}$ remains constant along the ϕ-axis. (**b**) For $\mathbf{u} = u^\phi \boldsymbol{\phi}$, differentiation along the ϕ-axis is similar because of the constant magnitude of the tangent vector $\boldsymbol{\phi}$ along the ϕ-axis

Suppose the coefficient u^ϕ remains constant along the θ-axis. Then, it implies that the vector $\mathbf{u}^\phi$ along ϕ-axis changes along the θ-axis:

$$\frac{\partial \mathbf{u}^\phi}{\partial \theta} \neq 0. \tag{3.61}$$

Alternatively, suppose that the vector $\mathbf{u}^\phi$ is constant along the θ direction:

$$\|\mathbf{u}^\phi\| = \text{constant along the } \theta\text{-axis}. \tag{3.62}$$

In this case, the coefficient u^ϕ must vary accordingly along the θ-axis:

$$\frac{\partial u^\phi}{\partial \theta} \neq 0. \tag{3.63}$$

Thus, the covariant differentiation of $\mathbf{u}^\phi$ along the θ-axis must be corrected by adding the term $\Gamma^\phi_{\theta\phi} u^\phi$, or equivalently $\cot\theta\, u^\phi$, to the scalar differentiation:

$$\nabla_\theta \mathbf{u}^\phi = \left[\frac{\partial u^\phi}{\partial \theta} + \frac{\cos\theta}{\sin\theta} u^\phi \right] \widehat{\boldsymbol{\phi}}. \tag{3.64}$$

Note that the direction of this covariant derivative remains aligned along the ϕ-axis. Consequently, the covariant differentiation of a general vector $\mathbf{u} = u^\theta \widehat{\boldsymbol{\theta}} + u^\phi \widehat{\boldsymbol{\phi}}$ generally produces components along both the θ and ϕ axes. This may initially seem odd because differentiation along one axis yields components in both directions, but it reflects the curvature of the domain.

Differentiation along the ϕ-axis is more complex, because the ϕ-axis tangent vector changes magnitude along the θ-axis. Compensation terms involving Christoffel symbols must be added to the scalar differentiation. Rearranging Eq. (3.58) groups terms according to the direction of the resulting components:

$$\nabla_{\phi}\mathbf{u} = \left[\frac{\partial u^{\theta}}{\partial \phi}\widehat{\boldsymbol{\theta}} + \Gamma^{\phi}_{\theta\phi}u^{\theta}\widehat{\boldsymbol{\phi}}\right] + \left[\frac{\partial u^{\phi}}{\partial \phi}\widehat{\boldsymbol{\phi}} + \Gamma^{\theta}_{\phi\phi}u^{\phi}\widehat{\boldsymbol{\theta}}\right]. \tag{3.65}$$

Consider the vector $\mathbf{u}^{\theta}$ that is aligned along with the θ-axis, as shown in Fig. 3.10a. When differentiating $\mathbf{u}^{\theta}$ with respect to ϕ-axis, the ϕ-axis' tangent vector magnitude changes along the θ-axis. Thus, the corresponding compensation term, or $\Gamma^{\phi}_{\theta\phi}u^{\theta}\widehat{\boldsymbol{\phi}} = \cot\theta u^{\theta}$, should be added to the scalar differentiation along the ϕ-axis. This compensation is the same as shown for the covariant differentiation of the vector $\mathbf{u}^{\phi}$ along the θ-axis. Thus, the covariant differentiation of the vector $\mathbf{u}^{\theta}$ is given as follows:

$$\nabla_{\phi}\mathbf{u} = \left[\frac{\partial u^{\theta}}{\partial \phi}\widehat{\boldsymbol{\theta}} + \frac{\cos\theta}{\sin\theta}u^{\theta}\widehat{\boldsymbol{\phi}}\right]. \tag{3.66}$$

Here, the first term is aligned along the θ-axis, and the second term is aligned along the ϕ-axis.

$$\nabla_{\phi}\mathbf{u} = \left[\frac{\partial u^{\phi}}{\partial \phi}\widehat{\boldsymbol{\phi}} - \frac{\sin 2\theta}{2}u^{\phi}\widehat{\boldsymbol{\theta}}\right]. \tag{3.67}$$

The first term corresponds to the expected scalar differentiation of the u^{ϕ} component along the ϕ-axis, aligned with the ϕ-axis. However, the second term is perhaps less intuitive: differentiation of a vector aligned along the ϕ-axis produces a component in the θ direction.

This can be explained by recognizing that the covariant derivative acts on a *vector field*. Differentiating all vectors in the field along the ϕ-axis yields scalar derivatives of u^{ϕ}, but it also captures changes in the magnitude of the ϕ-axis tangent vectors with respect to θ, as shown in Fig. 3.11.

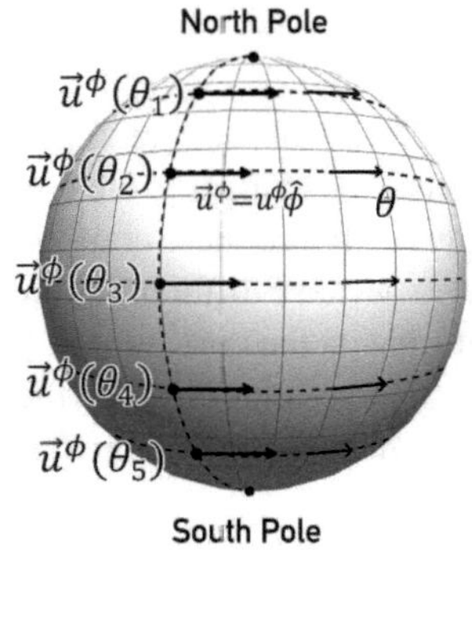

Fig. 3.11 Field-wise covariant derivative along the ϕ-axis. The magnitude of the tangent vector $\boldsymbol{\phi}$ decreases from the equator ($\theta = 0$) toward the poles ($\theta = \pm\pi/2$). Consequently, for a constant vector field $\mathbf{u} = u^{\phi}\boldsymbol{\phi}$, the component u^{ϕ} increases from the equator toward the poles

3.7 Covariant Derivatives via Orthonormal Moving Frames

The differentiation of a vector on a curved domain is well established and has been applied to various fields of science and engineering. However, critical challenges remain regarding the derivation and computation of the metric tensor g_{ij} and Christoffel symbols Γ^i_{jk} for realistic complex geometries. To derive these geometric components, a global and continuous coordinate system for the entire domain must first be constructed. Computing the accurate tangent vector lengths along a curved axis is one of the most challenging computational tasks, although very fine discretization may mitigate this difficulty. Nevertheless, finding a universal coordinate system for the entire domain is often extremely difficult, or even impossible, for realistic complex geometries.

Alternatively, orthonormal moving frames can be adopted to compute covariant derivatives without explicitly computing the metric tensor and Christoffel symbols. However, two main issues must be considered in advance regarding covariant derivatives in moving frames.

The first issue is that the Christoffel symbols, which represent the change of basis vectors as shown in Eq. (3.37), can only be computed using *coordinate frames*, also known as *holonomic frames*. A coordinate frame refers to an actual coordinate reference system associated with the domain. The coordinate system and its corresponding tangent vectors are required to compute the Christoffel symbols. However, moving frames with unit-length vectors on a curved domain are not holonomic frames. In general, the coordinate system of a domain does not have to be orthogonal nor composed of unit-length vectors. Orthonormal moving frames are holonomic only in Euclidean space. Thus, an alternative approach is necessary to represent the change of basis vectors on a curved domain.

The second issue concerns the accuracy of covariant derivatives computed using moving frames. Even with an alternative measure representing the change of basis, the covariant derivative in moving frames is fundamentally designed for locally Euclidean spaces. A locally Euclidean space implies that the magnitudes of tangent vectors along the curved axis should remain nearly constant.

Consider a moving frame $\mathbf{e}_i$ placed at a point P on the unit sphere, as illustrated in Fig. 3.12. The two tangent vectors of the moving frame lie on the tangent plane at the point P. Suppose the first tangent vector $\mathbf{e}_1$ is aligned along the θ-axis; then the second tangent vector $\mathbf{e}_2$ naturally aligns with the ϕ-axis because the spherical coordinate system is orthogonal. As shown in Fig. 3.12b, the grid points used for differentiating a vector lie on the tangent plane. However, the actual physical grid points for each axis lie on the curved surface. Therefore, differentiation *on* the moving frames has a different meaning from differentiation *along* the curved axes, even when projections along the surface normal direction are applied. This discrepancy leads to a deterioration of computational accuracy.

In the following sections, we examine these issues in greater depth and propose a new scheme for moving frames to address these two challenges, thereby enabling accurate covariant differentiation with moving frames (Chun 2023a).

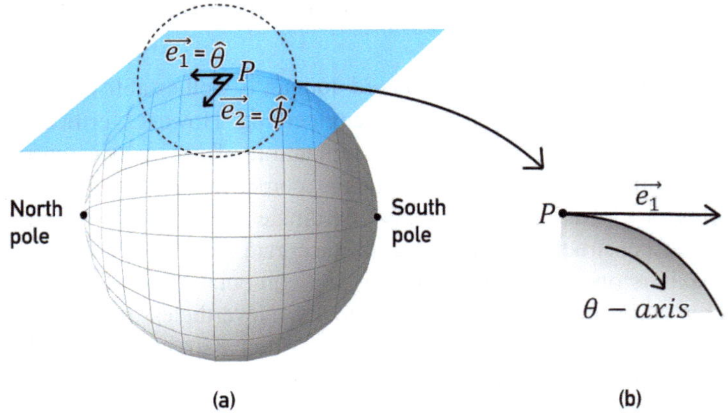

Fig. 3.12 Illustration of moving frames on a sphere aligned with the spherical coordinate system. The unit tangent vectors $\mathbf{e}_1$ and $\mathbf{e}_2$ are oriented along the θ- and ϕ-axis directions, respectively

3.7.1 Connection 1-Forms as Alternatives to Christoffel Symbols

Consider a sufficiently smooth two-dimensional curved surface, denoted by Ω. Let Ω_e be a tessellation of the surface Ω such that

$$\Omega = \cup \Omega_e \;\text{ and }\; \Omega_i \cap \Omega_j = \emptyset \;\text{ if } i \neq j, \tag{3.68}$$

where the empty set $\emptyset$ indicates that the common area between different elements is zero, except for shared lines or points. Figure 3.13a illustrates the tessellation of a sphere. Suppose that each element is locally Euclidean. Moving frames $\mathbf{e}_i$ are constructed at each point such that each tangent vector $\mathbf{e}_i$ $(1 \leq i \leq 2)$ is differentiable within each element. Let x^i denote the axis corresponding to the tangent vector $\mathbf{e}_i$. Figure 3.13b shows that a tessellated element on the sphere is a curved element such that the sum of its interior angles is less than 2π.

Consider a vector $\mathbf{u}$ on an element Ω_e. Expand the vector $\mathbf{u}$ in the tangent vectors $\mathbf{e}_i$, which are orthogonal to the surface normal vector, as follows:

$$\mathbf{u} = u^1\mathbf{e}_1 + u^2\mathbf{e}_2. \tag{3.69}$$

Recall that the moving frames are orthonormal. Thus, the magnitude of the curved axis's tangent vector is incorporated into the coefficients u^i. Differentiating Eq. (5.80) with respect to the axis x^j, we obtain

$$\frac{\partial \mathbf{u}}{\partial x^j} = \sum_{i=1}^{2} \left[(\nabla u^i \cdot \mathbf{e}_j)\mathbf{e}_i + u^i \frac{\partial \mathbf{e}_i}{\partial x^j} \right], \tag{3.70}$$

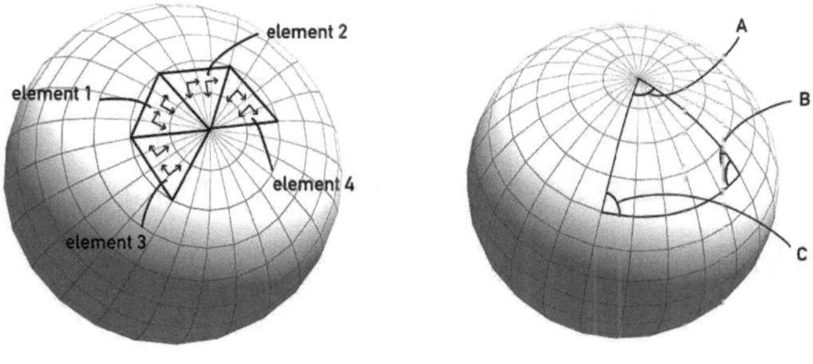

(a) (b)

Fig. 3.13 (**a**) Example of a triangular tessellation of the sphere. Moving frames are continuous and differentiable within each element, but not necessarily across element interfaces. (**b**) Element edges may be straight in general but can curved for effective tessellation. Illustrated is a curved triangular tessellation on a sphere where $\measuredangle A + \measuredangle B + \measuredangle C < 2\pi$

The first term in Eq. (3.70) is the directional derivative, as in Euclidean space. The second term corresponds to changes in the coordinate system and is analogous to the covariant derivative involving Christoffel symbols, as shown in Eq. (3.29). However, a different method of measuring the change of basis is required, since the orthonormal tangent vectors $\mathbf{e}_i$ do not form a holonomic basis. In this formulation, the magnitude of the tangent vector is incorporated into the coefficient u^i, rather than the basis vectors themselves.

The change of the moving frames can be measured by expanding the variation of each frame in terms of the moving frame itself (Cartan 2001a,b; Needham 2021). Consider a small change in the moving frame, $d\mathbf{e}_i$. Since this is a vector-valued 1-form, the variation has both direction and magnitude. Then, the change can be expanded as

$$d\mathbf{e}_1 = \omega_{11}\mathbf{e}_1 + \omega_{12}\mathbf{e}_2 + \omega_{13}\mathbf{e}_3, \tag{3.71}$$

$$d\mathbf{e}_2 = \omega_{21}\mathbf{e}_1 + \omega_{22}\mathbf{e}_2 + \omega_{23}\mathbf{e}_3, \tag{3.72}$$

$$d\mathbf{e}_3 = \omega_{31}\mathbf{e}_1 + \omega_{32}\mathbf{e}_2 + \omega_{33}\mathbf{e}_3, \tag{3.73}$$

where each ω_{ij} is a scalar-valued 1-form, known as the **connection 1-form**. The notation ω_{ij}, popular in modern literature (O'Neill 1997; Needham 2021), is equivalent to Cartan's classical notation ω^i_j (Cartan 2001a, 1937; Akivis and Rosenfeld 2011). The orthonormality of the moving frames implies the following properties for $1 \le i, j \le 3$:

$$\omega_{ii} = 0, \tag{3.74}$$

$$\omega_{ij} = -\omega_{ji}. \tag{3.75}$$

Thus, Eqs. (3.71)–(3.73) can be written in matrix form as

$$\begin{bmatrix} d\mathbf{e}_1 \\ d\mathbf{e}_2 \\ d\mathbf{e}_3 \end{bmatrix} = \begin{bmatrix} 0 & \omega_{12} & \omega_{13} \\ -\omega_{12} & 0 & \omega_{23} \\ -\omega_{13} & -\omega_{23} & 0 \end{bmatrix} \begin{bmatrix} \mathbf{e}_1 \\ \mathbf{e}_2 \\ \mathbf{e}_3 \end{bmatrix}, \tag{3.76}$$

or equivalently,

$$d[\mathbf{e}] = [\omega][\mathbf{e}], \tag{3.77}$$

where the brackets indicate matrices. This formulation arises from the first Cartan structure equation, representing how frame fields vary infinitesimally over space.

In the Cartesian coordinate system (x, y, z), each moving frame $\mathbf{e}_i$ is expanded as

$$\mathbf{e}_i = e_i^x \mathbf{x} + e_i^y \mathbf{y} + e_i^z \mathbf{z}, \qquad 1 \leq i \leq 3. \tag{3.78}$$

which in matrix form becomes

$$\begin{bmatrix} \mathbf{e}_1 \\ \mathbf{e}_2 \\ \mathbf{e}_3 \end{bmatrix} = \begin{bmatrix} e_1^x & e_1^y & e_1^z \\ e_2^x & e_2^y & e_2^z \\ e_3^x & e_3^y & e_3^z \end{bmatrix} \begin{bmatrix} \mathbf{x} \\ \mathbf{y} \\ \mathbf{z} \end{bmatrix}, \tag{3.79}$$

or we have

$$[\mathbf{e}] = [A][\mathbf{x}\ \mathbf{y}\ \mathbf{z}]^T, \tag{3.80}$$

where $[\mathbf{x}\ \mathbf{y}\ \mathbf{z}]$ represents the basis vectors of the Cartesian coordinates. The matrix $[A]$, called the *attitude matrix* (O'Neill 1997; Needham 2021; Piuze et al. 2015), consists of the coefficients of the moving frames. Since the moving frames are orthonormal, the matrix $[A]$ satisfies

$$[A]^{-1} = [A]^T. \tag{3.81}$$

Thus, Eq. (3.80) implies

$$[\mathbf{x}\ \mathbf{y}\ \mathbf{z}]^T = [A]^T [\mathbf{e}]. \tag{3.82}$$

In the following proposition, we propose that the covariant derivative can be equivalently computed with the connection 1-forms. The covariant derivative in orthonormal moving frames can avoid explicit use of the metric tensor or Christoffel symbols, as these are implicitly encoded in the derivatives of the frames themselves. For coordinate frames or special geometries like the sphere, the connection 1-forms and Christoffel symbols coincide. In general, they must be related via frame transformation laws:

Proposition 3.1 *The covariant derivative involving the Christoffel symbols Γ^i_{jk} as in Eq.* (3.29) *can be equivalently computed using the connection 1-forms ω_{ij}:*

$$\nabla_{\mathbf{e}_j}\mathbf{u} = \sum_{i=1}^{2}\left[\nabla u^i \cdot \mathbf{e}_j + \sum_{k=1}^{2}\omega_{ik}(\mathbf{e}_j)u^k\right]\mathbf{e}_i. \tag{3.83}$$

The covariant derivative along an arbitrary vector $\mathbf{v}$ *is given by*

$$\nabla_{\mathbf{v}}\mathbf{u} = v^1\nabla_{\mathbf{e}_1}\mathbf{u} + v^2\nabla_{\mathbf{e}_2}\mathbf{u}, \tag{3.84}$$

where $v^i = \mathbf{v}\cdot\mathbf{e}_i$. Thus, the covariant derivatives computed via Eqs. (3.29) *and* (3.83) *are equivalent.*

Proof By differentiating both sides of Eq. (3.80), we obtain

$$d[\mathbf{e}] = d[A][\mathbf{x}\ \mathbf{y}\ \mathbf{z}]^T. \tag{3.85}$$

Using Eqs. (3.82) and (3.85) becomes

$$d[\mathbf{e}] = d[A][A]^T[\mathbf{e}]. \tag{3.86}$$

By comparing Eq. (3.77) with Eq. (3.86), we obtain

$$[\omega] = d[A][A]^T. \tag{3.87}$$

In matrix form, Eq. (3.87) reads

$$\begin{bmatrix} 0 & \omega_{12} & \omega_{13} \\ -\omega_{12} & 0 & \omega_{23} \\ -\omega_{13} & -\omega_{23} & 0 \end{bmatrix} = \begin{bmatrix} de_1^x & de_1^y & de_1^z \\ de_2^x & de_2^y & de_2^z \\ de_3^x & de_3^y & de_3^z \end{bmatrix}\begin{bmatrix} e_1^x & e_2^x & e_3^x \\ e_1^y & e_2^y & e_3^y \\ e_1^z & e_2^z & e_3^z \end{bmatrix}. \tag{3.88}$$

Component-wise, for $1 \le i, j \le 3$, Eq. (3.88) becomes

$$\omega_{ij} = de_i^x e_j^x + de_i^y e_j^y + de_i^z e_j^z. \tag{3.89}$$

Note that (de^x, de^y, de^z) and ω_{ij} are all 1-forms. Thus, for a specific tangent vector $\mathbf{e}_k$, we have

$$\omega_{ij}(\mathbf{e}_k) = de_i^x(\mathbf{e}_k)e_j^x + de_i^y(\mathbf{e}_k)e_j^y + de_i^z(\mathbf{e}_k)e_j^z. \tag{3.90}$$

The components $de_i^x(\mathbf{e}_k)$, $de_i^y(\mathbf{e}_k)$, and $de_i^z(\mathbf{e}_k)$ can be computed by

$$de_i^x(\mathbf{e}_k) = \nabla e_i^x \cdot \mathbf{e}_k, \tag{3.91}$$

$$de_i^y(\mathbf{e}_k) = \nabla e_i^y \cdot \mathbf{e}_k, \tag{3.92}$$

$$de_i^z(\mathbf{e}_k) = \nabla e_i^z \cdot \mathbf{e}_k. \tag{3.93}$$

Define the Jacobian matrix

$$J(\mathbf{e}_i) = \begin{bmatrix} \nabla e_i^x \\ \nabla e_i^y \\ \nabla e_i^z \end{bmatrix}. \tag{3.94}$$

Then, the above relations can be compactly written as

$$d\mathbf{e}_i(\mathbf{e}_k) = \nabla J(\mathbf{e}_i) \cdot \mathbf{e}_k. \tag{3.95}$$

Thus, the component $\omega_{ij}(\mathbf{e}_k)$ can be expressed as

$$\omega_{ij}(\mathbf{e}_k) = [e_j^x \; e_j^y \; e_j^z] \begin{bmatrix} \nabla e_i^x \cdot \mathbf{e}_k \\ \nabla e_i^y \cdot \mathbf{e}_k \\ \nabla e_i^z \cdot \mathbf{e}_k \end{bmatrix} = \mathbf{e}_j \cdot (\nabla J(\mathbf{e}_i) \cdot \mathbf{e}_k). \tag{3.96}$$

Consequently, the derivative $\partial \mathbf{e}_i / \partial x^j$ in Eq. (3.70) is obtained as

$$\frac{\partial \mathbf{e}_i}{\partial x^j} = d[\mathbf{e}_i](\mathbf{e}_j) = d[A(\mathbf{e}_j)][A]^T[\mathbf{e}_i]. \tag{3.97}$$

By substituting Eq. (3.97) into Eq. (3.70), we recover the expression for the covariant derivative along $\mathbf{e}_j$ given in Eq. (3.83). □

The magnitude of the tangent vector of the axis is not required in Eq. (3.83), as it is implicitly incorporated into the coefficient u^i. The tangent vectors are needed only initially, to compute the vector coefficients along each axis. These vector coefficients can also be obtained directly from analytic expressions or experimental data.

Thus, the Christoffel symbols, being derivatives of the metric tensor, are not generally necessary. However, the connection 1-forms ω_{ij}, representing the derivatives of the moving frames, naturally emerge in Eq. (3.83). Therefore, the covariant derivative in Eq. (3.83) is conceptually similar to that in Eq. (3.4).

In discrete orthonormal systems, connection 1-forms can be directly computed via frame derivatives, bypassing symbolic differentiation of the metric tensor. The connection 1-form is computed by the directional derivative along the unit tangent vector $\mathbf{e}_j$, as shown in Eq. (3.96). In contrast, computing the Christoffel symbols requires first constructing a continuous curved coordinate system and then differentiating the metric tensor along the curved axes, as shown in Eq. (3.37).

Thus, the covariant derivative formulation using Eq. (3.83) can be readily applied to complex two-dimensional surfaces, including irregular and non-convex geometries, without requiring a metric tensor or Christoffel symbols.

3.8 Connecting Moving Frames and the Metric Tensor

Novel numerical schemes for a high-order direct covariant derivative are formulated on curved surfaces. Construct moving frames $\{\mathbf{e}_i\}$ on a smooth curved surface Ω such that $\mathbf{e}_1$ and $\mathbf{e}_2$ lie on the tangent plane, while $\mathbf{e}_3$ is aligned with the surface normal vector. The velocity vectors $\mathbf{u}$ and $\mathbf{v}$ on Ω are expanded in terms of the moving frames as follows:

$$\mathbf{u} = u^1\mathbf{e}_1 + u^2\mathbf{e}_2, \tag{3.98}$$

$$\mathbf{v} = v^1\mathbf{e}_1 + v^2\mathbf{e}_2. \tag{3.99}$$

The covariant derivative $\nabla_{\mathbf{v}}\mathbf{u}$ is transformed using the linearity of the covariant derivative:

$$\nabla_{\mathbf{v}}\mathbf{u} = \nabla_{v^1\mathbf{e}_1+v^2\mathbf{e}_2}\mathbf{u} = v^1\nabla_{\mathbf{e}_1}\mathbf{u} + v^2\nabla_{\mathbf{e}_2}\mathbf{u}. \tag{3.100}$$

This can be rewritten as

$$\nabla_{\mathbf{v}}\mathbf{u} = \sum_{i=1}^{2} v^i\nabla_{\mathbf{e}_i}\mathbf{u}. \tag{3.101}$$

Alternatively, it can be expressed as

$$\nabla_{\mathbf{v}}\mathbf{u} = \sum_{j=1}^{2} v^j\left[(\nabla_{\mathbf{e}_j}\mathbf{u}\cdot\mathbf{e}_1)\mathbf{e}_1 + (\nabla_{\mathbf{e}_j}\mathbf{u}\cdot\mathbf{e}_2)\mathbf{e}_2\right]. \tag{3.102}$$

Substituting Eq. (3.83) into Eq. (3.102), we obtain

$$\nabla_{\mathbf{v}}\mathbf{u} = \sum_{i,j=1}^{2} v^j\left[\nabla u^i\cdot\mathbf{e}_j + \sum_{k=1}^{2}\omega_{ik}(\mathbf{e}_j)u^k\right]\mathbf{e}_i. \tag{3.103}$$

Each covariant derivative along the unit tangent vectors is given by

$$\nabla_{\mathbf{e}_1}\mathbf{u}\cdot\mathbf{e}_1 = \nabla u^1\cdot\mathbf{e}_1 - \omega_{21}(\mathbf{e}_1)u^2, \tag{3.104}$$

$$\nabla_{\mathbf{e}_1}\mathbf{u}\cdot\mathbf{e}_2 = \nabla u^2\cdot\mathbf{e}_1 + \omega_{21}(\mathbf{e}_1)u^1, \tag{3.105}$$

$$\nabla_{\mathbf{e}_2}\mathbf{u}\cdot\mathbf{e}_1 = \nabla u^1\cdot\mathbf{e}_2 - \omega_{21}(\mathbf{e}_2)u^2, \tag{3.106}$$

$$\nabla_{\mathbf{e}_2}\mathbf{u}\cdot\mathbf{e}_2 = \nabla u^2\cdot\mathbf{e}_2 + \omega_{21}(\mathbf{e}_2)u^1. \tag{3.107}$$

For example, the covariant derivative of the unit tangent vector $\mathbf{e}_i$ is obtained by setting $v^i = u^j = 1$:

$$\nabla_{\mathbf{e}_i}\mathbf{e}_j = \sum_{k=1}^{2} \omega_{ij}(\mathbf{e}_k)\mathbf{e}_k = \omega_{ij}(\mathbf{e}_1)\mathbf{e}_1 + \omega_{ij}(\mathbf{e}_2)\mathbf{e}_2. \tag{3.108}$$

Explicitly:

$$\nabla_{\mathbf{e}_1}\mathbf{e}_1 = 0, \tag{3.109}$$

$$\nabla_{\mathbf{e}_1}\mathbf{e}_2 = \omega_{12}(\mathbf{e}_1)\mathbf{e}_1 + \omega_{12}(\mathbf{e}_2)\mathbf{e}_2, \tag{3.110}$$

$$\nabla_{\mathbf{e}_2}\mathbf{e}_1 = -\omega_{12}(\mathbf{e}_1)\mathbf{e}_1 - \omega_{12}(\mathbf{e}_2)\mathbf{e}_2, \tag{3.111}$$

$$\nabla_{\mathbf{e}_2}\mathbf{e}_2 = 0. \tag{3.112}$$

Note that Eqs. (3.109) and (3.112) indicate that the magnitudes of $\mathbf{e}_i$ remain constant along their own directions, which holds due to the orthonormality of the moving frames.

Suppose that a two-dimensional curved surface possesses orthonormal tangent vectors at each point. Then, the covariant derivative expressions in Eqs. (3.29) and (3.83) are equivalent:

$$\Gamma^i_{kj} = \omega_{ik}(\mathbf{e}_j). \tag{3.113}$$

However, for a general two-dimensional curved surface, Eq. (3.113) does not generally hold.

3.9 Verifying Equivalence on the Sphere

We now examine whether the two different formulations of the covariant derivative given in Eqs. (3.29) and (3.83) yield the same results. For simplicity, we consider the unit sphere, represented as

$$(x, y, z) = (\sin\theta\cos\phi, \sin\theta\sin\phi, \cos\theta), \tag{3.114}$$

where θ is the polar angle and ϕ is the azimuthal angle. The infinitesimal length is computed as

$$ds^2 = d\theta^2 + \sin^2\theta d\phi^2. \tag{3.115}$$

Let the tangent vectors be denoted by $\partial/\partial\theta$ and $\partial/\partial\phi$. A velocity vector $\mathbf{u}$ is expanded in the spherical coordinate system as

$$\mathbf{u} = \widetilde{u}^{\theta}\frac{\partial}{\partial\theta} + \widetilde{u}^{\phi}\frac{\partial}{\partial\phi}. \tag{3.116}$$

Note that $\|\partial/\partial\theta\| = 1$ and $\|\partial/\partial\phi\| = \sin\theta$. The metric tensor components are obtained via the inner product:

$$g_{\theta\theta} = 1, \quad g_{\phi\phi} = \sin^2\theta. \tag{3.117}$$

To construct orthonormal moving frames $\{\mathbf{e}_i\}$, align $\mathbf{e}_\phi$ and $\mathbf{e}_\theta$ along $\partial/\partial\phi$ and $\partial/\partial\theta$, respectively, and normalize them to unit length. This set is referred to as the **spherical moving frames**. Spherical moving frames refer to orthonormal vectors aligned with the spherical coordinate axes after normalization, typically used in radial settings. Similarly to Eq. (3.116), the vector $\mathbf{u}$ can be expanded in spherical moving frames as

$$\mathbf{u} = u^{\theta}\mathbf{e}_\theta + u^{\phi}\mathbf{e}_\phi, \tag{3.118}$$

with the following relations:

$$\begin{bmatrix}\mathbf{e}_\theta\\ \mathbf{e}_\phi\end{bmatrix} = \begin{bmatrix}\partial/\partial\theta,\\ (1/\sin\theta)\partial/\partial\phi\end{bmatrix}, \tag{3.119}$$

$$\begin{bmatrix}u^\theta\\ u^\phi\end{bmatrix} = \begin{bmatrix}\widetilde{u}^\theta,\\ \sin\theta\,\widetilde{u}^\phi\end{bmatrix}. \tag{3.120}$$

The covariant derivative along the θ-axis contains nontrivial Christoffel symbols and can be obtained as follows:

$$\nabla_\theta\mathbf{u} = \frac{\partial\widetilde{u}^\theta}{\partial\theta}\frac{\partial}{\partial\theta} + \left(\frac{\partial\widetilde{u}^\phi}{\partial\theta} + \Gamma^{\phi}_{\theta\phi}\widetilde{u}^\phi\right)\frac{\partial}{\partial\phi}, \tag{3.121}$$

where the Christoffel symbol $\Gamma^{\phi}_{\theta\phi}$ is computed as

$$\Gamma^{\phi}_{\theta\phi} = \sum_{m=1}^{2}\frac{g^{\phi m}}{2}\left(\frac{\partial g_{\theta m}}{\partial\phi} + \frac{\partial g_{\phi m}}{\partial\theta} - \frac{\partial g_{\theta\phi}}{\partial x^m}\right) = \frac{\cos\theta}{\sin\theta}. \tag{3.122}$$

Similarly, using Eq. (3.83), the covariant derivative along the unit tangent vector $\mathbf{e}_\theta$ is expressed as

$$\nabla_{\mathbf{e}_\theta}\mathbf{u} = (\nabla u^\theta\cdot\mathbf{e}_\theta)\mathbf{e}_\theta + \left(\nabla u^\phi\cdot\mathbf{e}_\theta\right)\mathbf{e}_\phi. \tag{3.123}$$

The corresponding connection 1-form $\omega_{\phi\theta}(\mathbf{e}_\phi)$ can be computed using Eq. (3.96). The unit tangent vectors of the spherical moving frames are

$$\begin{bmatrix} \mathbf{e}_r \\ \mathbf{e}_\theta \\ \mathbf{e}_\phi \end{bmatrix} = \begin{bmatrix} \sin\theta\cos\phi & \sin\theta\sin\phi & \cos\theta \\ \cos\theta\cos\phi & \cos\theta\sin\phi & -\sin\theta \\ -\sin\phi & \cos\phi & 0 \end{bmatrix} \begin{bmatrix} \mathbf{x} \\ \mathbf{y} \\ \mathbf{z} \end{bmatrix}.$$

Thus, the connection 1-form is

$$\omega_{\theta\phi}(\mathbf{e}_\phi) = \frac{1}{\sin\theta}(-\sin\phi\ \cos\phi\ 0)\begin{bmatrix} -\cos\theta\sin\phi \\ \cos\theta\cos\phi \\ 0 \end{bmatrix} = \frac{\cos\theta}{\sin\theta}. \tag{3.124}$$

Swapping the indices of ω_{ij} yields

$$\omega_{\theta\phi}(\mathbf{e}_\phi) = -\omega_{\phi\theta}(\mathbf{e}_\phi). \tag{3.125}$$

Comparing Eq. (3.121) with Eq. (3.123) leads to the following corollary:

Corollary 3.1 *For the unit sphere, the covariant derivatives of* **u** *along the* θ*-axis computed using the Christoffel symbol in Eq.* (3.121) *are identical to those computed using the connection 1-form in Eq.* (3.123).

Proof Substituting Eqs. (3.119) and (3.120) into Eq. (3.121), the term $\Gamma^{\phi}_{\theta\phi}\widetilde{u}^{\phi}$ cancels, yielding

$$\nabla_\theta \mathbf{u} = (\nabla u^\theta \cdot \mathbf{e}_\theta)\mathbf{e}_\theta + \left(\nabla u^\phi \cdot \mathbf{e}_\theta\right)\mathbf{e}_\phi, \tag{3.126}$$

Comparing with Eq. (3.123), we conclude

$$\left\|\nabla_\theta \mathbf{u} - \nabla_{\mathbf{e}_\theta}\mathbf{u}\right\| = 0. \tag{3.127}$$

□

Similarly, the covariant derivative along the ϕ-axis in spherical coordinates is given by

$$\nabla_\phi \mathbf{u} = \left(\frac{\partial \widetilde{u}^\theta}{\partial\phi} + \Gamma^{\theta}_{\phi\phi}\widetilde{u}^\phi\right)\frac{\partial}{\partial\theta} + \left(\frac{\partial \widetilde{u}^\phi}{\partial\phi} + \Gamma^{\phi}_{\theta\phi}\widetilde{u}^\theta\right)\frac{\partial}{\partial\phi}, \tag{3.128}$$

where $\Gamma^{\phi}_{\theta\phi}$ is given by Eq. (3.122). Similarly, $\Gamma^{\theta}_{\phi\phi}$ is computed as

$$\Gamma^{\theta}_{\phi\phi} = \sum_{m=1}^{2}\frac{g^{\theta m}}{2}\left(\frac{\partial g_{\phi m}}{\partial\phi} + \frac{\partial g_{\phi m}}{\partial\phi} - \frac{\partial g_{\phi\phi}}{\partial x^m}\right) = -\cos\theta\sin\theta. \tag{3.129}$$

Using Eq. (3.83), the covariant derivative along $\mathbf{e}_\phi$ is expressed as

$$\nabla_{\mathbf{e}_\phi}\mathbf{u} = \left(\nabla u^\theta \cdot \mathbf{e}_\phi + \omega_{\theta\phi}(\mathbf{e}_\phi)u^\phi\right)\mathbf{e}_\theta + \left(\nabla u^\phi \cdot \mathbf{e}_\phi + \omega_{\phi\theta}(\mathbf{e}_\phi)u^\theta\right)\mathbf{e}_\phi, \tag{3.130}$$

where $\omega_{\theta\phi}(\mathbf{e}_\phi)$ and $\omega_{\phi\theta}(\mathbf{e}_\phi)$ are given by Eq. (3.124).

This yields the following corollary:

Corollary 3.2 *For the unit sphere, the covariant derivatives of* **u** *along the* ϕ*-axis computed using the Christoffel symbol in Eq.* (3.128) *are equivalent to those computed using the connection 1-form in Eq.* (3.130), *up to a scaling by* $\sin\theta$.

Proof Substituting Eqs. (3.119) and (3.120) into Eq. (3.128) and applying the linearity of the covariant derivative, Eq. (3.128) transforms into

$$\nabla_\phi\mathbf{u} = \sin\theta\left[\left(\nabla u^\theta \cdot \mathbf{e}_\phi - \frac{\cos\theta}{\sin\theta}u^\phi\right)\mathbf{e}_\theta + \left(\nabla u^\phi \cdot \mathbf{e}_\phi + \frac{\cos\theta}{\sin\theta}u^\theta\right)\mathbf{e}_\phi\right]. \tag{3.131}$$

Comparing Eq. (3.131) with Eq. (3.130), we obtain

$$\left\|\frac{1}{\sin\theta}\nabla_\phi\mathbf{u} - \nabla_{\mathbf{e}_\phi}\mathbf{u}\right\| = 0. \tag{3.132}$$

□

Using Corollaries 3.1 and 3.2, we establish the following proposition:

Proposition 3.2 *On the unit sphere, the covariant derivative of* **u** *along an arbitrary vector* **v**, *computed using the Christoffel symbols in Eq.* (3.29), *is equivalent to that computed using the connection 1-forms in Eq.* (3.83).

Proof Expand the vector **v** as

$$\mathbf{v} = \tilde{v}^\theta\,\partial/\partial\theta + \tilde{v}^\phi\,\partial/\partial\phi = v^\theta\mathbf{e}_\theta + v^\phi\mathbf{e}_\phi, \tag{3.133}$$

where the relations between coefficients are $\widetilde{v^\theta} = v^\theta$ and $\widetilde{v^\phi} = v^\phi/\sin\theta$. The covariant derivative of **u** along **v** using the Christoffel symbol is

$$\nabla_{\tilde{v}^\theta\partial/\partial\theta+\tilde{v}^\phi\partial/\partial\phi}\mathbf{u}, \tag{3.134}$$

while using the connection 1-form, it is

$$\nabla_{v^\theta\mathbf{e}_\theta+v^\phi\mathbf{e}_\phi}\mathbf{u}. \tag{3.135}$$

The difference between Eqs. (3.134) and (3.135) is estimated as

$$\begin{aligned}&\left\|v^\theta\nabla_{\partial/\partial\theta}\mathbf{u} + (v^\phi/\sin\theta)\nabla_{\partial/\partial\phi}\mathbf{u} - v^\theta\nabla_{\mathbf{e}_\theta}\mathbf{u} + v^\phi\nabla_{\mathbf{e}_\phi}\mathbf{u}\right\|\\ &\quad\le \|v^\theta\|\left\|\nabla_{\partial/\partial\theta}\mathbf{u} - \nabla_{\mathbf{e}_\theta}\mathbf{u}\right\| + \|v^\phi\|\left\|(1/\sin\theta)\nabla_{\partial/\partial\phi}\mathbf{u} - \nabla_{\mathbf{e}_\phi}\mathbf{u}\right\| = 0.\end{aligned}$$

where we have used Corollaries 3.1 and 3.2. □

3.10 [MATLAB] Computing Connection 1-Forms

The following code, *ch3MATLABb.m*, computes the connection of the orthonormal moving frames aligned along the spherical coordinate system. Spectral element methods using Gauss–Lobatto–Legendre (GLL) nodes are employed for the differentiation of tangent vectors along each coordinate axis (Fischer 2023).

```
function ch3MATLABb(Nphi, Ntheta, shape)
% ch3MATLABb.m
%
% This function computes and validates the connection 1-form on a spherical
% coordinate grid using spectral element methods (SEM).
%
% Inputs:
%   Nphi   - Number of discretization points in the azimuthal (longitude) direction
%   Ntheta - Number of discretization points in the polar (latitude) direction
%   alpha  - Interpolation factor controlling frame deformation (optional, default: 0)
%
% Outputs:
%   Visualization of selected components of the connection 1-form on the sphere
%
% Usage example:
%   ch3MATLABb(20, 20, 0.1);

%% --- Initialization ---
close all; clc;

% Set default value for alpha if not provided
if nargin < 3
    alpha = 0.0;
end

%% --- Parameter and Grid Setup ---
SetupParameter;  % Ensure global parameters are consistent (radius, grid dimensions)

% Generate spectral grid points for spherical coordinates (phi, theta)
[phigrid, thetagrid, ~, ~] = Generategrid(Nphi, Ntheta);

% Create 2D meshgrid for spherical coordinates
[~, TH] = meshgrid(phigrid, thetagrid);

%% --- Construct Moving Frame Field ---
% Create grid points and tangent vectors
[grids,movingframes] = CreateGridTangentVec(Nphi, Ntheta, 1, shape);

%% --- Compute Connection 1-Form ---
% Compute the connection 1-form from the numerical derivatives of moving frames
wij_k = Computeconnection(Nphi, Ntheta, shape, movingframes);

%% --- Analytical Comparisons (Optional Validation) ---
% Analytical expressions for selected components
if(shape=="sphere")
    w21_1_exact = cos(TH') ./ sin(TH');   % omega_{21}^1 = cot(theta)
    w21_2_exact = zeros(Nphip,Nthetap);   % omega_{21}^2 = 0.0
    w31_1_exact = ones(Nphip,Nthetap);        % omega_{31}^1 = 1/r^2

elseif(shape=="pseudosphere")
    w21_1_exact = -1.*ones(Nphip,Nthetap);   % omega_{21}^1 = 1.0
```

```
    w21_2_exact = zeros(Nphip,Nthetap);   % omega_{21}^2 = 0.0
    w31_1_exact = 1./sqrt( sech(TH').*sech(TH') + (TH'-tanh(TH')).*(TH'-tanh(TH')) );          % omega_{31}^1 constant
end

% Compute numerical errors
error_w21_1 = norm(wij_k(:,:,2,1,1) - w21_1_exact, 'fro') / sqrt(numel(w21_1_exact));
error_w21_2 = norm(wij_k(:,:,2,1,2) - w21_2_exact, 'fro') / sqrt(numel(w21_2_exact));
error_w31_1 = norm(wij_k(:,:,3,1,1) - w31_1_exact, 'fro') / sqrt(numel(w31_1_exact));

% Display numerical errors
fprintf('L2 Error in \\omega_{21}^1: %e\n', error_w21_1);
fprintf('L2 Error in \\omega_{21}^2: %e\n', error_w21_2);
fprintf('L2 Error in \\omega_{31}^1: %e\n', error_w31_1);


%% --- Visualization of Selected Components of the Connection 1-Form ---

% Component omega_{21}^1 of connection 1-form
figure;
surf(grids(:,:,1), grids(:,:,2), grids(:,:,3), wij_k(:,:,2,1,1), 'EdgeColor', 'none');
title('Connection 1-form: $\omega_{21}^1$', 'Interpreter', 'latex');
xlabel('x'); ylabel('y'); zlabel('z');
axis equal; view(45,45);
lighting gouraud; camlight;
colormap(parula(256));
colorbar;

% Component omega_{31}^1 of connection 1-formfigure;
figure;
surf(grids(:,:,1), grids(:,:,2), grids(:,:,3), wij_k(:,:,3,1,1), 'EdgeColor', 'none');
title('Connection 1-form: $\omega_{31}^1$', 'Interpreter', 'latex');
xlabel('x'); ylabel('y'); zlabel('z');
axis equal; view(45,45);
lighting gouraud; camlight;
colormap(parula(256));
colorbar;


end
```

Listing 3.3 Setting up spherical coordinate system to compute metric tensor and Christoffel symbol

The following file, *Computeconnection.m*, computes the connection 1-form for Eqs. (3.90) or (3.96) by differentiating using spectral element methods (Figs. 3.14 and 3.15).

```
function wij_k = Computeconnection(Nphi, Ntheta, shape, movingframes)
% compute_connection.m
%
% This function computes the connection 1-forms using moving frames and
% spectral element method (SEM) derivatives on a spherical grid.
%
% Inputs:
%   Ntheta        - Number of discretization points in the polar direction
%   Nphi          - Number of discretization points in the azimuthal direction
%   movingframes  - Array containing moving frame vectors at each grid point
%                   Dimensions: [Nphi+1 x Ntheta+1 x mfdim x spacedim]
```

```
%
% Outputs:
%   wij_k - Connection 1-form at each grid point
%           Dimensions: [Nphi+1 x Ntheta+1 x mfdim x spacedim x mfdim]
%
% Usage example:
%   wij_k = compute_connection(20, 20, movingframes);

%% --- Parameter and Grid Setup ---
SetupParameter;  % Load global parameters such as radius, grid dimensions

% Generate SEM grids and differentiation matrices
[~, ~, Dbphi, Dbth] = Generategrid(Nphi, Ntheta);

% Compute metric tensor
[~,curvedcoord] = CreateGridTangentVec(Nphi, Ntheta, 0, shape);
[gij, ~] = Computemetric(Nphi, Ntheta, curvedcoord);

%% --- Allocate Arrays for Jacobians and Connection 1-forms ---
MFJacobian = zeros(Nphi+1, Ntheta+1, mfdim, spacedim, expdim);   % Derivatives of moving frame vectors
wij_k = zeros(Nphi+1, Ntheta+1, mfdim, spacedim, mfdim);    % Connection 1-forms

%% --- Compute Derivatives of Moving Frame Vectors ---
for i = 1:mfdim
    for k = 1:spacedim
        frame_vec = squeeze(movingframes(:,:,i,k));
        % Derivative w.r.t. azimuthal direction (phi), scaled by 1/sin(theta)
        MFJacobian(:,:,i,k,1) = (Dbphi * frame_vec) ./ sqrt(gij(:,:,1,1));
        % Derivative w.r.t. polar direction (theta)
        MFJacobian(:,:,i,k,2) = frame_vec * Dbth' ./ sqrt(gij(:,:,2,2));
    end
end

%% --- Compute Connection 1-Forms via Inner Products ---
for indi = 1:mfdim            % Output index i (form superscript)
    for indj = 1:spacedim     % Component direction j
        for indk = 1:expdim   % Input index k (form subscripts)
            fj = squeeze(movingframes(:, :, indj, :));              % Frame vector j
            tmp = squeeze(MFJacobian(:, :, indi, :, indk));         % Derivative of i-th frame w.r.t. k-th direction
            wij_k(:, :, indi, indj, indk) = sum(fj .* tmp, 3);      % Inner product across spatial components
        end
    end
end

end
```

Listing 3.4 Differentiating the tangent vectors by spectral element methods

3.11 Summary

This chapter developed a foundational framework for computing **covariant derivatives** on curved surfaces using both traditional coordinate-based approaches and a modern reformulation using **orthonormal moving frames**. Key concepts and computational methodologies were established.

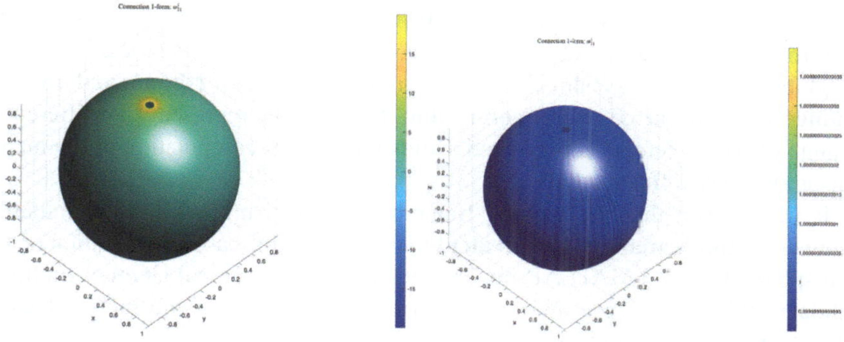

(a) (b)

Fig. 3.14 Computation of the connection 1-form of the orthonormal moving frames aligned along the spherical coordinate system using spectral element methods. (**a**) Component of ω_{11}^{2}, (**b**) Component of ω_{11}^{3}. The north pole is marked by the dark point

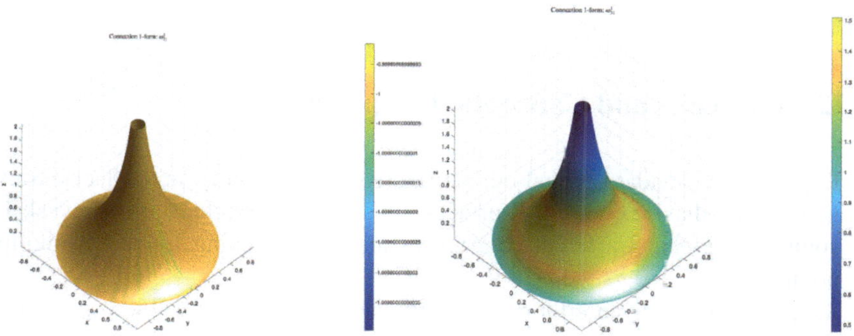

(a) (b)

Fig. 3.15 Computation of the connection 1-form of the orthonormal moving frames aligned along the pseudospherical coordinate system using spectral element methods. (**a**) Component of ω_{11}^{2}, (**b**) Component of ω_{11}^{3}

The covariant derivative of a vector field was expressed via **Christoffel symbols**, derived from the **metric tensor** and its derivatives. This formulation accurately captures the geometric effects of curvature but requires constructing a continuous, global coordinate system—an often impractical task for realistic geometries. By comparing **parallel transport**, **curved vs. Euclidean geometries**, and **directional derivatives** along surface coordinates (e.g., on the sphere), the chapter illustrated how basis vectors change and how these variations influence the differentiation of vectors.

On the unit sphere, the metric tensor and Christoffel symbols were explicitly computed in spherical coordinates. Covariant derivatives were illustrated through examples involving θ- and ϕ-axis-aligned vector fields. These examples emphasized the geometric significance of nontrivial Christoffel symbols like $\Gamma^{\theta}_{\phi\phi}$ and $\Gamma^{\phi}_{\theta\phi}$ and their role in expressing curvature-induced correction terms. To overcome limitations of global coordinate systems, a new approach based on **orthonormal moving frames** was introduced. Here, **connection 1-forms** ω_{ij} represent the differential change of basis vectors. This method avoids the explicit computation of the metric tensor and Christoffel symbols.

It was proven that the moving frame-based covariant derivative formulation using ω_{ij} is **mathematically equivalent** to the Christoffel-based formulation on the unit sphere. Detailed MATLAB implementations using **spectral element methods** validated this equivalence through numerical computation of the **metric tensor**, **Christoffel symbols**, and **connection 1-forms**.

This chapter lays the mathematical and computational foundation for applying **moving frame techniques** to partial differential equations on arbitrary curved surfaces. The methods developed here facilitate accurate and efficient differentiation even on geometries that defy traditional coordinate representations, making them essential tools for the numerical modeling of PDEs in complex domains.

3.12 Exercises and Conceptual Applications

1. Explain why standard partial derivatives (e.g., $\partial u/\partial x$) may not reflect intrinsic geometric behavior on curved surfaces. Illustrate your explanation with a sketch comparing global Cartesian coordinates and local moving frames at multiple points.
2. Suppose you are given orthonormal frame fields $\{\mathbf{e}_1, \mathbf{e}_2, \mathbf{e}_3\}$ defined on a curved surface. Derive the connection 1-forms ω_{ij} such that:

$$d\mathbf{e}_i = \sum_j \omega_{ij}\mathbf{e}_j.$$

   How does this compare to computing Christoffel symbols using the metric tensor?
3. Given a scalar field u defined on a curved surface and a moving frame $\{\mathbf{e}_1, \mathbf{e}_2\}$, express the gradient of u in terms of the directional derivatives along $\mathbf{e}_1$ and $\mathbf{e}_2$. How would you approximate these numerically?
4. Let $\mathbf{v} = v^1\mathbf{e}_1 + v^2\mathbf{e}_2$ be a vector field in a moving frame basis. Derive an expression for the covariant derivative $\nabla_{\mathbf{e}_i}\mathbf{v}$ in terms of the components v^j and the connection 1-forms ω_{jk}.
5. What does it mean for a frame to be holonomic? Show that if the commutator $[\mathbf{e}_i, \mathbf{e}_j] \neq 0$, then the frame is non-holonomic. Why does this matter in defining covariant derivatives?
6. Implement a finite difference approximation of the covariant derivative of a scalar field on a spherical surface using moving frames. What challenges arise from pole singularities or non-uniform frame alignment?

Chapter 4
Differential Operators in Moving Frames

It is a capital mistake to theorize before one has data. Insensibly one begins to twist facts to suit theories, instead of theories to suit facts.

—Sherlock Homes, "A Scandal in Bohemia"

This chapter presents test problems demonstrating that LOC (Local Orthonormal Coordinate) moving frames with the connection 1-form (ω_{ij}) can compute the covariant derivative with sufficient accuracy. The proposed scheme is applicable to any 2D curved surface, including irregular or non-convex geometries. For convenience in evaluating convergence error, we adopt the unit sphere with the following metric (Chun and Oh 2022). The invariant square of an infinitesimal length, known as the **interval**, is given by:

$$ds^2 = d\theta^2 + \sin^2\theta d\phi^2. \tag{4.1}$$

A steady flow modified from the classical steady zonal flow on the unit sphere is constructed (Williamson et al. 1992). The velocity vector of the zonal flow is multiplied by $\sin\theta$ to avoid division by zero at the poles ($\theta = 0, \pi$). For a vector field on the sphere, consider the following expansion in moving frames:

$$\mathbf{u} = u^\theta \mathbf{e}_\theta + u^\phi \mathbf{e}_\phi = u^x \mathbf{x} + u^y \mathbf{y} + u^z \mathbf{z}, \tag{4.2}$$

The components of the vector in spherical coordinates are derived as follows:

$$u^\theta = -u_0 \sin\theta \sin\phi \sin\alpha, \tag{4.3}$$

$$u^\phi = u_0 \sin\theta(\cos\theta\cos\alpha + \sin\theta\cos\phi\sin\alpha), \tag{4.4}$$

S. Chun, *Moving Frames for the Numerical Solution of Partial Differential Equations in Complex Domains*, Springer Asia Pacific Mathematics Series 7,
https://doi.org/10.1007/978-981-95-4495-0_4

where α denotes the angle between the velocity flow and the north pole. Let α be arbitrary. The components in Cartesian coordinates are given by:

$$u^x = u^\theta \cos\theta \cos\phi - u^\phi \sin\phi, \tag{4.5}$$

$$u^y = u^\theta \cos\theta \sin\phi + u^\phi \cos\phi, \tag{4.6}$$

$$u^z = -u^\theta \sin\theta. \tag{4.7}$$

The exact covariant derivatives along the spherical coordinate axes are expressed as:

$$\nabla_{\boldsymbol{\theta}} \mathbf{u} = \nabla_\theta u^\theta \mathbf{e}_\theta + \nabla_\theta u^\phi \mathbf{e}_\phi, \tag{4.8}$$

$$\nabla_{\boldsymbol{\phi}} \mathbf{u} = \left[\nabla_\phi u^\theta - \frac{\cos\theta}{\sin\theta} u^\phi\right] \mathbf{e}_\theta + \left[\nabla_\phi u^\phi + \frac{\cos\theta}{\sin\theta} u^\theta\right] \mathbf{e}_\phi, \tag{4.9}$$

with directional derivatives computed as:

$$\nabla_\theta u^\theta = -u_0 \cos\theta \sin\phi \sin\alpha, \tag{4.10}$$

$$\nabla_\theta u^\phi = u_0(\cos 2\theta \cos\alpha + \sin 2\theta \cos\phi \sin\alpha), \tag{4.11}$$

$$\nabla_\phi u^\theta = -u_0 \cos\phi \sin\alpha, \tag{4.12}$$

$$\nabla_\phi u^\phi = -u_0 \sin\theta \sin\phi \sin\alpha. \tag{4.13}$$

Two cases are considered: $\alpha = 0$ and $\alpha = \pi/2$. The corresponding velocity vector distributions are illustrated in Fig. 4.1. Table 4.1 presents the L_2 norm of the covariant derivative error, showing nearly exponential convergence as p increases. The LOCSPH moving frame scheme outperforms the spherical frame

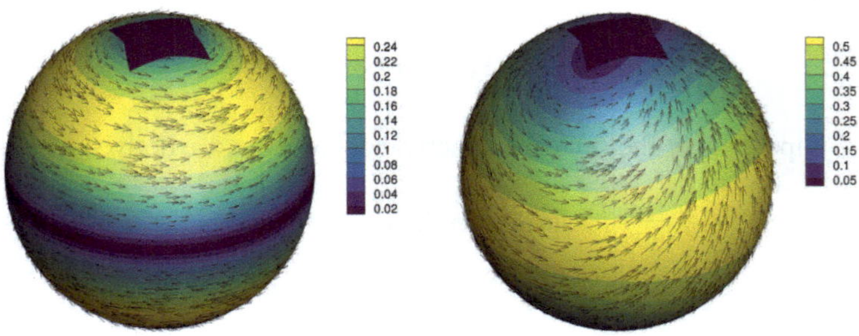

(a) $\alpha = 0$ (b) $\alpha = \pi/2$

Fig. 4.1 Distributions of velocity vectors for various angles α in the test problem. Contours show the corresponding velocity magnitude. Mesh elements near the poles are omitted to prevent singularities in the spherical coordinate system. (**a**) $\alpha = 0$. (**b**) $\alpha = \pi/2$

Table 4.1 Covariant derivative error in the L_2 norm for the vector field in Eqs. (4.3) and (4.4), evaluated along θ and ϕ directions. *SPH* spherical moving frames, *LOCSPH* local orthonormal coordinate spherical moving frames

	p	$\nabla_{\mathbf{e}_\theta}$ SPH	$\nabla_{\mathbf{e}_\theta}$ LOCSPH	$\nabla_{\mathbf{e}_\phi}$ SPH	$\nabla_{\mathbf{e}_\phi}$ LOCSPH
$\alpha = 0$	3	4.59×10^{-3}	1.51×10^{-4}	1.63×10^{-3}	2.43×10^{-4}
	5	9.62×10^{-4}	8.14×10^{-7}	2.31×10^{-4}	1.09×10^{-6}
	7	1.95×10^{-4}	3.22×10^{-9}	3.91×10^{-5}	4.05×10^{-9}
	9	4.09×10^{-5}	1.78×10^{-11}	5.87×10^{-6}	1.46×10^{-11}
$\alpha = \frac{\pi}{2}$	3	8.77×10^{-4}	3.19×10^{-3}	3.38×10^{-3}	4.66×10^{-3}
	5	1.84×10^{-4}	7.30×10^{-4}	6.75×10^{-4}	8.89×10^{-4}
	7	3.11×10^{-5}	1.40×10^{-4}	1.68×10^{-4}	2.19×10^{-4}
	9	5.94×10^{-6}	3.09×10^{-5}	3.46×10^{-5}	4.53×10^{-5}

(SPH) scheme for $\alpha = 0$, while for $\alpha = \pi/2$, SPH achieves better accuracy. These variations are attributed to the relative alignment accuracy of grid points along the respective moving frames. Despite mesh differences, both methods exhibit exponential convergence, and the difference becomes negligible for sufficiently large p.

For subsequent test problems, a more complex velocity field is introduced. This field resembles the Rossby–Haurwitz wave, commonly used as a benchmark in shallow water equations:

$$\mathbf{v} = v_\phi \mathbf{e}_\phi + v_\theta \mathbf{e}_\theta, \tag{4.14}$$

where the vector components are derived as follows:

$$v_\phi = \gamma \sin\theta + K \sin^3\theta (4\cos^2\theta - \sin^2\theta) \cos 4\phi. \tag{4.15}$$

$$v_\theta = -4K \sin^3\theta \cos\theta \sin 4\phi, \tag{4.16}$$

where $\gamma = K = 7.848 \times 10^{-6}$ sec^{-1}. Equations (4.15) and (4.16) are normalized such that the radius of the earth corresponds to the unit length.

4.1 Gradient

The gradient of a scalar field on a 2D curved surface should be independent of the coordinate system. Locally, the gradient points in the direction of maximal increase with respect to the intrinsic geometry of the surface. In the spherical coordinate system (θ, ϕ), the gradient of the scalar component v_ϕ in Eq. (4.16) is given by:

$$\nabla v_\phi = \frac{\partial v_\phi}{\partial \theta}\frac{\partial}{\partial \theta} + \frac{1}{\sin^2\theta}\frac{\partial v_\phi}{\partial \phi}\frac{\partial}{\partial \phi}, \tag{4.17}$$

Table 4.2 Error between Eqs. (4.17) and (4.20) on the unit sphere. Here, $h = 0.4$ and 498 elements are used. Elements near the poles are removed due to singularities in the spherical coordinate system

	p				
	3	4	5	6	7
Error	3.23×10^{-4}	1.51×10^{-5}	6.54×10^{-7}	3.87×10^{-8}	1.77×10^{-9}

where each term is computed as follows:

$$\frac{\partial v_\phi}{\partial \theta} = \omega \cos\theta + K \sin^2\theta[3\cos\theta(4\cos^2\theta - \sin^2\theta) - 10\sin^2\theta\cos\theta]\cos 4\phi, \tag{4.18}$$

$$\frac{\partial v_\phi}{\partial \phi} = -4K \sin\theta(4\cos^2\theta - \sin^2\theta)\sin 4\phi, \tag{4.19}$$

where ω and K are defined in Eqs. (4.14) and (4.15). The gradient in spherical or LOC moving frames should yield the same result and can be written as:

$$\nabla v_\phi = (\nabla v_\phi \cdot \mathbf{e}_1)\mathbf{e}_1 + (\nabla v_\phi \cdot \mathbf{e}_2)\mathbf{e}_2. \tag{4.20}$$

Computationally, Eq. (4.20) should converge to Eq. (4.17) as spatial resolution increases. Table 4.2 confirms exponential convergence of the difference between the gradient computed via covariant differentiation and that using LOC moving frames for v_ϕ as p increases.

4.1.1 [MATLAB] Computing Gradient

This section illustrates how to compute the gradient of a scalar function on curved surfaces using orthonormal moving frames. The MATLAB script *ch4MATLABa.m* carries out this computation, where the gradient is evaluated in terms of the local basis vectors $\mathbf{e}_\phi$ and $\mathbf{e}_\theta$.

The necessary moving frames are generated for both the sphere and pseudosphere using the helper script *CreateGridTangentVec.m*, which constructs the local tangent vectors at each grid point based on the surface geometry.

We consider the scalar test function:

$$f(\phi, \theta) = \sin\theta \cos\phi. \tag{4.21}$$

The analytical form of the gradient of f, expressed in the moving frame basis $\{\boldsymbol{\phi}, \boldsymbol{\theta}\}$, is given by:

$$\nabla f \cdot \boldsymbol{\phi} = -\sin\theta \sin\phi, \tag{4.22}$$

$$\nabla f \cdot \boldsymbol{\theta} = \cos\theta \cos\phi. \tag{4.23}$$

These expressions serve as reference values for validating the numerical computations.

```
function ch4MATLABa(Nphi, Ntheta, shape)

% ch4MATLABa - Computes directional derivatives using moving frames efficiently
%
% Syntax: ch4MATLABa(Nphi, Ntheta, alpha)
%
% Inputs:
%    Nphi   - Number of discretization points in phi-direction
%    Ntheta - Number of discretization points in theta-direction
%    alpha  - Optional orientation parameter for moving frames (default: 0.0)
%
% Outputs:
%    Visual plots of directional derivatives

%% Initialization and Grid Setup
SetupParameter;
[phigrid, thetagrid, ~, ~] = Generategrid(Nphi, Ntheta);
[TH, PHI] = meshgrid(thetagrid, phigrid);

%% Compute Moving Frames and Curved Coordinates
[grids,movingframes] = CreateGridTangentVec(Nphi, Ntheta, 1, shape);
[~,curvedcoord] = CreateGridTangentVec(Nphi, Ntheta, 0, shape);

%% Scalar Field and Analytical Gradient Calculation
inarray = sin(TH') .* cos(PHI');
gradphi = -sin(TH') .* sin(PHI');
gradtheta = cos(TH') .* cos(PHI');

gradexact = zeros(Nphip, Nthetap, spacedim);
for k = 1:spacedim
    gradexact(:,:,k) = gradphi .* squeeze(curvedcoord(:,:,1,k)) + gradtheta .* squeeze(curvedcoord(:,:,2,k));
end

%% Directional Derivatives via Moving Frames
dirderiv = zeros(Nphip, Nthetap, expdim);
for j = 1:expdim
      dirderiv(:,:,j) = Computedirderiv(squeeze(movingframes(:,:,j,:)), inarray, shape);
end

%% Gradient Reconstruction from Directional Derivatives
MFgrad = zeros(Nphip, Nthetap, spacedim);
for k = 1:spacedim
    MFgrad(:,:,k) = dirderiv(:,:,1) .* squeeze(movingframes(:,:,1,k)) + ...
                    dirderiv(:,:,2) .* squeeze(movingframes(:,:,2,k));
end

%% Error Norm Calculation
MFgraddiff = MFgrad - gradexact;
MFgraderr = norm(MFgraddiff(:)) / numel(MFgraddiff(:));
fprintf('Gradient error norm: %.4e\n', MFgraderr);
```

```

%% Visualization of Directional Derivatives
figure;
surf(grids(:,:,1), grids(:,:,2), grids(:,:,3), MFgrad(:,:,1), 'EdgeColor', 'none');
title('$\nabla␣f␣\cdot␣\mathbf{e}_1$', 'Interpreter', 'latex');
xlabel('x'); ylabel('y'); zlabel('z');
axis equal; view(45,45);
lighting gouraud; camlight;
colormap(parula(256));
colorbar;

figure;
surf(grids(:,:,1), grids(:,:,2), grids(:,:,3), MFgrad(:,:,2), 'EdgeColor', 'none');
title('$\nabla␣f␣\cdot␣\mathbf{e}_2$', 'Interpreter', 'latex');
xlabel('x'); ylabel('y'); zlabel('z');
axis equal; view(45,45);
lighting gouraud; camlight;
colormap(parula(256));
colorbar;

end
```

Listing 4.1 Computation of the gradient of the scalar test function f using moving frames (MATLAB script: ch4MATLABa.m)

To evaluate the directional derivatives numerically, we use the MATLAB file *Computedirderiv.m*, which computes the derivative of the scalar function f along each local frame direction. The implementation is based on spectral element methods to achieve high-order accuracy on curved geometries (Turner 2017).

```
function dirderiv = Computedirderiv(movingframe, inarray, shape)
% MFdirderiv - Computes directional derivative along a moving frame
%
% Syntax: dirderiv = MFdirderiv(polacoord, movingframe, inarray)
%
% Inputs:
%   movingframe  - [Nphi x Ntheta x spacedim] Moving frame vector field
%   inarray      - [Nphi x Ntheta] Scalar field defined on spherical grid
%   shape        - Shape of the domain: sphere or pseudosphere
%
% Outputs:
%   dirderiv     - [Nphi x Ntheta] Directional derivative of scalar field along moving frame

%% --- Grid and Parameter Setup ---
[Nphip, Nthetap, ~] = size(movingframe);  % Extract grid and spatial dimensions
Nphi = Nphip - 1; Ntheta = Nthetap - 1;

% --- Generate spherical coordinate grids using spectral element method (SEM) ---
[~, ~, Dbphi, Dbtheta] = Generategrid(Nphi, Ntheta);

% Compute metric tensor
[~,curvedcoord] = CreateGridTangentVec(Nphi, Ntheta, 0, shape);
[gij, ~] = Computemetric(Nphi, Ntheta, curvedcoord);

% --- Compute gradients in phi and theta directions with proper scaling ---
Grad_phi = (Dbphi * inarray) ./ sqrt(squeeze(gij(:,:,1,1)));        % Phi-direction gradient scaled by 1/sin(theta)
Grad_theta = inarray * Dbtheta'./ sqrt(squeeze(gij(:,:,2,2)));       % Theta-direction gradient (no scaling required)

```

```
% --- Project gradients onto moving frame vectors ---
dirderiv = zeros(Nphip, Nthetap);  % Initialize directional derivative array
for i = 1:Nphip
    for j = 1:Nthetap
        mf = squeeze(movingframe(i,j,:));    % Moving frame vector at node (i,j)
        ep = squeeze(curvedcoord(i,j,1,:))./ sqrt(gij(i,j,1,1));    % Phi-direction basis vector at node (i,j)
        et = squeeze(curvedcoord(i,j,2,:))./ sqrt(gij(i,j,2,2));    % Theta-direction basis vector at node (i,j)

        % Compute directional derivative as projection of gradient onto moving frame
        dirderiv(i,j) = Grad_phi(i,j) * dot(mf, ep) + Grad_theta(i,j) * dot(mf, et) ;
    end
end
end
```

Listing 4.2 Computation of the directional derivatives using spectral element methods (MATLAB script: `Computedirderiv.m`)

The following figures visualize the components of the gradient on the sphere (Fig. 4.2) and the pseudosphere (Fig. 4.3), evaluated along the directions of the local tangent vectors $\mathbf{e}_\phi$ and $\mathbf{e}_\theta$. These plots confirm the alignment between the analytical gradients and the numerically computed directional derivatives.

4.2 Divergence

In Euclidean space, divergence is computed as $\nabla \cdot \mathbf{v} = \partial_i v^i$. On curved domains, it requires geometric correction via metric tensor components or covariant derivatives. In the spherical coordinate system on the unit sphere, the divergence of the velocity

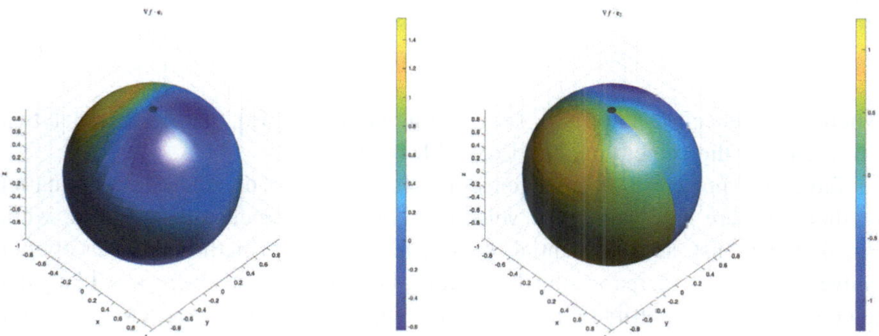

(a) Directional derivative $\nabla f \cdot \mathbf{e}_\phi$ (b) Directional derivative $\nabla f \cdot \mathbf{e}_\theta$

Fig. 4.2 Gradient components of $f(\phi, \theta) = \sin\theta \cos\phi$ along the local directions $\mathbf{e}_\phi$ and $\mathbf{e}_\theta$ on the unit sphere. (**a**) Evaluation of $\nabla f \cdot \mathbf{e}_\phi$ and (**b**) evaluation of $\nabla f \cdot \mathbf{e}_\theta$

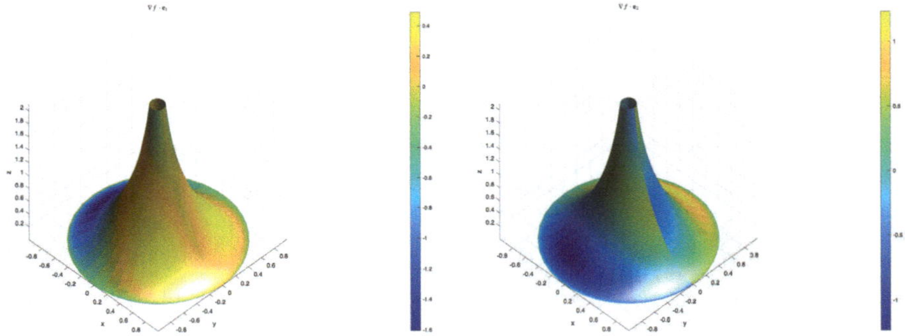

(a) Directional derivative $\nabla f \cdot \mathbf{e}_\phi$ (b) Directional derivative $\nabla f \cdot \mathbf{e}_\theta$

Fig. 4.3 Gradient components of $f(\phi, \theta) = \sin\theta \cos\phi$ along the local directions $\mathbf{e}_\phi$ and $\mathbf{e}_\theta$ on the unit pseudosphere. (**a**) Evaluation of $\nabla f \cdot \mathbf{e}_\phi$ and (**b**) evaluation of $\nabla f \cdot \mathbf{e}_\theta$

vector $\mathbf{v}$ from Eq. (4.14) is given by the covariant formulation:

$$\nabla \cdot \mathbf{v} = \frac{1}{\sin\theta}\left(\frac{\partial v_\phi}{\partial \phi} + \frac{\partial}{\partial \theta}(v_\theta \sin\theta)\right). \tag{4.24}$$

Alternatively, in orthonormal moving frames, the divergence is computed by:

$$\nabla \cdot \mathbf{u} = \sum_{i=1}^{2} \nabla_{\mathbf{e}_i} \mathbf{u} \cdot \mathbf{e}_i = \nabla_{\mathbf{e}_1} \mathbf{u} \cdot \mathbf{e}_1 + \nabla_{\mathbf{e}_2} \mathbf{u} \cdot \mathbf{e}_2, \tag{4.25}$$

Note that the direction of the unit tangent vector $\mathbf{e}_1$ is arbitrary and may be discontinuous across element interfaces. Using the covariant derivatives in Eqs. (3.104)–(3.107), we obtain:

$$\nabla \cdot \mathbf{u} = \nabla u^1 \cdot \mathbf{e}_1 + \omega_{21}(\mathbf{e}_1) u^2 + \nabla u^2 \cdot \mathbf{e}_2 + \omega_{21}(\mathbf{e}_2) u^1, \tag{4.26}$$

where the connection 1-form $\omega_{ij}(\mathbf{e}_k)$ is computed using Eq. (3.96). The analytical divergence of the velocity field in Eq. (4.14) is zero.

Table 4.3 presents h-convergence results for three divergence computation methods. Figure 4.4a shows the velocity field and divergence distribution, while Fig. 4.4b demonstrates exponential convergence. The LOC method achieves optimal convergence. As p increases, the discrepancy between methods increases, likely due to geometric error contributions. However, computational efficiency remains similar across methods.

Table 4.3 h-convergence of divergence on the unit sphere using three methods for $p = 5$

	$h = 0.1867$ $N_e = 1918$	$h = 0.2623$ $N_e = 970$	$h = 0.3445$ $N_e = 498$	$h = 0.5027$ $N_e = 278$
Covariant	1.93×10^{-5}	6.29×10^{-5}	2.68×10^{-4}	1.23×10^{-3}
(order)		3.47×10^{0}	5.32×10^{0}	4.03×10^{0}
Spherical MF	2.38×10^{-5}	7.23×10^{-5}	3.02×10^{-4}	1.39×10^{-3}
(order)		3.27×10^{0}	5.25×10^{0}	4.04×10^{0}
LOC MF	4.26×10^{-6}	2.39×10^{-5}	8.83×10^{-5}	7.29×10^{-4}
(order)		5.08×10^{0}	4.79×10^{0}	5.58×10^{0}

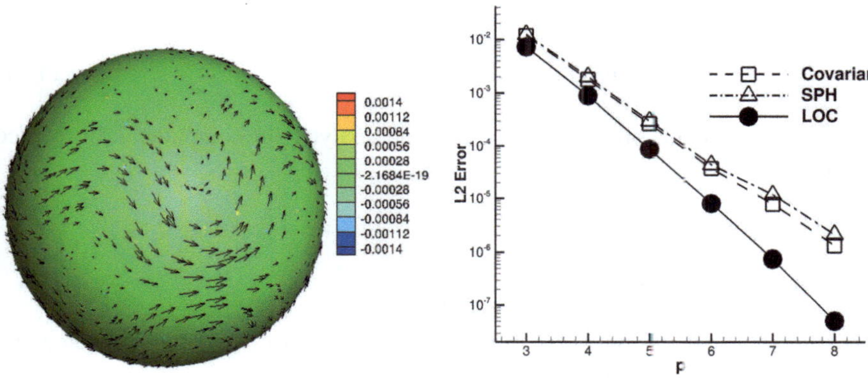

(a) $\nabla \cdot \mathbf{v}$ (b) Error convergence

Fig. 4.4 (**a**) Distribution of the velocity vector $\mathbf{v}$ (arrow) of the test problem and its divergence $\nabla \cdot \mathbf{v}$ (contour) by LOC moving frames for $p = 5$. (**b**) The p-convergence of L_2 error is shown for $h = 0.3445$. *Covariant* covariant formulation, *SPH* spherical moving frames, *LOC* local moving frames

4.2.1 [MATLAB] Computing Divergence

This subsection demonstrates how to compute the covariant divergence of a velocity vector field $\mathbf{v}$ defined on curved surfaces, using orthonormal moving frames. The MATLAB script *ch4MATLABb.m* performs the divergence calculation on both spherical and pseudospherical geometries. The required moving frames are constructed in the associated utility script, shared with previous gradient computations.

We define the angular velocity components u_ϕ and u_θ as:

$$u_\phi = \cos\phi, \tag{4.27}$$

$$u_\theta = \sin\theta, \tag{4.28}$$

which are used to construct the full 3D velocity vector $\mathbf{v} = (v_x, v_y, v_z)$ in Cartesian coordinates, mapped from the angular components via the moving frame basis.

For the sphere, the components of $\mathbf{v}$ become:

$$v_x = -u_\phi \sin\phi - u_\theta \cos\theta \cos\phi, \tag{4.29}$$

$$v_y = u_\phi \cos\phi - u_\theta \cos\theta \sin\phi, \tag{4.30}$$

$$v_z = u_\theta \sin\theta. \tag{4.31}$$

For the pseudosphere, which exhibits hyperbolic geometry, the components are:

$$v_x = -u_\phi \sin\phi - u_\theta \operatorname{sech}\theta \cos\phi, \tag{4.32}$$

$$v_y = u_\phi \cos\phi - u_\theta \operatorname{sech}\theta \sin\phi, \tag{4.33}$$

$$v_z = u_\theta \tanh\theta. \tag{4.34}$$

These velocity fields are constructed to yield analytically computable divergence expressions:

- On the sphere: $\nabla \cdot \mathbf{v} = -\dfrac{\sin\phi}{\sin\theta}$.
- On the pseudosphere: $\nabla \cdot \mathbf{v} = -\dfrac{\sin\phi}{\operatorname{sech}\theta} + \dfrac{\cos\theta}{\tanh\theta}$.

These analytic expressions serve as reference solutions for validating the numerical results obtained using moving frames.

```
function ch4MATLABb(Nphi, Ntheta, shape)
% ch4MATLABb - Computes divergence of a velocity field using moving frames
    on a sphere
%
% Syntax:
%    ch4MATLABb(Nphi, Ntheta, alpha)
%
% Description:
%    This function computes the divergence of a synthetic velocity field on
    a spherical surface
%    using covariant derivative expressions in a moving frame coordinate
    system.
%
% Inputs:
%    Nphi   - Number of discretization points in \phi-direction (longitude)
%    Ntheta - Number of discretization points in \theta-direction (
    colatitude)
%    alpha  - Optional orientation parameter for moving frames (in radians,
    default = 0.0)
%
% Outputs:
%    A surface plot of the computed divergence on the spherical grid
%    Error norm printed to the console (for verification against analytic
    value)
%
% Note:
%    \phi (phi) denotes azimuthal angle (longitude), \theta (theta) denotes
    polar angle (colatitude)

alpha = 0.0;

%% --- Parameter and Grid Setup ---
SetupParameter;  % Load parameters including radius R, space dimension, etc.
```

```
% Generate spectral-element-type grids in (phi, theta) domain
[phigrid, thetagrid, ~, ~] = Generategrid(Nphi, Ntheta);

% Create meshgrid for 2D evaluation: PHI (azimuthal), TH (colatitudinal)
[PHI, TH] = meshgrid(phigrid, thetagrid);

%% Compute Moving Frames and Curved Coordinates
[grids,movingframes] = CreateGridTangentVec(Nphi, Ntheta, 1, shape);

%% --- Define Synthetic Velocity Field ---
% Initialize velocity field in 3D: (Nphi x Ntheta x 3)
velocity = zeros(Nphip, Nthetap, spacedim);

% Define component fields in spherical coordinates
uphi = cos(PHI');                % \phi-component, constant
uth  = sin(TH');  % \theta-component, includes 1/sin(\theta) singularity at poles

% Map to Cartesian components based on spherical unit vectors
% These are velocity components expressed in the global Cartesian basis

% Sphere
if(shape=="sphere")
    velocity(:,:,1) = -uphi .* sin(PHI') + uth .* cos(TH') .* cos(PHI');  % x-component
    velocity(:,:,2) =  uphi .* cos(PHI') + uth .* cos(TH') .* sin(PHI');  % y-component
    velocity(:,:,3) = -uth .* sin(TH');                                   % z-component

    ExactDiv = -sin(PHI')./sin(TH');

    % Pseudosphere
elseif(shape=="pseudosphere")
    velocity(:,:,1) = -uphi .* sin(PHI') - uth .* sech(TH') .* cos(PHI');  % x-component
    velocity(:,:,2) =  uphi .* cos(PHI') - uth .* sech(TH') .* sin(PHI');  % y-component
    velocity(:,:,3) =  uth .* tanh(TH');

    ExactDiv = -sin(PHI')./sech(TH') + sin (TH') + cos(TH')./tanh(TH');
end

%% --- Compute Covariant Divergence ---
% Use moving frame divergence computation
MFDiv = Computecovdivg(Nphi, Ntheta, movingframes, velocity, shape);

%% --- Compute Error Norm ---
% Compare with exact divergence = 0 for this synthetic field
MFDivdiff = MFDiv - ExactDiv; % Exact divergence is zero
MFDiverr = norm(MFDivdiff(:)) / sqrt(numel(MFDivdiff(:)));  % L2 norm of error
fprintf('Divergence␣error␣norm:␣%.4e\n', MFDiverr)

%% --- Visualization ---
% Surface plot of computed divergence on the sphere
figure;
surf(grids(:,:,1), grids(:,:,2), grids(:,:,3), MFDiv, 'EdgeColor', 'none'); % Smooth 3D surface
title('Error␣for␣␣$\nabla␣\cdot␣\mathbf{v}$␣using␣moving␣frame', 'Interpreter', 'latex');  % LaTeX title
xlabel('x'); ylabel('y'); zlabel('z');
axis equal; view(45,45);
lighting gouraud; camlight;
colormap(parula(256));
colorbar;
```

```
end
```

Listing 4.3 Computation of the covariant divergence of the velocity field **v** using moving frames (MATLAB script: `ch4MATLABb.m`)

To enhance numerical accuracy and geometric fidelity, the MATLAB script *Computecovdivg.m* uses spectral element methods combined with precomputed connection coefficients to evaluate the covariant divergence. This approach ensures compatibility with the underlying curved geometry.

```
function Div = Computecovdivg(Nphi, Ntheta, movingframes, velocity, shape)
% ComputeCovDiv - Computes covariant divergence of a vector field in moving frames
%
% Syntax:
%   Div = ComputeCovDiv(Nphi, Ntheta, velocity, alpha)
%
% Description:
%   This function calculates the divergence of a vector field defined on a
%   spherical domain using moving frames. It uses the covariant derivative,
%   accounting for rotational effects through the connection 1-form.
%
% Inputs:
%   Nphi     - Number of grid points in \phi (azimuthal) direction
%   Ntheta   - Number of grid points in \theta (polar) direction
%   velocity - 3D velocity field (Nphi x Ntheta x 3) in Cartesian coordinates
%   alpha    - Optional rotation angle for moving frame orientation
%
% Output:
%   Div - Covariant divergence of the input velocity field on the sphere

%% --- Parameter and Grid Setup ---
SetupParameter;  % Load parameters including radius R, space dimension, etc.

%% --- Convert Cartesian Velocity to Local Moving Frame Coordinates ---
% Convert 3D velocity vector field into components (v1, v2) in moving frame basis
inarray = CartesianToMovingframes(movingframes, velocity);

%% --- Compute Directional Derivatives along Moving Frame Vectors ---
% dirderiv(:,:,k,j) = \partial (v_k)/\partial (e_j) where:
%   k: component index of vector field in moving frame
%   j: directional derivative index (e.g., along e1, e2)
dirderiv = zeros(Nphip, Nthetap, mfdim, expdim);
for k = 1:mfdim
    for j = 1:expdim
        % Compute directional derivative of component k along direction j
        mf = squeeze(movingframes(:,:,j,:));
        f = squeeze(inarray(:,:,k));
        dirderiv(:,:,k,j) = Computedirderiv(mf, f, shape);
    end
end

%% --- Compute Connection 1-Form ---
% \omega ij_k(i,j,m,n) \sim \omega_{ij}(e_m)_n
% This encodes rotation of frame vectors along the surface for covariant correction
wij_k = Computeconnection(Nphi, Ntheta, shape, movingframes);

%% --- Assemble Covariant Divergence Expression ---
% Divergence in moving frame: \nabla \cdot \vec{v} = D_{e1}(v1) + \omega_{21}(e1)*v2 + D_{e2}(v2) + \omega_{21}(e2)*v1
Div = dirderiv(:,:,1,1) - wij_k(:,:,2,1,1) .* inarray(:,:,2) ...
```

```
    + dirderiv(:,:,2,2) + wij_k(:,:,2,1,2) .* inarray(:,:,1);

end
```

Listing 4.4 Covariant divergence of **v** computed via spectral element methods using connection coefficients (MATLAB script: `Computecovdivg.m`)

The following visualizations in Fig. 4.5 compare the divergence field $\nabla \cdot \mathbf{v}$ computed numerically on both the spherical and pseudospherical surfaces. The patterns confirm agreement with the analytical expressions, validating the spectral-metric formulation in curved moving frames.

4.3 Curl

Consider the projection of the curl onto the surface normal vector **k**, denoted by $\mathbf{k} \cdot \nabla \times \mathbf{v}$. For the Rossby–Haurwitz wave velocity field in Eq. (4.14), the analytic curl projection is:

$$\mathbf{k} \cdot (\nabla \times \mathbf{v}) = -2\omega \cos\theta + 30K \sin^4\theta \cos\theta \cos 4\phi. \tag{4.35}$$

The covariant curl operator in spherical coordinates is given by:

$$\mathbf{k} \cdot (\nabla \times \mathbf{v}) = \frac{1}{\sin\theta} \left(\frac{\partial v_\theta}{\partial \phi} - \frac{\partial}{\partial \theta}(v_\phi \sin\theta) \right). \tag{4.36}$$

Alternatively, using orthonormal moving frames, the projected curl is:

$$\mathbf{k} \cdot (\nabla \times \mathbf{u}) = \nabla_{\mathbf{e}_1} \mathbf{u} \cdot \mathbf{e}_2 - \nabla_{\mathbf{e}_2} \mathbf{u} \cdot \mathbf{e}_1. \tag{4.37}$$

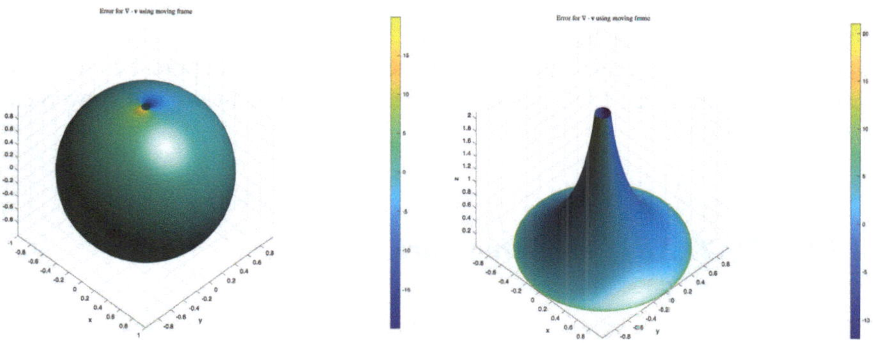

(a) $\nabla \cdot \mathbf{v}$ on the sphere (b) $\nabla \cdot \mathbf{v}$ on the pseudosphere

Fig. 4.5 Divergence of the velocity field **v**, i.e., $\nabla \cdot \mathbf{v}$, on curved surfaces using moving frames, consistent with the analytic solutions. (**a**) Unit sphere; (**b**) unit pseudosphere

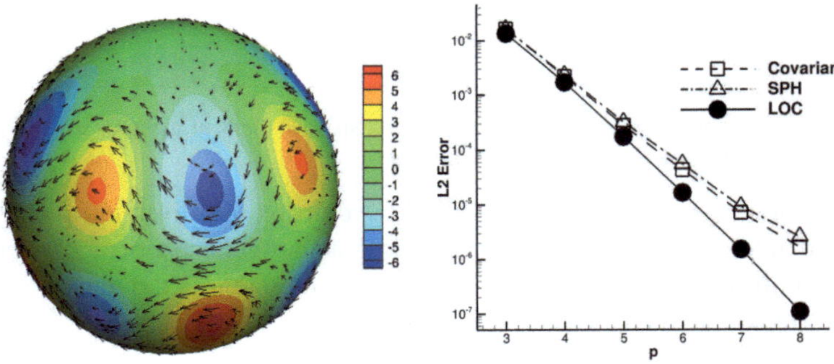

(a) $\mathbf{k} \cdot (\nabla \times \mathbf{v})$ (b) Error convergence

Fig. 4.6 (**a**) Distribution of the velocity vector $\mathbf{v}$ (arrows) and the curl projection $\mathbf{k} \cdot (\nabla \times \mathbf{v})$ (contours) using LOC moving frames for $p = 5$. (**b**) p-convergence of the error for $h = 0.3445$. *Covariant* covariant formulation, *SPH* spherical moving frames, *LOC* local moving frames

Table 4.4 h-convergence of curl on the unit sphere using three methods for $p = 5$

	$h = 0.1867$ $N_e = 1918$	$h = 0.2623$ $N_e = 970$	$h = 0.3445$ $N_e = 498$	$h = 0.5027$ $N_e = 278$
Covariant (order)	1.96×10^{-5}	6.83×10^{-5} 3.4746×10^{0}	2.77×10^{-4} 5.3218×10^{0}	1.41×10^{-3} 4.0315×10^{0}
Spherical MF (order)	1.92×10^{-5}	7.46×10^{-5} 3.2655×10^{0}	3.18×10^{-4} 5.2488×10^{0}	1.72×10^{-3} 4.0375×10^{0}
LOC MF (order)	7.67×10^{-6}	4.07×10^{-5} 5.0775×10^{0}	1.73×10^{-4} 4.7865×10^{0}	1.08×10^{-3} 5.5840×10^{0}

Using covariant derivatives, this becomes:

$$\mathbf{k} \cdot (\nabla \times \mathbf{u}) = \nabla u^2 \cdot \mathbf{e}_1 + \omega_{21}(\mathbf{e}_1)u^1 - \nabla u^1 \cdot \mathbf{e}_2 + \omega_{21}(\mathbf{e}_2)u^2. \tag{4.38}$$

Figure 4.6a, b show the velocity field and curl convergence. As with divergence, LOC moving frames exhibit the highest accuracy and optimal convergence. Table 4.4 confirms that LOC moving frames achieve near-theoretical convergence orders.

4.3.1 [MATLAB] Computing Curl

This subsection demonstrates how to compute the curl of a velocity field $\mathbf{v}$ on curved surfaces using orthonormal moving frames. The MATLAB script *ch4MATLABc.m* performs this operation on both spherical and pseudospherical geometries. The required moving frames are generated using the setup previously introduced in the same script.

The angular velocity components u_ϕ and u_θ are defined as:

$$u_\phi = \sin\theta, \tag{4.39}$$

$$u_\theta = \cos\phi, \tag{4.40}$$

which are used to construct the full 3D velocity vector $\mathbf{v} = (v_x, v_y, v_z)$, expressed in Cartesian coordinates based on the local moving frame directions.

On the sphere, the components of $\mathbf{v}$ are given by:

$$v_x = -u_\phi \sin\phi - u_\theta \cos\theta \cos\phi, \tag{4.41}$$

$$v_y = u_\phi \cos\phi - u_\theta \cos\theta \sin\phi, \tag{4.42}$$

$$v_z = u_\theta \sin\theta. \tag{4.43}$$

On the pseudosphere (a negatively curved surface), the vector field components are adjusted to match the pseudospherical geometry:

$$v_x = -u_\phi \sin\phi - u_\theta \operatorname{sech}\theta \cos\phi, \tag{4.44}$$

$$v_y = u_\phi \cos\phi - u_\theta \operatorname{sech}\theta \sin\phi, \tag{4.45}$$

$$v_z = u_\theta \tanh\theta. \tag{4.46}$$

These velocity fields are designed to yield known analytic expressions for the **normal component of the curl**, i.e., $\mathbf{k} \cdot (\nabla \times \mathbf{v})$, where $\mathbf{k}$ is the unit normal vector to the surface. The resulting values are:

- On the sphere: $\mathbf{k} \cdot (\nabla \times \mathbf{v}) = -\dfrac{\sin\phi}{\sin\theta}$.
- On the pseudosphere: $\mathbf{k} \cdot (\nabla \times \mathbf{v}) = -\dfrac{\sin\phi}{\operatorname{sech}\theta} - \sin\theta - \dfrac{\cos\theta}{\tanh\theta}$.

```
function ch4MATLABc(Nphi, Ntheta, shape)
% ch4MATLABcb - Computes curl of a velocity field using moving frames on a
    sphere
%
% Syntax:
%    ch4MATLABb(Nphi, Ntheta, alpha)
%
% Description:
%    This function computes the divergence of a synthetic velocity field on
    a spherical surface
%    using covariant derivative expressions in a moving frame coordinate
    system.
%
% Inputs:
%    Nphi   - Number of discretization points in \phi-direction (longitude)
%    Ntheta - Number of discretization points in \theta-direction (
    colatitude)
%    alpha  - Optional orientation parameter for moving frames (in radians,
    default = 0.0)
%
% Outputs:
```

```
%    A surface plot of the computed divergence on the spherical grid
%    Error norm printed to the console (for verification against analytic value)
%
% Note:
%    \phi (phi) denotes azimuthal angle (longitude), \theta (theta) denotes polar angle (colatitude)

%% --- Parameter and Grid Setup ---
SetupParameter;  % Load parameters including radius R, space dimension, etc.

% Generate spectral-element-type grids in (phi, theta) domain
[phigrid, thetagrid, ~, ~] = Generategrid(Nphi, Ntheta);

% Create meshgrid for 2D evaluation: PHI (azimuthal), TH (colatitudinal)
[PHI, TH] = meshgrid(phigrid, thetagrid);

%% Compute Moving Frames and Curved Coordinates
[grids,movingframes] = CreateGridTangentVec(Nphi, Ntheta, 1, shape);

%% --- Define Synthetic Velocity Field ---
% Initialize velocity field in 3D: (Nphi x Ntheta x 3)
velocity = zeros(Nphip, Nthetap, spacedim);

% Define component fields in spherical coordinates
uphi = sin(TH');              % \phi-component, constant
uth  = cos(PHI');    % \theta-component, includes 1/sin(\theta) singularity at poles

% Map to Cartesian components based on spherical unit vectors
% These are velocity components expressed in the global Cartesian basis

% Sphere
if(shape=="sphere")
    velocity(:,:,1) = -uphi .* sin(PHI') + uth .* cos(TH') .* cos(PHI');  % x-component
    velocity(:,:,2) =  uphi .* cos(PHI') + uth .* cos(TH') .* sin(PHI');  % y-component
    velocity(:,:,3) = -uth .* sin(TH');                                    % z-component

    ExactCurl = -sin(PHI')./sin(TH');

    % Pseudosphere
elseif(shape=="pseudosphere")
    velocity(:,:,1) = -uphi .* sin(PHI') - uth .* sech(TH') .* cos(PHI');  % x-component
    velocity(:,:,2) =  uphi .* cos(PHI') - uth .* sech(TH') .* sin(PHI');  % y-component
    velocity(:,:,3) =  uth .* tanh(TH');

    ExactCurl = -sin(PHI')./sech(TH') - sin(TH') - cos(TH')./tanh(TH');
end                                       % z-component

%% --- Compute Covariant Divergence ---
% Use moving frame divergence computation
MFCurl = Computecovcurl(Nphi, Ntheta, movingframes, velocity, shape);

%% --- Compute Error Norm ---
% Compare with exact divergence = 0 for this synthetic field
MFCurldiff = MFCurl - ExactCurl;  % Exact divergence is zero
MFCurlerr = norm(MFCurldiff(:)) / sqrt(numel(MFCurldiff(:)));  % L2 norm of error
fprintf('Curl error norm: %.4e\n', MFCurlerr)

%% --- Visualization ---
% Surface plot of computed divergence on the sphere
```

```
figure;
surf(grids(:,:,1), grids(:,:,2), grids(:,:,3), MFCurl, 'EdgeColor', 'none');
        % Smooth 3D surface
title('$\nabla␣\times␣\mathbf{v}$␣using␣moving␣frame', 'Interpreter', 'latex
    '); % LaTeX title
xlabel('x'); ylabel('y'); zlabel('z');
axis equal; view(45,45);
lighting gouraud; camlight; % Lighting effects for better 3D shading
colormap(parula(256));
colorbar;

end
```

Listing 4.5 Computation of the covariant curl of the velocity field **v** using moving frames (MATLAB script: ch4MATLABc.m)

For improved numerical accuracy, the script *Computecovcurl.m* evaluates the curl using spectral element methods and connection coefficients derived from the moving frames. This allows correct geometric behavior even on non-Euclidean domains and captures rotational effects consistent with the underlying curvature.

```
function Curl = Computecovcurl(Nphi, Ntheta, movingframes, velocity, shape)
% ComputeCovDiv - Computes covariant divergence of a vector field in moving
    frames
%
% Syntax:
%  Curl = ComputeCovDiv(Nphi, Ntheta, velocity, alpha)
%
% Description:
%   This function calculates the curl of a vector field defined on a
%   spherical domain using moving frames. It uses the covariant derivative,
%   accounting for rotational effects through the connection 1-form.
%
% Inputs:
%   Nphi     - Number of grid points in \phi (azimuthal) direction
%   Ntheta   - Number of grid points in \theta (polar) direction
%   velocity - 3D velocity field (Nphi x Ntheta x 3) in Cartesian
    coordinates
%   alpha    - Optional rotation angle for moving frame orientation
%
% Output:
%   Curl - Covariant curl of the input velocity field on the sphere

%% --- Parameter and Grid Setup ---
SetupParameter; % Load parameters including radius R, space dimension, etc.

%% --- Convert Cartesian Velocity to Local Moving Frame Coordinates ---
% Convert 3D velocity vector field into components (v1, v2) in moving frame
    basis
inarray = CartesianToMovingframes(movingframes, velocity);

%% --- Compute Directional Derivatives along Moving Frame Vectors ---
% dirderiv(:,:,k,j) = \partial (v_k)/\partial (e_j) where:
%   k: component index of vector field in moving frame
%   j: directional derivative index (e.g., along e1, e2)
dirderiv = zeros(Nphip, Nthetap, mfdim, expdim);
for k = 1:mfdim
    for j = 1:expdim
        % Compute directional derivative of component k along direction j
        mf = squeeze(movingframes(:,:,j,:));
        f = squeeze(inarray(:,:,k));
        dirderiv(:,:,k,j) = Computedirderiv(mf, f, shape);
    end
end

```

```
%% --- Compute Connection 1-Form ---
% \omega ij_k(i,j,m,n) \sim \omega_{ij}(e_m)_n
% This encodes rotation of frame vectors along the surface for covariant
    correction
wij_k = Computeconnection(Nphi, Ntheta, shape, movingframes);

%% --- Assemble Covariant Divergence Expression ---
% Curl in moving frame: \nabla \cdot \vec{v} = D_{e1}(v2) + \omega_{21}(e1)*
    v1 - D_{e2}(v1) + \omega_{21}(e2)*v2

Curl = dirderiv(:,:,2,1) + wij_k(:,:,2,1,1) .* inarray(:,:,1) ...
     - dirderiv(:,:,1,2) + wij_k(:,:,2,1,2) .* inarray(:,:,2);

end
```

Listing 4.6 Covariant curl calculation using spectral element methods and connection 1-forms (MATLAB script: `Computecovcurl.m`)

The following visualizations display the computed curl field on the sphere and pseudosphere. The numerical results match the expected analytical values, confirming the accuracy of the implementation based on moving frames (Fig. 4.7).

4.4 Application: Helmholtz–Hodge Decomposition

This section illustrates the application of the proposed scheme to a complex surface. As an extension of the covariant derivative formulation, a numerical scheme for the Helmholtz–Hodge decomposition (HHD) is implemented on curved surfaces. For a two-dimensional curved surface Ω with either Neumann boundary conditions or no boundary, the HHD decomposes a vector field $\mathbf{v}$ into three orthogonal components (Bhatia et al. 2013):

$$\mathbf{v} = \nabla u + \nabla \times \mathbf{R} + \mathbf{h}, \tag{4.47}$$

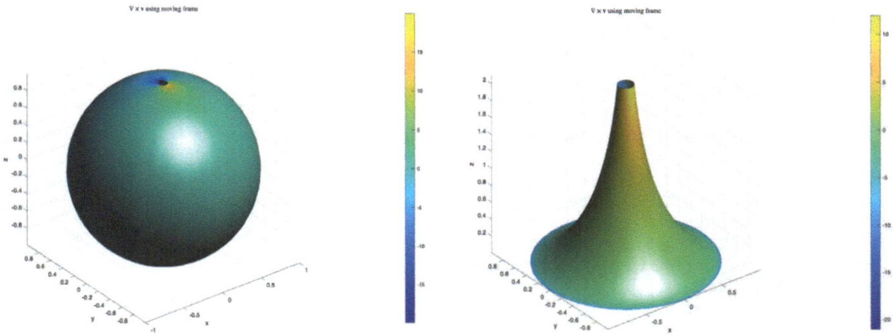

(a) $\mathbf{k} \cdot (\nabla \times \mathbf{v})$ on the sphere

(b) $\mathbf{k} \cdot (\nabla \times \mathbf{v})$ on the pseudosphere

Fig. 4.7 Projected curl of the velocity field $\mathbf{v}$, i.e., $\nabla \times \mathbf{v}$, on curved surfaces using moving frames, consistent with the analytic solutions. (**a**) Unit sphere; (**b**) unit pseudosphere where ∇u is the curl-free (irrotational) component, $\nabla \times \mathbf{R}$ is the divergence-free (incompressible) component, and $\mathbf{h}$ is a harmonic vector satisfying

$$\nabla^2 \mathbf{h} = \mathbf{0}. \tag{4.48}$$

The irrotational component is computed by applying the divergence operator to Eq. (4.47). The incompressible component is rewritten using the linear operator J, which rotates a vector

$$\mathbf{v} = v^1 \mathbf{e}_1 + v^2 \mathbf{e}_2, \tag{4.49}$$

as:

$$J\mathbf{v} = -v^2 \mathbf{e}_1 + v^1 \mathbf{e}_2. \tag{4.50}$$

Applying the divergence operator to Eq. (4.47) yields:

$$\nabla^2 u = \nabla \cdot \boldsymbol{\xi} - \frac{1}{A} \int \nabla \cdot \boldsymbol{\xi} dx, \tag{4.51}$$

$$\nabla^2 v = -\nabla \cdot J\boldsymbol{\xi} + \frac{1}{A} \int \nabla \cdot J\boldsymbol{\xi} dx, \tag{4.52}$$

where A denotes the area of Ω, and the additional terms account for the Neumann or boundaryless nature of the domain. The harmonic component $\mathbf{h}$ is obtained by subtracting the irrotational and incompressible components from $\mathbf{v}$:

$$\nabla \cdot \mathbf{h} = \frac{1}{A} \int \nabla \cdot \boldsymbol{\xi} dx, \tag{4.53}$$

$$\nabla \cdot J\mathbf{h} = \frac{1}{A} \int \nabla \cdot J\boldsymbol{\xi} dx. \tag{4.54}$$

Equations (4.53) and (4.54) are constants over the domain. The harmonic vector satisfies:

$$\nabla^2 \mathbf{h} = \nabla(\nabla \cdot \mathbf{h}) + \nabla \times (\nabla \times \mathbf{h}) = 0. \tag{4.55}$$

On a curved surface, the terms $\nabla \cdot \boldsymbol{\xi}$ and $\nabla \cdot (J\boldsymbol{\xi})$ in Eqs. (4.51) and (4.52) are computed using Eq. (4.26). Inaccurate computation of these terms can misidentify the source structure of the flow field. The Helmholtz equations are solved using the **Nektar++** spectral/hp library (Cantwell et al. 2015), implemented for both continuous and discontinuous Galerkin methods.

For computational validation, a curl-free vector field was constructed using spherical moving frames, multiplied by $\sin\theta$. A divergence-free vector field was derived from the Rossby–Haurwitz vector in Eq. (4.14). Figures 4.8 and 4.9 show

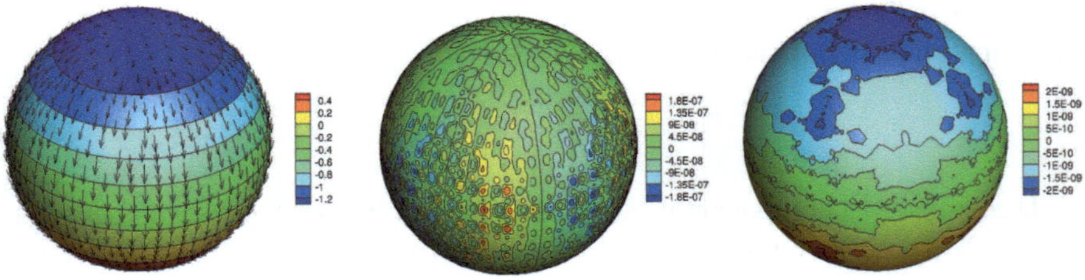

(a) Irrotational (b) Incompressible (c) Harmonic

Fig. 4.8 Helmholtz–Hodge decomposition of a curl-free vector: spherical moving frames multiplied by $\sin\theta$. (**a**) Irrotational component with potential u, (**b**) incompressible component with potential v, (**c**) harmonic component with potential U, where $\mathbf{h} = \nabla U$. $h = 0.2$, $p = 10$

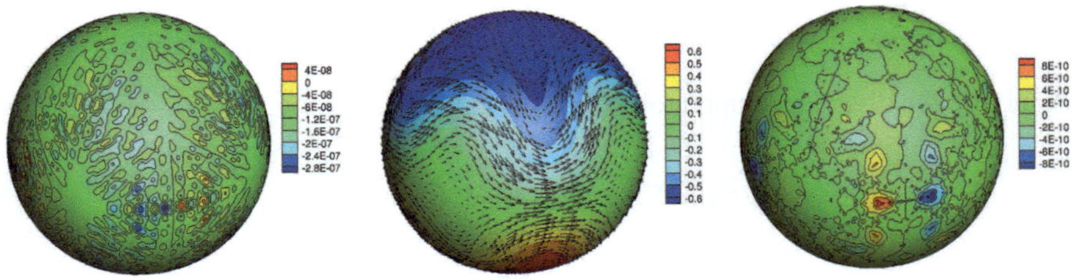

(a) Irrotational (b) Incompressible (c) Harmonic

Fig. 4.9 Helmholtz–Hodge decomposition of a divergence-free vector (Rossby–Haurwitz field). (**a**) Irrotational potential u, (**b**) incompressible potential v, (**c**) harmonic potential U, where $\mathbf{h} = \nabla U$. $h = 0.2$, $p = 10$

Table 4.5 Error in the Helmholtz–Hodge decomposition on the unit sphere ($p = 5$). Asterisks (*) and (**) indicate comparison to $(1/A) \int \nabla \cdot \boldsymbol{\xi} \, dx$ and $(1/A) \int \nabla \cdot (J\boldsymbol{\xi}) \, dx$, respectively

	$\nabla \times \nabla u$	$\nabla \cdot J \nabla v$	$\nabla \cdot \mathbf{h}^*$	$\nabla \cdot J\mathbf{h}^{**}$	$\nabla^2 \mathbf{h}$
Curl-less	2.44×10^{-9}	2.77×10^{-9}	6.60×10^{-6}	6.51×10^{-6}	1.79×10^{-3}
Divergence-less	1.95×10^{-9}	2.40×10^{-9}	6.16×10^{-6}	5.51×10^{-6}	1.63×10^{-3}

the HHD results for the curl-free and divergence-free test fields. The decomposition errors for each component are shown in Table 4.5.

To demonstrate HHD on geometrically complex domains, two curved surfaces are considered: a human atrial surface and the Stanford bunny. Initial vector fields were generated using a reaction–diffusion wave model (e.g., the two-variable Aliev–Panfilov model). Waves were initiated from a point source and propagated outward, with moving frames aligned along the gradient of the action potential. Although both geometries are topologically closed ($\beta_0 = 1$, no holes), the excised initial and final activation regions create effective holes, raising the Betti number to one. Therefore, the harmonic field is at most one-dimensional.

Figures 4.10 and 4.11 show the decomposition of the propagation field into three components on the atrium and the bunny, respectively. For the atrium, the following

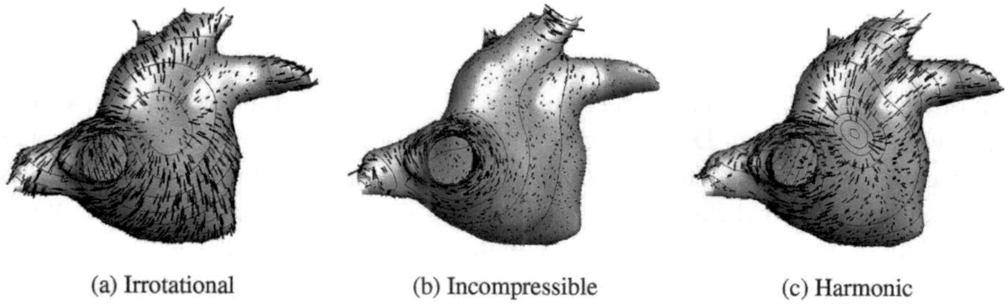

Fig. 4.10 Helmholtz–Hodge decomposition using aligned moving frames on an atrial surface. (**a**) Irrotational component with potential u, (**b**) incompressible component with potential v, (**c**) harmonic component with potential U

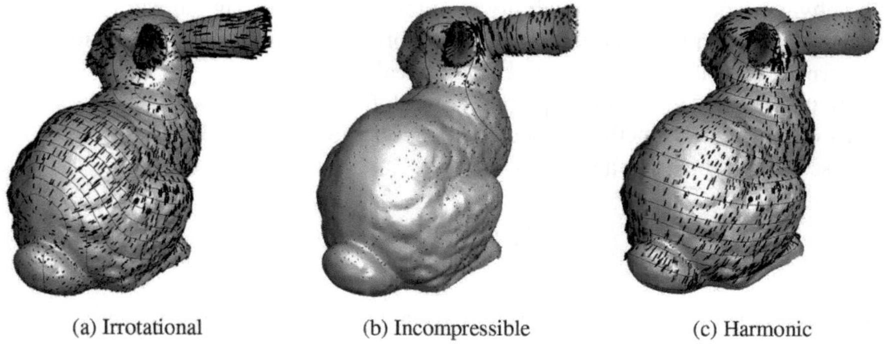

Fig. 4.11 Helmholtz–Hodge decomposition using aligned moving frames on the Stanford Bunny. (**a**) Irrotational component with potential u, (**b**) incompressible component with potential v, (**c**) harmonic component with potential U

errors were computed using a discontinuous Helmholtz solver:

$$\|\nabla \times \nabla u\| = 4.71 \times 10^{-13}, \tag{4.56}$$

$$\|\nabla \cdot J\nabla v\| = 1.91 \times 10^{-12}. \tag{4.57}$$

For the bunny, solved using a continuous Helmholtz solver:

$$\|\nabla \times \nabla u\| = 4.52 \times 10^{-11}, \tag{4.58}$$

$$\|\nabla \cdot J\nabla v\| = 1.46 \times 10^{-9}. \tag{4.59}$$

Although the vector Laplacian is non-negligible near boundary regions (atrium) or non-smooth junctions (bunny), the proposed scheme still recovers a smooth harmonic potential U.

4.5 Weak Formulation

Another form of the covariant derivative arises from an integral formulation, known as a weak formulation. Depending on the choice of test function, various branches of weak formulations have been proposed. The most common test functions are continuous and differentiable polynomials defined over each element. Weak formulations form the foundation of popular numerical methods such as the finite element and finite volume methods. While the advantages of weak formulations remain a subject of discussion, an increasing number of computational simulations adopt them for numerical schemes.

Weak formulations avoid direct differentiation of scalar or vector functions. Instead, derivatives are transferred to test functions through integration by parts, thereby reducing the differentiability requirements for the original variables. However, the solution obtained through a weak formulation may not strictly satisfy the governing equations—it is referred to as a **weak solution**.

In this section, we outline a particular weak formulation known as the discontinuous Galerkin (DG) method. Consider a conservation law governing the advection of a scalar variable u by a velocity vector $\mathbf{v}$:

$$\frac{\partial u}{\partial t} + (\nabla \cdot \mathbf{v})u = 0. \tag{4.60}$$

Multiplying Eq. (4.60) by a test function φ and integrating over a sufficiently smooth domain Ω gives the weak form:

$$\int_{\Omega} \frac{\partial u}{\partial t} \varphi dx + \int_{\Omega} (\nabla \cdot \mathbf{v}) u \varphi dx = 0. \tag{4.61}$$

Let Ω^h be a tessellation of Ω into N non-overlapping elements Ω_e^h $(1 \leq e \leq N)$. Table 4.6 illustrates representative curvilinear meshes adopted for the weak formulation with moving frames for the spherical test problems. The subscript h denotes the characteristic edge length, representing the largest edge in each element:

$$\Omega^h = \bigcup_{e}^{N} \Omega_e^h, \quad \Omega_i^h \bigcap \Omega_j^h = \emptyset, \quad \text{if } i \neq j. \tag{4.62}$$

Table 4.6 Comparison of curvilinear meshes used in spherical test problems

	ProjMesh	Gmsh	NekMesh
Type	Projection	Lower-order mesh	High-order mesh
Optimal vertex distribution	No	Yes	Yes
Optimal high-order node distribution	No	No	Yes
Geometric approximation error	Yes	Yes	Negligible

We define the solution space $\mathscr{S}_h^p$ as a finite-dimensional space of discontinuous piecewise polynomials over Ω^h:

$$\mathscr{S}_h^p = \{u^h \in P^p(\Omega_e^h), \quad \Omega_e^h \in \Omega_h\}, \tag{4.63}$$

meaning that u^h is a polynomial of degree p on each element Ω_e^h. Choosing basis functions $\Phi_n \in \mathscr{S}_h^p$, the solution and test functions are expanded as:

$$u^h = \sum_n \hat{u}_n^h \Phi_n, \tag{4.64}$$

$$\varphi^h = \sum_n \hat{\varphi}_n^h \Phi_n, \tag{4.65}$$

where $\hat{u}_n^h$ and $\hat{\varphi}_n^h$ are the coefficients of u^h and φ^h, respectively. The basis can be interpreted in either the **modal space** or **nodal space**. Substituting Eqs. (4.64)–(4.65) into Eq. (4.61) yields:

$$\int_{\Omega_e^h} \frac{\partial \hat{u}^h}{\partial t} \hat{\varphi}^h dx + \int_{\Omega_e^h} (\nabla \cdot \mathbf{v}) \hat{u}^h \varphi^h dx = 0. \tag{4.66}$$

Applying integration by parts:

$$\int_{\Omega_e^h} \frac{\partial \hat{u}^h}{\partial t} \hat{\varphi}^h dx - \int_{\Omega_e^h} \left(\nabla \hat{\varphi}^h \cdot \mathbf{v}\right) \hat{u}^h dx + \int_{\partial \Omega_e^h} \tilde{u}^h \hat{\varphi}^h \left(\mathbf{v} \cdot \mathbf{n}\right) ds = 0, \tag{4.67}$$

where $\mathbf{n}$ is the outward unit normal vector on $\partial \Omega_e^h$, and $\tilde{u}^h$ is the numerical flux—an approximation at element interfaces (see Fig. 4.12).

4.5.1 Weak Formulation of Gradient

The direct computation of a gradient does not explicitly require the metric tensor. Therefore, the geometric characteristics of the domain do not influence the accuracy of computing the gradient. Consider the gradient of a scalar field f expanded in terms of a local moving frame $\{\mathbf{e}_i\}$ on a curved surface Ω:

$$\nabla f = (\nabla f \cdot \mathbf{e}_1)\mathbf{e}_1 + (\nabla f \cdot \mathbf{e}_2)\mathbf{e}_2 = \sum_{i=1}^{2} g_i \mathbf{e}_i. \tag{4.68}$$

To derive the weak formulation for ∇f, we multiply Eq. (4.68) by a vector-valued test function $\boldsymbol{\omega}_i = \varphi \mathbf{e}_i$ and integrate:

$$\int_{\Omega_e^h} g_i \varphi dx = \int_{\Omega_e^h} (\nabla f \cdot \mathbf{e}_i) \varphi dx. \tag{4.69}$$

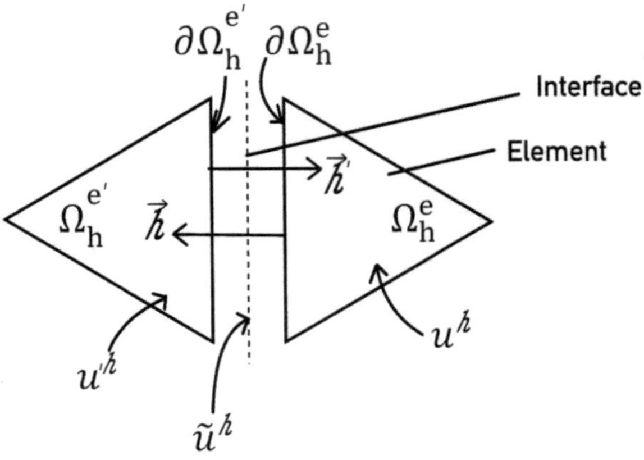

Fig. 4.12 Illustration of two neighboring elements in the DG weak formulation. Ω_e^h and Ω'^h_e denote a curved element and its neighboring element, respectively. u^h and u'^h are the numerical solutions in Ω_e^h and Ω'^h_e. $\mathbf{n}$ represents the outward normal vector on the element interface, denoted by $\partial\Omega_e^h$. A similar description is applied to $\mathbf{n}'$, which is the opposite vector to $\mathbf{n}$

Using the vector identity

$$\nabla \cdot (f\varphi \mathbf{e}_i) = (\nabla f \cdot \mathbf{e}_i)\varphi + (\nabla\varphi \cdot \mathbf{e}_i) f + (\nabla \cdot \mathbf{e}_i) f\varphi, \tag{4.70}$$

we rewrite Eq. (4.69) as:

$$\int_{\Omega_e^h} g_i \varphi dx = -\left[\int_{\Omega_e^h} (\nabla\varphi \cdot \mathbf{e}_i) f dx + \int_{\Omega_e^h} (\nabla \cdot \mathbf{e}_i) f\varphi dx\right] + \int_{\Omega_e^h} (\nabla \cdot (f\varphi\mathbf{e}_i))\varphi dx \tag{4.71}$$

By the divergence theorem, we obtain the weak formulation of the gradient:

$$\int_{\Omega_e^h} g_i \varphi dx = -\left[\int_{\Omega_e^h} (\nabla\varphi \cdot \mathbf{e}_i) f dx + \int_{\Omega_e^h} (\nabla \cdot \mathbf{e}_i) f\varphi dx\right] + \int_{\partial\Omega_e^h} (\mathbf{n} \cdot \mathbf{e}_i)\tilde{f}\varphi ds, \tag{4.72}$$

where $\tilde{f}$ denotes the numerical flux—an approximation of f on element boundaries. Common choices include upwind or central fluxes. As spatial resolution increases, the solution becomes less sensitive to the choice of flux. Notably, the geometric term $(\nabla \cdot \mathbf{e}_i)$ arises naturally in the weak form, as seen in Eq. (4.72).

4.5.1.1 Case I

As an example, consider the polar coordinate system on the Euclidean plane. Let $f(r, \theta) = \frac{1}{2}r^2$, where r is the radial coordinate and θ is the angular coordinate. Then the gradient is:

$$\nabla f = r\mathbf{e}_r = \hat{\mathbf{r}}. \tag{4.73}$$

Thus, $g_r = r$. For this coordinate system, the divergence of the unit basis vectors is:

$$\nabla \cdot \mathbf{e}_r = \frac{1}{r}, \tag{4.74}$$

$$\nabla \cdot \mathbf{e}_\theta = 0. \tag{4.75}$$

Substituting Eqs. (4.73), (4.74), and (4.75) into Eq. (4.72), we obtain:

$$\int_{\Omega_e^h} r\varphi dx = -\left[\int_{\Omega_e^h} (\nabla\varphi \cdot \mathbf{e}_r)\frac{r^2}{2} dx + \int_{\Omega_e^h} r\varphi dx\right] + \int_{\partial\Omega_e^h} (\mathbf{n} \cdot \mathbf{e}_r)\frac{r^2}{2}\varphi ds. \tag{4.76}$$

Canceling the term $\int_{\Omega_e^h} r\varphi\, dx$ on both sides gives:

$$\int_{\Omega_e^h} (\nabla\varphi \cdot \mathbf{d}_r) dx = \int_{\partial\Omega_e^h} (\mathbf{n} \cdot \mathbf{d}_r)\varphi ds, \tag{4.77}$$

where $\mathbf{d}_r = \frac{r^2}{2}\mathbf{e}_r$. Equation (4.77) states that the directional derivative of φ along $\mathbf{d}_r$ is equal to the boundary flux in the same direction. This implies that φ is a conservative field along $\mathbf{e}_r$, which is expected because $\mathbf{e}_r$ is itself conservative:

$$\nabla \times \mathbf{d}_r = \mathbf{0}. \tag{4.78}$$

4.5.1.2 Case II

Now consider $f(\theta) = \cos\theta$ on the same 2D plane. The gradient is:

$$\nabla f = -\sin\theta \mathbf{e}_\theta = -\sin\theta\hat{\boldsymbol{\theta}}. \tag{4.79}$$

So $g_\theta = -\sin\theta$. Substituting into Eq. (4.72) gives:

$$\left[\int_{\Omega_e^h} (\nabla\varphi \cdot \mathbf{d}_\theta) dx\right] = \int_{\partial\Omega_e^h} (\mathbf{n} \cdot \mathbf{d}_\theta) ds + \int_{\Omega_e^h} \sin\theta\varphi dx, \tag{4.80}$$

where $\mathbf{d}_\theta = \cos\theta\, \mathbf{e}_\theta$. Equation (4.80) shows that the gradient of φ along $\mathbf{d}_\theta$ depends on the integration path. This non-conservativeness is confirmed since the curl of $\mathbf{d}_\theta$ is non-zero:

$$\nabla \times \mathbf{d}_\theta = \frac{\sin\theta}{r}\mathbf{z} \neq \mathbf{0}. \tag{4.81}$$

When $\mathbf{e}_i$ is a conservative field, the geometric term $\int_{\Omega_e^h} (\nabla \cdot \mathbf{e}_i)\, f\varphi\, dx$ is generally non-zero, but the resulting weak gradient is path-independent. Conversely, when $\mathbf{e}_i$ is not conservative, this term vanishes and the weak formulation becomes path-dependent.

4.5.2 Weak Formulation of Divergence

Consider a scalar variable u and its associated velocity vector $\mathbf{v}$ defined on a smooth surface element Ω_e^h. The divergence of the product $u\mathbf{v}$ is given by:

$$\nabla \cdot (u\mathbf{v}) = f. \tag{4.82}$$

Expanding the left-hand side, we obtain:

$$\nabla u \cdot \mathbf{v} + u \nabla \cdot \mathbf{v} = f. \tag{4.83}$$

Equation (4.83) shows that the divergence of $u\mathbf{v}$ is the sum of the directional derivative of u along $\mathbf{v}$ and the divergence of $\mathbf{v}$ scaled by u. Expanding the velocity vector $\mathbf{v}$ in the moving frame basis as $\mathbf{v} = v^1\mathbf{e}_1 + v^2\mathbf{e}_2$, we get:

$$\sum_{i=1}^{2} v^i \nabla u \cdot \mathbf{e}_i + u \left(\nabla \cdot \sum_{j=1}^{2} v^j \mathbf{e}_j \right) = f. \tag{4.84}$$

Expanding the divergence term, we obtain:

$$\sum_{i=1}^{2} \left[\nabla (u v^i) \cdot \mathbf{e}_i + u v^i \nabla \cdot \mathbf{e}_i \right] = f. \tag{4.85}$$

Suppose $\mathbf{e}_1$ is a curl-free unit tangent vector and $\mathbf{e}_2$ is a divergence-free unit tangent vector, as illustrated in Fig. 4.13. If u and v^i are approximately constant, then the divergence of $u\mathbf{v}$ becomes approximately zero when $v^1 = 0$. This corresponds to a field with no sources or sinks. Conversely, if $v^1 \neq 0$, the divergence of $u\mathbf{v}$ is generally non-zero.

Multiplying Eq. (4.82) by a test function φ and integrating over the curved element Ω_e^h yields:

$$\int_{\Omega_e^h} \nabla \cdot (u\mathbf{v})\varphi dx = \int_{\Omega_e^h} f\varphi dx. \tag{4.86}$$

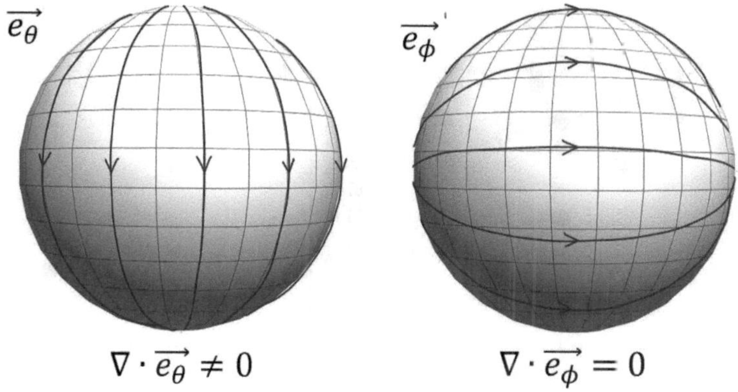

Fig. 4.13 Illustration of divergence on the sphere: the unit tangent vector $\mathbf{e}_\phi$, aligned with the ϕ-axis, has zero divergence ($\nabla \cdot \mathbf{e}_\phi = 0$), whereas the unit tangent vector $\mathbf{e}_\theta$ yields nonzero divergence ($\nabla \cdot \mathbf{e}_\theta \neq 0$)

Applying integration by parts, Eq. (4.88) becomes:

$$-\int_{\Omega_e^h} \nabla\varphi \cdot u\mathbf{v}dx + \int_{\partial\Omega_e^h} (\mathbf{n}\cdot\mathbf{v})u^*\varphi ds = \int_{\Omega_e^h} f\varphi dx. \tag{4.87}$$

where the superscript $*$ indicates that the corresponding quantity is an approximate solution at the element interfaces, often called the **numerical trace**. Substituting the expansion of $\mathbf{v}$ in the moving frame basis into Eq. (4.87), we obtain:

$$-\int_{\Omega_e^h} (\nabla\varphi \cdot \mathbf{e}_i)uv^i dx + \int_{\partial\Omega_e^h} (\mathbf{n}\cdot\mathbf{e}_i)u^*v^i\varphi ds = \int_{\Omega_e^h} f\varphi dx. \tag{4.88}$$

If $f = 0$, then Eq. (4.88) implies a conservation law: the gradient of φ along $\mathbf{e}_i$ depends only on the endpoints, regardless of the path. If $f \neq 0$, it implies that a source or sink must be included along $\mathbf{e}_i$. Note that the divergence of the moving frame vectors $\mathbf{e}_i$ does not influence the weak form divergence of $u\mathbf{v}$.

The same test problems used for direct derivatives in Eqs. (4.25) are reused here with $u = 1$. The velocity vector is given by the Rossby–Haurwitz wave defined in Eqs. (4.15)–(4.16). Three mesh types are tested: ProjMesh, Gmsh (Geuzaine and Remacle 2009), and NekMesh (Sherwin and Peiró 2002; Turner et al. 2016). ProjMesh starts with the edges of a unit cube subdivided into equal-length segments, then projects the resulting vertices onto the sphere. Gmsh is an open-source 3D mesh generator using second-order polynomial-curved edges and a frontal meshing algorithm. NekMesh is an open-source high-order 3D mesh generator. ProjMesh and Gmsh are low-order meshes with noticeable internal point errors, while NekMesh produces high-order meshes with negligible internal point error.

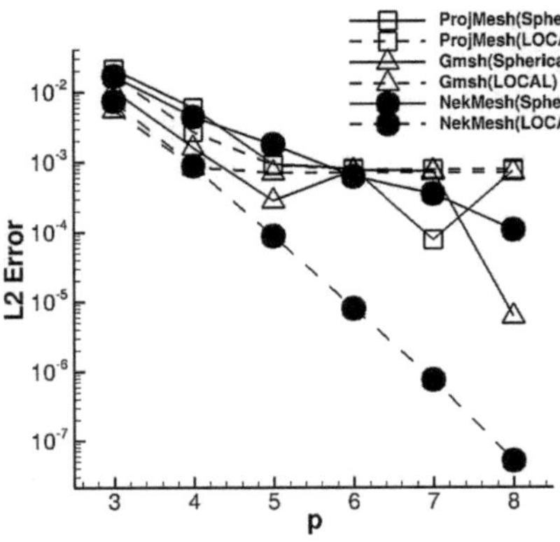

(a) By direct formulation

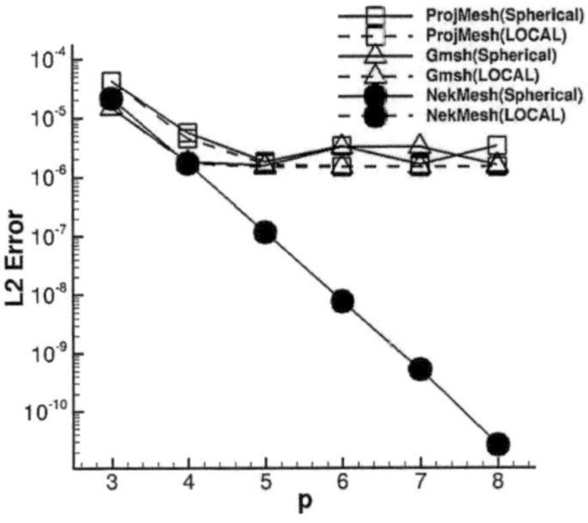

(b) By weak formulation

Fig. 4.14 Error convergence of the divergence operator using direct and weak formulations. $h = 0.4$. *Solid line* spherical MF, *dashed line* local MF, *square* ProjMesh, *triangle* Gmsh, *filled circle* NekMesh. (**a**) By direct formulation. (**b**) By weak formulation

Figure 4.14 displays convergence errors of both direct and weak divergence formulations across the three mesh types. For ProjMesh and Gmsh, error convergence stagnates as polynomial order p increases. NekMesh, however, demonstrates exponential convergence at high p, indicating that its geometric approximation error is significantly smaller than its discretization error. Note that spherical and local moving frames yield nearly identical accuracy for NekMesh.

4.5.3 Weak Formulation of Curl

Consider the curl operator of a vector field $\mathbf{v}$ defined on a curved element Ω_e^h. Assume that the curl is projected onto the surface normal vector $\mathbf{k}$ of Ω_e^h:

$$(\nabla \times \mathbf{v}) \cdot \mathbf{k} = f, \qquad \mathbf{v} \in \Omega_e^h. \tag{4.89}$$

Construct moving frames such that the unit tangent vectors $\mathbf{e}_1$ and $\mathbf{e}_2$ lie on the tangent plane. Consequently, $\mathbf{e}_3$ aligns with the surface normal vector $\mathbf{k}$. The velocity vector $\mathbf{v}$ is then expanded as:

$$\mathbf{k} = \mathbf{e}_3, \tag{4.90}$$

$$\mathbf{v} = v^1\mathbf{e}_1 + v^2\mathbf{e}_2. \tag{4.91}$$

Substituting Eqs. (4.90) and (4.91) into Eq. (4.89) gives:

$$\sum_{i=1}^{2}(\nabla \times v^i \mathbf{e}_i) \cdot \mathbf{e}_3 = f. \tag{4.92}$$

Using vector calculus identities, we obtain:

$$\sum_{i=1}^{2}\Big[\nabla \cdot ((-1)^{i+1} v^i \mathbf{e}_{3-i}) - v^i \mathbf{e}_i \cdot (\nabla \times \mathbf{e}_3)\Big] = f. \tag{4.93}$$

The first term represents the divergence of $\mathbf{v} \times \mathbf{e}_3$, while the second term accounts for geometric effects, specifically the projection of the curl of the surface normal vector onto $\mathbf{e}_i$. Multiplying Eq. (4.93) by a test function φ and integrating over Ω_e^h yields the weak formulation:

$$\sum_{i=1}^{2}\int_{\Omega_e^h}\Big[(-1)^{i+1}\nabla\varphi \cdot \mathbf{e}_{3-i} + \mathbf{e}_i \cdot (\nabla \times \mathbf{e}_3)\Big] v^i \varphi dx + \int_{\partial\Omega_e^h} \mathbf{n} \cdot (\mathbf{v}^* \times \mathbf{e}_3)\varphi ds = \int_{\Omega_e^h} f\varphi dx, \tag{4.94}$$

where $\mathbf{v}^*$ denotes the numerical flux, i.e., an approximate solution at element interfaces. Equation (4.94) describes conservation laws for the quantity $\mathbf{v} \times \mathbf{e}_3$, with an additional geometric source term $\mathbf{e}_i \cdot (\nabla \times \mathbf{e}_3)$. Using vector identities, this term expands to:

$$\mathbf{e}_i \cdot (\nabla \times \mathbf{e}_3) = \mathbf{e}_3 \cdot (\nabla \times \mathbf{e}_i) + (-1)^{i+1}\nabla \cdot \mathbf{e}_{3-i}. \tag{4.95}$$

Equation (4.95) shows that a non-negligible curl of $\mathbf{e}_i$ and non-trivial divergence of $\mathbf{e}_{3-i}$ contribute additional source terms to the curl equation in Eq. (4.89).

Figure 4.15 shows the convergence behavior for both direct and weak formulations. ProjMesh and Gmsh exhibit stagnation for higher polynomial orders ($p \geq 5$). Only NekMesh, which avoids significant internal grid point error, shows exponential convergence across all p values.

4.5.4 Laplacian Operator in Moving Frames

The Laplacian operator is a second-order differential operator defined as the divergence of the gradient of a scalar function. For a scalar function u, the Laplacian is given by:

$$\nabla^2 u = \nabla \cdot \nabla u = f. \tag{4.96}$$

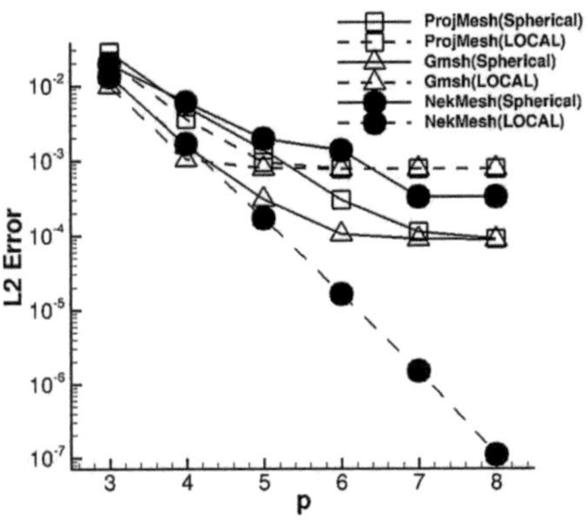

(a) By direct formulation

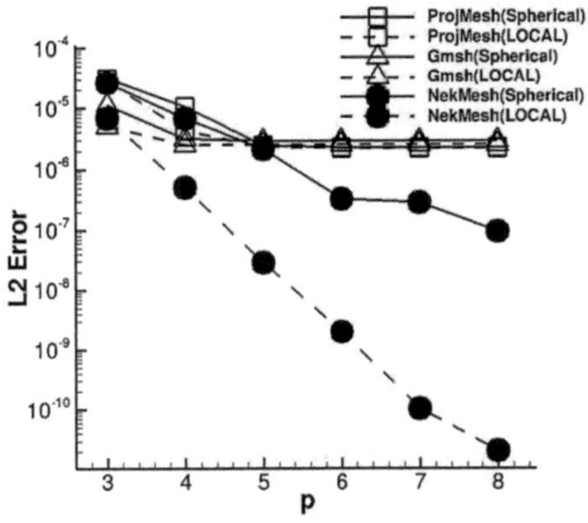

(b) By weak formulation

Fig. 4.15 Error convergence of the curl operator using direct and weak formulations. $h = 0.4$. *Solid line* spherical MF, *dashed line* local MF, *square* ProjMesh, *triangle* Gmsh, *filled circle* NekMesh. (**a**) By direct formulation. (**b**) By weak formulation

The Laplacian appears in many partial differential equations, including Poisson's equation, diffusion equations, and wave equations. In curved geometry, it generalizes to an elliptic operator known as the **Laplace–Beltrami** operator. For instance, in spherical coordinates (r, θ, ϕ), the Laplacian of a scalar function f restricted to the unit sphere becomes:

$$\nabla^2 f = \left(\frac{1}{\sin^2\theta}\right)\frac{\partial^2 f}{\partial\varphi^2} + \left(\frac{\cos\theta}{\sin\theta}\right)\frac{\partial f}{\partial\theta} + \frac{\partial^2 f}{\partial\theta^2}. \tag{4.97}$$

For example, let

$$f = \omega \sin^2\theta \cos 2\varphi, \tag{4.98}$$

Then, using Eq. (4.97), the Laplacian of f on the sphere is:

$$\nabla^2 f = -4\omega\cos 2\varphi + 2\omega\cos^2\theta\cos 2\varphi + 2\omega(2\cos^2\theta - 1)\cos 2\varphi. \tag{4.99}$$

For a direct computation of the Laplacian using moving frames, one can apply the identity $\nabla^2 f = \nabla\cdot\nabla f$, with the gradient first expanded in orthonormal moving frames as:

$$\mathbf{q} = \nabla f = q_1\mathbf{e}_1 + q_2\mathbf{e}_1. \tag{4.100}$$

Substituting Eq. (4.100) into Eq. (4.96) yields:

$$\nabla^2 f = \nabla\cdot[q_1\mathbf{e}_1 + q_2\mathbf{e}_2] = \sum_{i=1}^{2}[\nabla q_i\cdot\mathbf{e}_i + q_i(\nabla\cdot\mathbf{e}_i)]. \tag{4.101}$$

Fig. 4.16 Examples of nontrivial divergence of the first tangent vector $\mathbf{e}_1$, i.e., $\nabla \cdot \mathbf{e}_1$

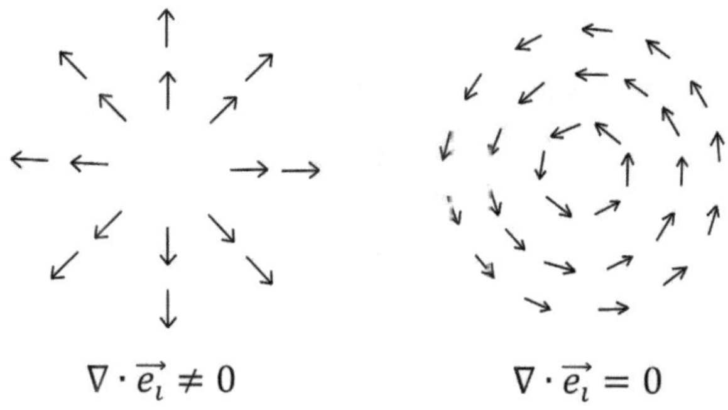

This expression reveals that the divergence of the frame vectors $\nabla \cdot \mathbf{e}_i$ contributes to the Laplacian operator. If $\mathbf{e}_i$ has nonzero divergence and $q_i \neq 0$, then $\nabla^2 f \neq 0$. However, even when $\nabla \cdot \mathbf{e}_i$ is nontrivial, as shown in Fig. 4.16, the Laplacian can vanish under the following condition:

$$-\frac{\nabla q_i}{q_i} = (\nabla \cdot \mathbf{e}_i)\mathbf{e}_i. \tag{4.102}$$

On the other hand, the Laplacian becomes nonzero if q_i varies along $\mathbf{e}_i$, producing a nonzero second derivative, even if $\mathbf{e}_i$ is divergence-free. In summary, the Laplacian is nonzero when:

1. The second derivative of f along $\mathbf{e}_i$ is nonzero.
2. The divergence of $\mathbf{e}_i$ is nonzero and q_i is nonzero and does not satisfy Eq. (4.102).

For the weak formulation, the local discontinuous Galerkin (LDG) scheme with moving frames can be applied (Chun 2014a). Consider the mixed formulation of Eq. (4.96):

$$\mathbf{q} = \nabla u, \tag{4.103}$$

$$\nabla \cdot \mathbf{q} = f. \tag{4.104}$$

The vector $\mathbf{q}$ is expanded in moving frames as:

$$\mathbf{q} = q_1\mathbf{e}_1 + q_2\mathbf{e}_2, \tag{4.105}$$

where the subscript i in q_i denotes that $\mathbf{q}$ is a covariant vector. Multiplying Eqs. (4.103) and (4.104) by a scalar test function φ or a vector test function $\boldsymbol{\omega} = \varphi\mathbf{e}_i$ and integrating over Ω_e^h yields:

$$\int_{\Omega_e^h} \mathbf{q} \cdot \boldsymbol{\omega} d\mathbf{x} = \int_{\Omega_e^h} \nabla u \cdot \boldsymbol{\omega} d\mathbf{x}, \tag{4.106}$$

$$\int_{\Omega_e^h} \nabla \cdot \mathbf{q}\varphi d\mathbf{x} = \int_{\Omega_e^h} f\varphi d\mathbf{x}. \tag{4.107}$$

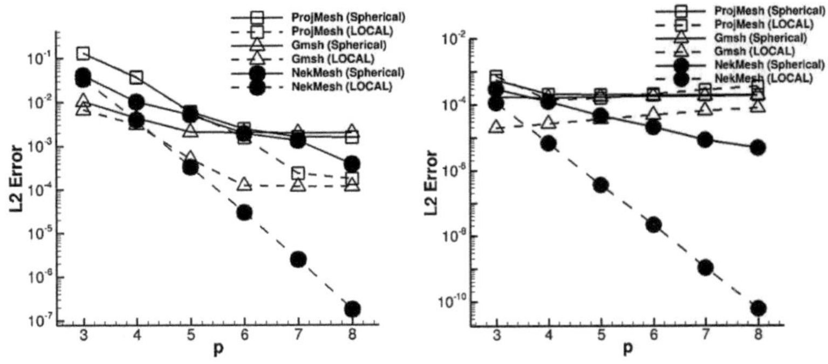

(a) By direct formulation (b) By weak formulation

Fig. 4.17 Error convergence of the Laplacian operator using direct and weak formulations. $h = 0.4$. *Solid line* spherical MF, *dashed line* local MF, *square* ProjMesh, *triangle* Gmsh, *filled circle* NekMesh. (**a**) By direct formulation. (**b**) By weak formulation

Using integration by parts, we obtain the weak formulation:

$$\int_{\Omega} f\varphi dx = \sum_{i=1}^{2}\left[\int_{\Omega} q_i(\nabla\varphi \cdot \mathbf{e}_i)dx + \int_{\partial\Omega} q_i^*(\mathbf{n}\cdot\mathbf{e}_i)\varphi ds\right], \tag{4.108}$$

$$\int_{\Omega} q_i\varphi dx = -\int_{\Omega}(\nabla\varphi\cdot\mathbf{e}_i)udx - \int_{\Omega}(\nabla\cdot\mathbf{e}_i)\varphi udx + \int_{\partial\Omega}(\mathbf{n}\cdot\mathbf{e}_i)\varphi f^* ds, \tag{4.109}$$

where q_i^* and f^* represent numerical fluxes at element interfaces.

Figure 4.17 illustrates convergence behavior similar to that observed in the divergence and curl tests. For ProjMesh and Gmsh, convergence stagnates for $p \geq 5$ (direct formulation) and for all p (weak formulation). The poor convergence in the weak formulation highlights the substantial influence of internal point errors on Laplacian computation over curved elements. In contrast, NekMesh, with negligible internal point error, shows near-ideal exponential convergence for all p.

4.6 Effects of Mesh Error on Covariant Derivative

The computation of covariant derivatives is typically performed on curved elements. These elements have curved edges or faces, as illustrated in Fig. 4.18, resulting in a spatially varying Jacobian. Using regular elements with straight edges to approximate a curved geometry requires many smaller elements, making curved elements a more efficient choice in modern scientific computing. However, any discrepancy between the curved mesh and the exact geometry introduces **mesh error**, which may influence the accuracy of numerical schemes. Since generating a perfectly accurate mesh for a given domain is often challenging or even impossible,

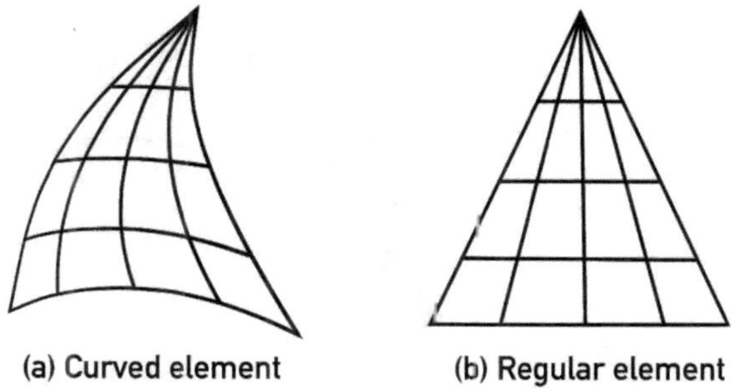

Fig. 4.18 Illustrations of (**a**) a curved element with curved edges and (**b**) a regular element with straight edges. In curved elements the Jacobian varies within the element, while in regular elements it is constant

such errors are inevitable. This section examines the possible effects of mesh error on the numerical computation of covariant derivatives.

Consider the divergence of a velocity vector **v** on a smooth curved element Ω_e^h. Multiply the divergence by a polynomial test function φ and integrate over Ω_e^h, yielding:

$$\int_{\Omega_e^h} \nabla \cdot \mathbf{v}\varphi dx = \int_{\Omega_e^h} f\varphi ds. \tag{4.110}$$

Let **k** denote the surface normal vector. The velocity vector is tangent to the surface, and thus orthogonal to **k**:

$$\mathbf{v}(\mathbf{x}) \cdot \mathbf{k}(\mathbf{x}) = 0. \qquad \mathbf{x} \in \Omega_e^h. \tag{4.111}$$

In practice, computational meshes may not satisfy this orthogonality due to geometric approximation errors, known as **mesh error**. Mesh error can be subcategorized as **vertex error**, arising at mesh vertices, and **internal point error**, occurring elsewhere. For example, assume that a velocity field in spherical coordinates lies exactly on a sphere and takes the form:

$$\mathbf{v} = v_\theta \boldsymbol{\theta} + v_\phi \boldsymbol{\phi}. \tag{4.112}$$

The velocity vector is orthogonal to **k** in the exact geometry. However, a mesh error may perturb the direction of **k**, introducing a small deviation ε. For instance, if **k** tilts slightly toward the θ-axis, we obtain a perturbed normal vector (See Fig. 4.19):

$$\mathbf{k}^{\varepsilon} = (1-\varepsilon)\mathbf{k} + \varepsilon\boldsymbol{\theta}, \qquad \text{for a small } \varepsilon. \tag{4.113}$$

Consider an exact smooth domain Ω^0 without mesh error, partitioned as:

$$\Omega^0 = \cup_e \Omega_e^0 \quad \text{where} \quad \Omega_i^0 \cap \Omega_j^0 = \emptyset, \quad \text{when } i \neq j. \tag{4.114}$$

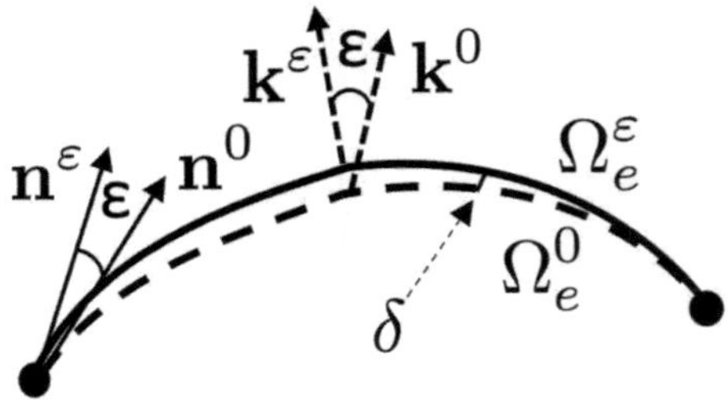

Fig. 4.19 Geometric components of a curved line Ω_e^0 and its deformed configuration Ω_e^ε displaced by a distance δ. The surface normal vectors are denoted by $\mathbf{k}^0$ and $\mathbf{k}^\varepsilon$, and the edge normal vectors are denoted by $\mathbf{n}^0$ and $\mathbf{n}^\varepsilon$ for Ω_e^0 and Ω_e^ε, respectively

For a velocity vector $\mathbf{v}^0$ defined on a curved element Ω_e^0, the divergence theorem states:

$$\int_{\Omega_e^0} \nabla^0 \cdot \mathbf{v}^0 d\mathbf{x} = \int_{\partial\Omega_e^0} \mathbf{n}^0 \cdot \mathbf{v}^0 ds, \tag{4.115}$$

where superscript 0 denotes quantities defined on the exact domain. Let Ω_e^ε be a perturbed element whose surface normal differs from the exact one by an angle ε, as in Eq. (4.113). Then the divergence theorem still holds:

$$\int_{\Omega_e^\varepsilon} \nabla^\varepsilon \cdot \mathbf{v}^\varepsilon d\mathbf{x} = \int_{\partial\Omega_e^\varepsilon} \mathbf{n}^\varepsilon \cdot \mathbf{v}^\varepsilon ds. \tag{4.116}$$

On the slightly deformed surface Ω_e^ε, the conservation of sources and sinks remains valid. In the absence of an ambient space, all vectors are projected onto the domain Ω_e^ε, and the projection is effectively instantaneous.

Although the divergence theorems in Eqs. (4.115) and (4.116) hold independently, this does not imply equivalence of the computed divergence values:

$$\int_{\Omega_e^\varepsilon} \nabla^\varepsilon \cdot \mathbf{v}^\varepsilon d\mathbf{x} \neq \int_{\Omega_e^0} \nabla^0 \cdot \mathbf{v}^0 d\mathbf{x}. \tag{4.117}$$

Domain geometry affects surface divergence. For instance, consider a spherical shell of radius r with metric:

$$ds^2 = r^2 d\theta^2 + r^2 \sin^2\theta d\phi^2, \tag{4.118}$$

corresponding to the Cartesian mapping:

$$x = r \sin\theta \cos\phi, \tag{4.119}$$

$$y = r \sin\theta \sin\phi, \tag{4.120}$$

$$z = r\cos\theta, \tag{4.121}$$

where ϕ and θ are the azimuthal and polar angles, respectively. Let $\mathbf{v}^0$ be a velocity field expanded in the spherical coordinate basis:

$$\mathbf{v}^0 = v_\theta \boldsymbol{\theta} + v_\phi \boldsymbol{\phi}. \tag{4.122}$$

where $\boldsymbol{\theta}$ and $\boldsymbol{\phi}$ are unit tangents. Then the surface divergence is:

$$\nabla^0 \cdot \mathbf{v}^0 = \frac{1}{r\sin\theta}\left(\frac{\partial v_\phi}{\partial \phi} + \frac{\partial}{\partial \theta}(v_\theta \sin\theta)\right). \tag{4.123}$$

In a perturbed domain Ω_e^ε, assume deformation only along the θ-component. The velocity field becomes:

$$\mathbf{v}^\varepsilon = v_{\theta'} \boldsymbol{\theta}' + v_\phi \boldsymbol{\phi}. \tag{4.124}$$

where the deformed unit vector $\boldsymbol{\theta}'$ satisfies:

$$\boldsymbol{\theta}' = \sin\varepsilon \mathbf{r} + \cos\varepsilon \boldsymbol{\theta}, \tag{4.125}$$

The velocity projection error $\|v_\theta - v_{\theta'}\|$ is of order $\mathscr{O}(\varepsilon^2)$. The radial deformation varies with θ':

$$r' = r + \sin\varepsilon(\theta'), \tag{4.126}$$

resulting in:

$$x = r'\sin\theta'\cos\phi, \tag{4.127}$$

$$y = r'\sin\theta'\sin\phi, \tag{4.128}$$

$$z = r'\cos\theta'. \tag{4.129}$$

The divergence on the deformed surface is then:

$$\nabla^\varepsilon \cdot \mathbf{v}^\varepsilon = \frac{1}{r'\sin\theta'}\left[\frac{\cos\varepsilon}{r'}\frac{\partial \varepsilon(\theta')}{\partial \theta'} v_{\theta'}\sin\theta' + \left(\frac{\partial v_\phi}{\partial \phi} + \frac{\partial}{\partial \theta'}(v_{\theta'}\sin\theta')\right)\right]. \tag{4.130}$$

Assuming ε is small and $\theta \approx \theta'$, the difference between Eqs. (4.123) and (4.130) is approximated by:

$$\|\nabla^\varepsilon \cdot \mathbf{v}^\varepsilon - \nabla^0 \cdot \mathbf{v}^0\| \approx \frac{v_{\theta'}}{r^2}\frac{\partial \varepsilon(\theta')}{\partial \theta'}. \tag{4.131}$$

Equation (4.131) shows how mesh error alters divergence. Perturbations in the direction of flow (e.g., θ) can significantly impact divergence, whereas errors in the orthogonal direction (e.g., ϕ) do not. Additionally, deformation of the surface area further amplifies divergence discrepancies.

4.7 Summary

This chapter developed a comprehensive numerical framework for computing **covariant derivatives** on two-dimensional curved surfaces using **local orthonormal coordinate (LOC) moving frames**. By replacing traditional coordinate-based differential operators with frame-based formulations, the chapter demonstrated how differential operators such as **gradient, divergence, curl**, and **Laplacian** can be computed with improved geometric adaptability and numerical accuracy.

We began by formulating **covariant derivatives** for vector fields in spherical and Cartesian coordinates and verified accuracy through comparison with exact expressions. Exponential convergence was observed using LOC moving frames on various mesh configurations, even in cases where spherical frames suffered from singularities near poles.

The operators were then individually redefined and numerically validated:

- **Gradient operators** were reformulated in terms of moving frames, avoiding explicit metric tensors. The convergence tests verified their robustness and efficiency.
- **Divergence** was analyzed both in direct and weak forms, highlighting the role of the frame divergence term $\nabla \cdot \mathbf{e}_i$. LOC frames provided consistently accurate results across a range of polynomial orders.
- **Curl operators** were reformulated through projections onto the surface normal and verified using the Rossby–Haurwitz velocity field. Results showed that incorporating frame geometry yields superior accuracy.
- **Laplacian operators**, expressed as the divergence of the gradient, were explored both analytically and numerically. A weak formulation using the local discontinuous Galerkin (LDG) method was introduced and validated.

An important application presented was the **Helmholtz–Hodge decomposition (HHD)**, which decomposes a vector field into irrotational, incompressible, and harmonic components. The method was validated on standard and complex geometries, including the human atrium and Stanford bunny. Results showed high fidelity in reconstructing all components, including accurate harmonic fields despite geometric complexity.

The chapter also highlighted the impact of **mesh error**, particularly from curved vs. regular elements. Analytical derivations and numerical experiments confirmed that internal point errors significantly affect the accuracy of covariant operators. High-order curvilinear meshes (e.g., from NekMesh) minimized these errors and allowed exponential convergence in all test problems.

Together, these developments affirm that **moving frame-based methods** are not only geometrically faithful but also computationally robust. They offer a **unified, mesh-compatible, and accurate** strategy for implementing differential operators on curved domains, laying the foundation for advanced PDE solvers in scientific and engineering applications.

4.8 Exercises and Conceptual Applications

1. At a point on a curved surface, an orthonormal moving frame $\{\mathbf{e}_1, \mathbf{e}_2, \mathbf{e}_3\}$ satisfies:

$$\mathbf{e}_1 = \mathbf{e}_2 \times \mathbf{e}_3, \tag{4.132}$$

$$\mathbf{e}_2 = \mathbf{e}_3 \times \mathbf{e}_1, \tag{4.133}$$

$$\mathbf{e}_3 = \mathbf{e}_1 \times \mathbf{e}_2. \tag{4.134}$$

   a. Prove that this system is positively oriented.
   b. Show that any deviation from this triple-cross condition results in either a non-orthogonal or negatively oriented frame.

2. Given the structure equations for a moving frame:

$$d\mathbf{x} = \sum_i \omega_i \mathbf{e}_i, \tag{4.135}$$

$$d\mathbf{e}_i = \sum_j \omega_{ij} \mathbf{e}_j, \tag{4.136}$$

   a. Explain the geometric meaning of the 1-forms ω_i and ω_{ij}.
   b. For a spherical surface, derive expressions for ω_{ij} in terms of the coordinates (θ, ϕ).

3. In numerical simulations on curved surfaces:

   a. Describe how moving frames help reduce geometric complexity in approximating differential operators.
   b. Outline an algorithm to compute the connection 1-forms ω_{ij} using discrete frame fields on a triangulated surface.

4. Let the frames $\{\mathbf{e}_1^p, \mathbf{e}_2^p, \mathbf{e}_3^p\}$ and $\{\mathbf{e}_1^q, \mathbf{e}_2^q, \mathbf{e}_3^q\}$ be constructed at two neighboring grid points p and q.

   a. Describe how to measure the angular difference between frames using rotation matrices.

b. If the angle between normals $\mathbf{e}_3^p$ and $\mathbf{e}_3^q$ is small, show that the rotation matrix approximates the identity plus a skew-symmetric matrix.

5. Consider the scalar diffusion equation on a curved surface:

$$\frac{\partial u}{\partial t} = \nabla \cdot (D \nabla u), \tag{4.137}$$

where D is the diffusivity tensor.

a. Write the weak formulation of this equation using a local moving frame field.
b. Explain how the basis transformation via moving frames improves the accuracy and stability of the numerical scheme.

Chapter 5
Applications to PDEs on Curved Surfaces

Provando e riprovando (Verify the one and disprove the other)

—Galileo Galilei

After introducing the methods for computing covariant differential operators in the preceding chapters, we now incorporate moving frames into the numerical solution of partial differential equations (PDEs). We are particularly interested in solving PDEs defined on complex domains. By a **complex domain**, we mean a non-Euclidean domain, which typically requires a **curved coordinate system** for reference. Depending on the number of basis vectors, this domain may be a two-dimensional (2D) curved surface or a three-dimensional (3D) curved volume. Additionally, the medium may be anisotropic—meaning that it exhibits direction-dependent properties—referred to as an **anisotropic medium**, where wave propagation speeds differ along specific directions. See Fig. 5.1 for illustrations.

PDEs generally consist of differential operators acting on scalar or vector fields. One application of moving frames in PDE solvers is the expansion of vectors—such as gradients—in terms of orthonormal moving frames. Moving frames constructed on a complex domain reflect both the geometry and the medium properties of the domain. Thus, expressing vector quantities in moving frames naturally adapts the PDEs to the underlying domain. The **orthonormality** of moving frames simplifies the computation of vector components, eliminating the need to explicitly involve the metric tensor or Christoffel symbols.

Although moving frames can be applied to PDEs in any domain, including Euclidean domains with homogeneous and isotropic media, their advantages are most pronounced in complex geometries or anisotropic environments. The moving frame-based numerical scheme supports both direct and weak formulations of PDEs. In this chapter, we primarily focus on the weak formulation, as it aligns more comfortably with the authors' methodology. The direct formulation follows naturally from the numerical treatment of covariant derivatives introduced in Chap. 4.

S. Chun, *Moving Frames for the Numerical Solution of Partial Differential Equations in Complex Domains*, Springer Asia Pacific Mathematics Series 7,
https://doi.org/10.1007/978-981-95-4495-0_5

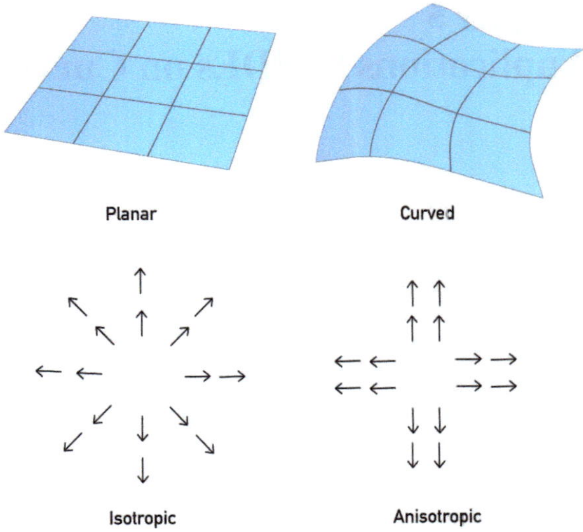

Fig. 5.1 Top row: Illustrations of planar and curved domains, where the coordinate axes are linear and curved, respectively. Bottom row: Illustrations of isotropic and anisotropic media. In isotropic media, propagation is uniform in all directions, whereas in anisotropic media, propagation depends on direction

We will consider several representative PDEs, including conservation laws, diffusion equations, shallow water equations, and Maxwell's equations. Conservation laws involve divergence operators, diffusion equations involve Laplacians, shallow water equations combine divergence and gradient operators, and Maxwell's equations include curl operators. In the remainder of this chapter, we explore how such PDEs can be formulated in a weak form using moving frames on a general curved surface denoted by Ω. The domain Ω may represent a spherical shell with constant Gaussian curvature, an irregular surface with varying positive curvature, or a non-convex surface with both positive and negative curvature. Solving PDEs numerically on such general curved surfaces is one of the most significant computational achievements of the moving frame framework.

Before proceeding, we must clarify certain conditions on the surface Ω to ensure the applicability of moving frames in numerical PDE solvers. Most importantly, moving frames must exist and be **uniquely** defined at every grid point. This condition is mathematically expressed by the concept of a **locally Euclidean** domain, which loosely requires that the neighborhood around each point be sufficiently flat. An example of a non-locally Euclidean domain is a sharply pointed geometry where the endpoint is not differentiable, as shown in Fig. 5.2.

$$\Omega = \cup \Omega_e^h, \quad \Omega_i^h \cap \Omega_j^h = \emptyset \text{ if } i \neq j. \tag{5.1}$$

Each curved element may be a regular element with straight edges or a curvilinear element with curved edges. Throughout this book, we assume all elements are curvilinear with sufficiently smooth edges. Curvilinear elements are generally more computationally efficient and accurate than regular elements in terms of element count and domain approximation quality.

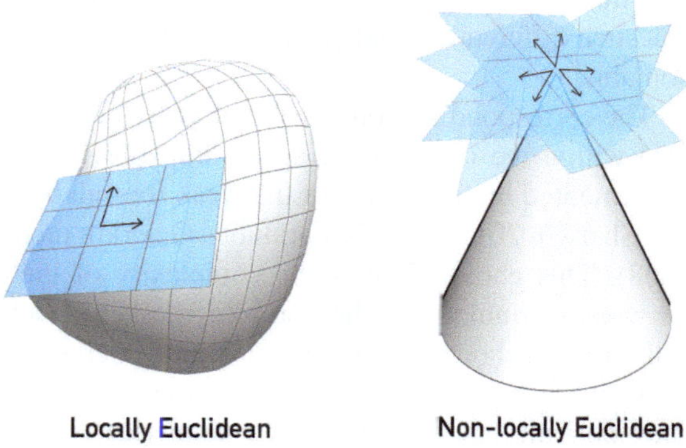

Fig. 5.2 Examples of locally Euclidean and non-locally Euclidean domains. In a locally Euclidean surface, a unique tangent plane exists at every point, whereas in a non-locally Euclidean geometry, the tangent plane is not uniquely defined at certain points

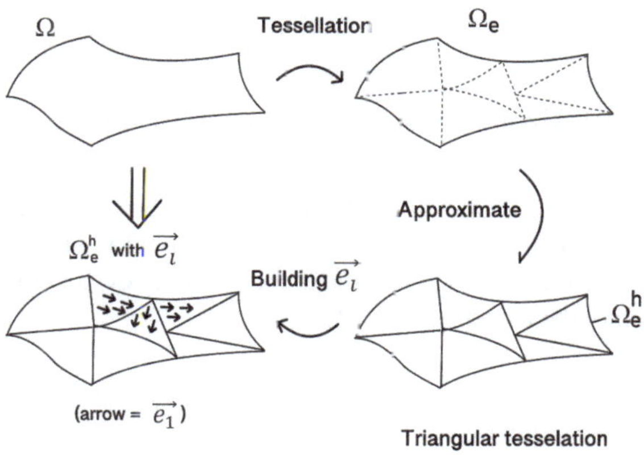

Fig. 5.3 Process of mesh construction with moving frames ($\mathbf{e}_i$) at grid points in a multidimensional domain Ω. The domain is tessellated into triangular elements (Ω_e^h), and continuous, differentiable first tangent vectors ($\mathbf{e}_1$) are constructed within each element. The other tangent vectors ($\mathbf{e}_i$, $2 \leq i \leq 3$)are constructed by an orthonormalization scheme

A computational tessellation, which approximates the domain by Ω_e^h, is also referred to as a **mesh**. The deviation in grid point locations between the mesh Ω_e^h and the original domain Ω is called the **mesh error**. In practice, this mesh error is typically smaller than the discretization error. However, it is not entirely negligible—particularly in long-time integrations—where it can accumulate. For simplicity, we assume that the mesh error is negligible in this chapter.

On the domain Ω, the moving frame $\mathbf{e}_i$ is constructed as follows. At each grid point, the third unit vector $\mathbf{e}_3$ is aligned with the surface normal vector $\mathbf{k}$. The other two unit vectors, $\mathbf{e}_1$ and $\mathbf{e}_2$, lie on the tangent plane, orthogonal to the normal vector $\mathbf{k}$ (or equivalently, to $\mathbf{e}_3$). This specific configuration is a special case where the moving frames are adapted to the geometry of the domain and are therefore referred to as **adapted moving frames** (Needham 2021).

Moreover, the adapted moving frames must be differentiable within each element Ω_e^h. However, this does not imply that they must be differentiable across the entire domain. In fact, differentiability across element interfaces is not required. This configuration is called **element-wise differentiable adapted moving frames**, as previously discussed in Chap. 2. See Fig. 5.3 for an overview of the mesh construction process incorporating moving frames. Throughout this chapter, we assume that such differentiable adapted moving frames are appropriately constructed on the domain.

The corresponding computational code for each PDE is provided as C++ source files within the open-source spectral/hp library Nektar++ (Cantwell et al. 2015), implemented in the context of the weak formulation using spectral/hp (Karniadakis and Sherwin 2013) and discontinuous Galerkin methods (Hesthaven and Warburton 2008). This choice reflects the complexity of the numerical scheme inherent in the weak formulation. The complex meshes—such as the sphere, pseudosphere, and atrium—shown in this section were generated using the open-source 3D finite element mesh generator `Gmsh` (Geuzaine and Remacle 2009). For higher accuracy and compatibility with high-order numerical schemes, high-order meshes can be generated using `NekMesh` (Green et al. 2024).

5.1 Conservation Laws on Curved Surfaces

Conservation laws in PDEs describe the preservation of a physical quantity, denoted by u, as it is transported by a velocity field $\mathbf{v}$ within a smooth domain Ω. Consider the following mathematical formulation of a conservation law on a sufficiently smooth curved surface Ω:

$$\frac{\partial u}{\partial t} + \nabla \cdot (u\mathbf{v}) = 0, \qquad \mathbf{v} \in \Omega. \tag{5.2}$$

Equation (5.2) states that the time evolution of the quantity u is governed by the divergence of the flux $u\mathbf{v}$, which acts as a source or sink for u. Integrating over the domain Ω and applying the divergence theorem yields the equivalent integral form:

$$\int_{\Omega} \frac{\partial u}{\partial t} dx + \int_{\partial\Omega} (\mathbf{n} \cdot \mathbf{v}) u dA = 0, \tag{5.3}$$

where $\mathbf{n}$ is the outward-pointing normal vector to the boundary $\partial\Omega$. Equation (5.3) indicates that changes in u over time are due to fluxes across the domain boundary. The quantity u is conserved unless there is an influx or efflux through the boundary.

The conservation law in Eq. (5.2) involves a single divergence operator. When $\nabla \cdot (u\mathbf{v}) = 0$, the PDE becomes time-independent and describes a steady-state condition. In Euclidean space, assuming $\mathbf{v} = (v_x, v_y, v_z)$, the divergence of the flux $u\mathbf{v}$ is computed component-wise as:

$$\nabla \cdot (u\mathbf{v}) = \frac{\partial u v_x}{\partial x} + \frac{\partial u v_y}{\partial y} + \frac{\partial u v_z}{\partial z}. \tag{5.4}$$

However, the divergence operator behaves differently on a curved domain. Even when a vector field has constant components along a curved coordinate line, the divergence may still be nontrivial due to geometric effects induced by the curvature of the coordinate axes. Suppose that an adapted set of moving frames is constructed on a two-dimensional curved domain Ω. Then, the velocity vector $\mathbf{v}$ can be expressed as:

$$\mathbf{v} = v^1\mathbf{e}_1 + v^2\mathbf{e}_2, \qquad \mathbf{e}^i \in \Omega. \tag{5.5}$$

Multiplying Eq. (5.2) by a test function φ and integrating over an approximated curved element Ω_e^h, we obtain:

$$\int_{\Omega_e^h} \frac{\partial u}{\partial t}\varphi dx + \int_{\Omega_e^h} \nabla \cdot (u\mathbf{v})\varphi dx = 0. \tag{5.6}$$

Assuming the mesh approximates the curved domain sufficiently well, we apply the divergence theorem to obtain:

$$\int_{\Omega_e^h} \frac{\partial u}{\partial t}\varphi d\mathbf{x} - \int_{\Omega_e^h} (\nabla\varphi \cdot \mathbf{v})\, u d\mathbf{x} + \int_{\partial\Omega_e^h} (\mathbf{v}\cdot\mathbf{n})\, \tilde{u}\varphi dS = 0, \tag{5.7}$$

where the tilde notation, referred to as the *numerical flux*, indicates that the corresponding quantity is an approximation at the interface between elements. For example, $\tilde{u}$ represents an approximate value of u at the interface, since u may be discontinuous across element boundaries. A simple approach is to average the values of u across the interface, which is permissible in the discontinuous Galerkin (DG) method. More accurate schemes, however, assign different $\tilde{u}$ values on each edge, depending on conservation constraints or the direction of information propagation. In essence, the numerical flux modifies the flux term to improve the stability and accuracy of the divergence operator. Substituting Eq. (5.5) into Eq. (5.7) gives:

$$\int_{\Omega_e^h} \frac{\partial u}{\partial t}\varphi d\mathbf{x} - \sum_{i=1}^{2}\left[\int_{\Omega_e^h} (\nabla\varphi \cdot \mathbf{e}_i)\, v^i u d\mathbf{x} - \int_{\partial\Omega_e^h} (\mathbf{e}_i \cdot \mathbf{n})\, v^i \tilde{u}\varphi dS\right] = 0. \tag{5.8}$$

Interestingly, Eq. (5.8) can be further simplified by rescaling the moving frames, i.e., altering the magnitude of the unit tangent vectors $\mathbf{e}_i$. While $\mathbf{e}_i$ are typically of unit length, we define scaled tangent vectors $\mathbf{d}_i$ by:

$$\mathbf{d}_i = v^i\mathbf{e}_i, \qquad 1 \le i \le 2, \tag{5.9}$$

as illustrated in Fig. 5.4. These vectors satisfy:

$$\mathbf{d}_i \cdot \mathbf{d}_j = \delta^i_j, \qquad \|\mathbf{d}_i\| = v^i. \tag{5.10}$$

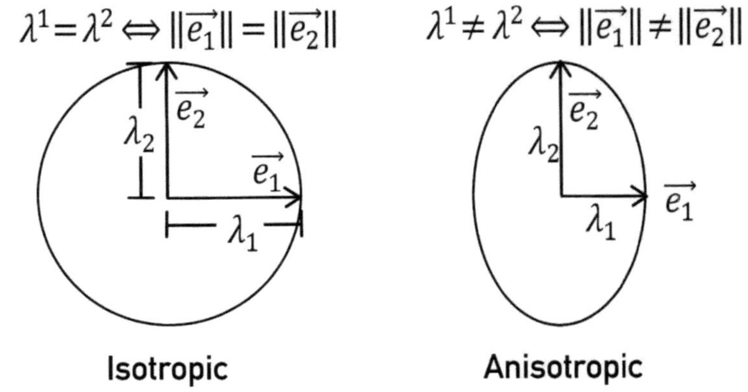

Fig. 5.4 Rescaled moving frames $\mathbf{e}_i$ in isotropic and anisotropic media. In an isotropic medium, the lengths of $\mathbf{e}_1$ and $\mathbf{e}_2$, denoted by λ^1 and λ^2, are equal. In an anisotropic medium, they are generally not equal, i.e., $\lambda^1 \neq \lambda^2$

Here, only the magnitude of the tangent vectors changes; their directions remain the same. Substituting into Eq. (5.8), we obtain:

$$\int \frac{\partial u}{\partial t}\varphi d\mathbf{x} - \sum_{i=1}^{2}\left[\int \left(\nabla\varphi \cdot \mathbf{d}^i\right) u d\mathbf{x} - \int_{\partial\Omega_e} \left(\mathbf{d}^i \cdot \mathbf{n}\right) \tilde{u}\varphi dS\right] = 0. \tag{5.11}$$

If the velocity field is time-independent, then the scaled frames $\mathbf{d}_i$ are also time-independent and can be precomputed once, reducing computational cost. Compared to Eq. (5.7), Eq. (5.11) avoids multiplying v^i at every time step. However, this approach can break down if the velocity vanishes, i.e., $v^i = 0$. In that case:

$$\|\mathbf{d}_i\| = 0, \quad 1 \leq i \leq 2, \tag{5.12}$$

and the spatial terms in Eq. (5.11) vanish, yielding:

$$\int \frac{\partial u}{\partial t}\varphi d\mathbf{x} = 0, \tag{5.13}$$

which implies that u is constant in time. Thus, the region where u evolves over time is restricted to where the velocity (and hence the frame length) is nonzero.

Note that Eq. (5.11) contains no explicit geometric terms, such as metric tensors, even though Eq. (5.2) originally involves divergence on a curved surface. In Eq. (5.11), the term $\nabla\varphi \cdot \mathbf{d}^i$ is a dot product between a covariant vector ($\nabla\varphi$) and a contravariant vector ($\mathbf{d}^i$), in which the metric tensor cancels out. The term $\mathbf{d}^i \cdot \mathbf{n}$ represents a boundary integral, which incorporates geometric information implicitly through the Jacobian of the curved edges, not through the metric tensor. Consequently, Eq. (5.11) is effectively free of explicit metric tensor terms.

5.1.1 [NEKTAR++] Computing Conservation Laws on Curved Surfaces

The numerical schemes defined in Eqs. (5.8) or (5.11) have been implemented within the spectral/hp element framework of `Nektar++`, featuring essential geometric adjustments to accurately represent conservation laws on curved surfaces. The primary modification involves replacing the standard Euclidean derivative operator `IProductWRT DerivBase` with the directional derivative `IProductWRT DirectionalDerivBase`. This allows differential operations to be precisely aligned with local geometry through the use of moving frames.

Moreover, the edge normal vector **n** is represented explicitly in the local moving frame basis rather than global Cartesian coordinates. This approach ensures that flux computations and differential operations are coherent with the curved geometry. The numerical flux evaluation via *GetFwdBwdTracePhys* supports scenarios where the moving frames may be discontinuous across element boundaries, emphasizing the flexibility and accuracy of the method.

Below is an annotated implementation illustrating these key geometric and numerical considerations within `Nektar++`:

```
/**
 * @brief Computes the weak Discontinuous Galerkin (DG) formulation of a
     directional advection term
 *        using moving frames. Applies to a scalar/vector field over a
     curved surface or domain.
 *
 * @param InField  Input field values at each node (physical space).
 * @param OutField Output field values after applying weak directional
     advection operator.
 */
void MMFAdvection::WeakDGDirectionalAdvection(
    const Array<OneD, const Array<OneD, NekDouble>> &InField,
    Array<OneD, Array<OneD, NekDouble>> &OutField)
{
    // Number of modal coefficients (per variable)
    int ncoeffs         = GetNcoeffs();

    // Total number of trace points (interface DoFs)
    int nTracePointsTot = GetTraceNpoints();

    // Number of solution variables (e.g., components)
    int nvariables      = m_fields.size();

    // Allocate a container for input fields (already in physical space)
    Array<OneD, Array<OneD, NekDouble>> physfield(nvariables);

    // Assign input field values directly
    for (int i = 0; i < nvariables; ++i)
    {
        physfield[i] = InField[i];
    }

    // Allocate weak directional derivatives (in modal space) for each
       moving frame direction
    Array<OneD, Array<OneD, NekDouble>> WeakDeriv(m_shapedim);

    // Loop over all solution variables
    for (int i = 0; i < nvariables; ++i)
    {
```

```
        // Compute weak derivatives along each moving frame direction
        for (int j = 0; j < m_shapedim; ++j)
        {
            WeakDeriv[j] = Array<OneD, NekDouble>(ncoeffs, 0.0);

            // Compute inner product of grad(u) with moving frame e^j in weak form
            // Projects gradient of physfield[i] along m_movingframes[j]
            m_fields[0]->IProductWRTDirectionalDerivBase(
                m_movingframes[j], physfield[i], WeakDeriv[j]);
        }

        // --- Numerical Flux Computation for DG Method ---

        // Containers for upwind flux, forward and backward trace values
        Array<OneD, NekDouble> flux(nTracePointsTot, 0.0);
        Array<OneD, NekDouble> Fwd(nTracePointsTot);
        Array<OneD, NekDouble> Bwd(nTracePointsTot);

        // Compute forward and backward trace values at element interfaces
        m_fields[i]->GetFwdBwdTracePhys(physfield[i], Fwd, Bwd);

        // Compute upwind numerical flux using dot product of velocity and normal vector
        // m_traceVn is the normal velocity at trace points
        m_fields[i]->GetTrace()->Upwind(m_traceVn, Fwd, Bwd, flux);

        // Initialize the output array for variable i
        OutField[i] = Array<OneD, NekDouble>(ncoeffs, 0.0);

        // Loop over shape dimensions (e.g., 2D: phi and theta)
        for (int j = 0; j < m_shapedim; ++j)
        {
            // Compute (n \cdot MF) * flux for both Fwd and Bwd traces
            Vmath::Vmul(nTracePointsTot, &flux[0], 1, &m_ncdotMFFwd[j][0], 1,
                        &Fwd[0], 1);
            Vmath::Vmul(nTracePointsTot, &flux[0], 1, &m_ncdotMFBwd[j][0], 1,
                        &Bwd[0], 1);

            // Negate weak derivative for integration
            Vmath::Neg(ncoeffs, WeakDeriv[j], 1);

            // Integrate numerical flux across faces using Fwd and Bwd values
            // Note: This generalizes the case where (n \cdot MF)_FWD \neq -(n \cdot MF)_BWD
            m_fields[i]->AddFwdBwdTraceIntegral(Fwd, Bwd, WeakDeriv[j]);

            // Mark field as no longer in physical space
            m_fields[i]->SetPhysState(false);

            // Accumulate directional contributions into output modal coefficients
            Vmath::Vadd(ncoeffs, &WeakDeriv[j][0], 1, &OutField[i][0], 1,
                        &OutField[i][0], 1);
        }
    }
}
```

Listing 5.1 Computation of conservation laws on a curved surface using moving frames. This `Nektar++` implementation applies weak DG formulations and directional derivatives aligned explicitly with surface geometry, ensuring accurate flux computations and differential consistency

This carefully designed implementation ensures robust handling of conservation laws by explicitly integrating geometric effects into the numerical schemes, thereby significantly enhancing accuracy and stability on complex curved geometries.

5.1.2 Numerical Test Problem for Conservation Laws

To test the convergence of the numerical scheme described by Eq. (5.8), we consider a benchmark problem adapted from Chun (2012). A smooth cosine bell is initialized on a spherical surface and advected along a great circle. After a full 360° rotation, the bell returns to its original position and shape, allowing for a direct comparison to the initial condition to measure numerical error. The scalar field u and the velocity field $\mathbf{v}(\mathbf{x})$ evolve according to the following conservation law:

$$\frac{\partial u(\mathbf{x})}{\partial t} + \nabla \cdot (\mathbf{v}(\mathbf{x})u(\mathbf{x})) = 0, \qquad \mathbf{x} \in \Omega \tag{5.14}$$

The velocity field is defined by:

$$\mathbf{v}(\mathbf{x}) = (-\tilde{u}\sin\varphi - \tilde{v}\sin\theta\cos\varphi,\ \tilde{u}\cos\varphi - \tilde{v}\sin\theta\sin\varphi,\ \tilde{v}\cos\theta), \tag{5.15}$$

where the velocity component is defined as:

$$\tilde{u} = \omega_f(\cos\theta\cos\alpha + \sin\theta\cos\varphi\sin\alpha), \tag{5.16}$$

$$\tilde{v} = -\omega_f\sin\varphi\sin\alpha, \tag{5.17}$$

where φ and θ denote the longitude and latitude, respectively. The parameter ω_f is the angular frequency, and α specifies the inclination of the axis of rotation from the north pole. For a sphere of radius R, $\omega_f = 2\pi R$ corresponds to one full revolution per unit time.

The initial scalar field is a cosine bell defined as:

$$u(\mathbf{x}, t = 0) = \begin{cases} \frac{1}{2}(1 + \cos(\pi \,\mathrm{dist}(\theta, \varphi)/r_\ell)), & \text{if } \mathrm{dist}(\theta, \varphi) < r_\ell, \\ 0.0, & \text{if } \mathrm{dist}(\theta, \varphi) \geq r_\ell, \end{cases} \tag{5.18}$$

where r_ℓ is the radius of the cosine bell, and the distance function is given by:

$$\mathrm{dist}(\theta, \varphi) = R\cos^{-1}\left[\sin\theta_c\sin\theta + \cos\theta_c\cos\theta\cos(\varphi - \varphi_c)\right], \tag{5.19}$$

where (θ_c, φ_c) denote the center of the bell. The initial locations of θ_c and φ_c can be chosen arbitrarily and do not affect the computational results. The parameters are summarized in Table 5.1.

Figure 5.5 illustrates the convergence behavior of the numerical scheme for the conservation law in Eq. (5.8), applied using spherical moving frames on the sphere.

Table 5.1 Simulation parameters for the conservation law test problem

	Ω_f	Δt	r_ℓ	α	θ_c	φ_c
	$\frac{2\pi}{3}$	$\frac{1}{1000}$	$\frac{7\pi}{64}$	0	0	0

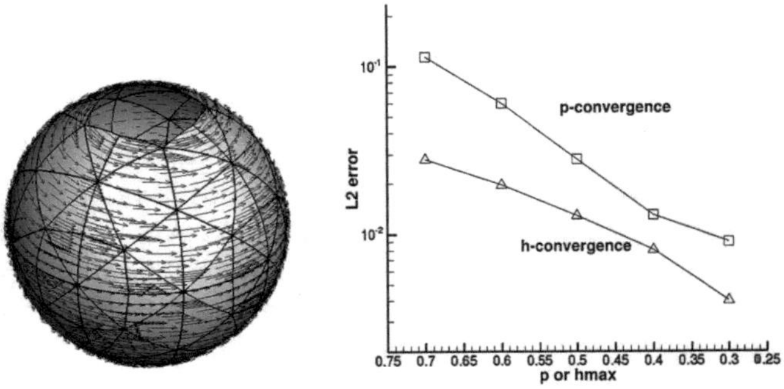

(a) Moving frames ($\mathbf{e}_1$) (b) Convergence

Fig. 5.5 (**a**) The cosine bell at $T = 0$ on a spherical shell, with elements near the poles removed to avoid singularities. The moving frames are aligned with the latitudinal axis and are referred to as *spherical moving frames*. (**b**) Exponential h- and p-convergence errors of the numerical scheme for conservation laws

Spherical moving frames introduce singularities at the poles, leading to the non-differentiability of the tangent vectors. Consequently, elements near the poles are excluded from the simulation. In this example, the first tangent vector is aligned along the latitudinal direction (φ), as shown in Fig. 5.5a. Figure 5.5b demonstrates spectral convergence with respect to p (the polynomial degree of the solution) and h (the maximum edge length of elements).

Similar convergence behavior is observed when using *local* moving frames. In this case, element truncation is not required because local moving frames do not generate geometric singularities. Figure 5.6 shows the initial condition, the distribution of the first moving frames, and both p- and h-convergence results for this setup.

Moreover, the moving frame-based numerical scheme can be applied to general curved surfaces. As an example, consider the irregular surface defined by:

$$2x^2 + y^4 + y^2 + z^4 + z^2 = 2. \tag{5.20}$$

An irregular surface refers to one where the Gaussian curvature is spatially varying, as shown in Fig. 5.7a. On this surface, we solve the same conservation law with a modified velocity field defined as:

$$\widetilde{\mathbf{v}}(\mathbf{x}) = \Omega_{\mathscr{P}}(\mathbf{x})\mathbf{v}_{\mathscr{P}}(\mathbf{x}), \tag{5.21}$$

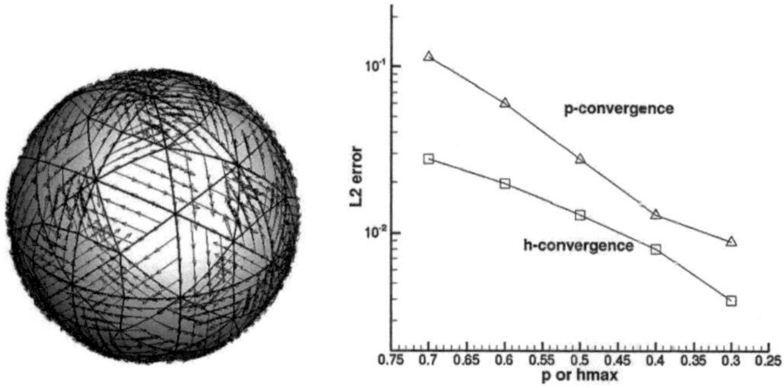

(a) Moving frames ($\mathbf{e}^1$) (b) Convergence

Fig. 5.6 (**a**) Similar setup as Fig. 5.5, but without truncation of elements around the poles. The moving frames are discontinuous across element interfaces and are referred to as *local moving frames*. (**b**) Exponential h- and p-convergence errors of the numerical scheme for conservation laws using local moving frames

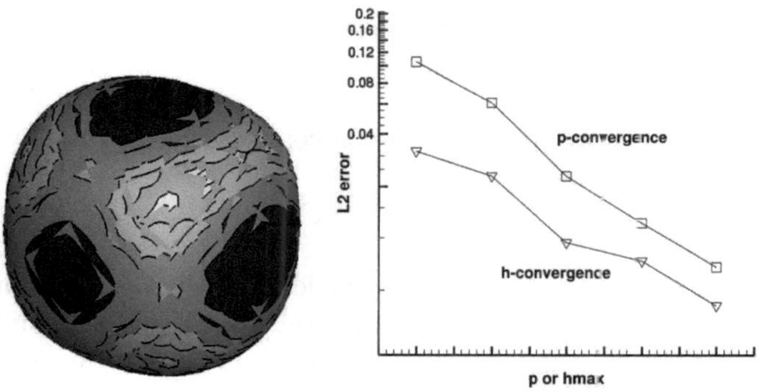

(a) Gaussian curvature (b) Convergence

Fig. 5.7 (**a**) An irregular closed surface with spatially varying Gaussian curvature. (**b**) Exponential h- and p-convergence errors of the numerical scheme for conservation laws on the irregular surface using local moving frames

where the subscript $\mathscr{P}$ denotes projection onto the tangent plane of the surface. This velocity vector, which is proportional to the circumferential length, preserves the shape of the cosine bell after a full rotation. The circumferential length at a given z is computed as follows:

Step 1: For each z, compute points (x_i, y_i) satisfying Eq. (5.20) N.
for a large number (5.22)

$$\text{Step 2: Evaluate } \Omega_{\mathscr{P}}(\mathbf{x}) = \sum_{i=1}^{N} \left[(x_{i+1} - x_i)^2 + (y_{i+1} - y_i)^2 \right]. \quad (5.23)$$

(a) Gaussian curvature

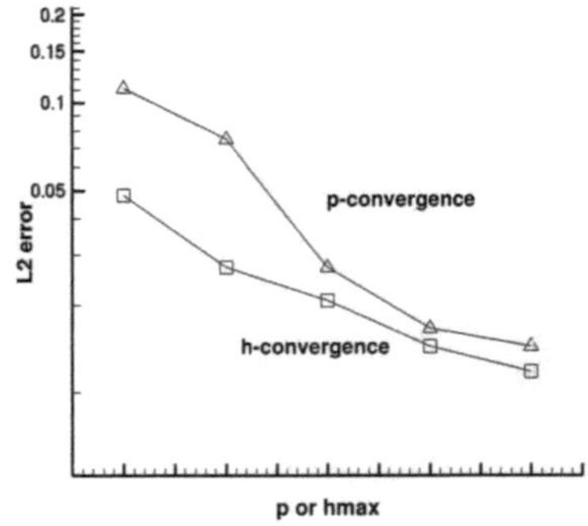

(b) Convergence

Fig. 5.8 (**a**) A non-convex closed surface with spatially varying Gaussian curvature. (**b**) Exponential h- and p-convergence errors of the numerical scheme for conservation laws on the non-convex surface using local moving frames

The rest of the test problem setup follows that of the spherical shell. Exponential convergence in both h and p is confirmed on this irregular surface, as shown in Fig. 5.7b.

Comparable results are obtained for a non-convex surface defined by:

$$2x^2 + 2(y^4 - y^2) + z^4 + z^2 = 2. \tag{5.24}$$

Figure 5.8a shows the distribution of rapidly varying Gaussian curvature on this surface. The angular frequency is calculated similarly to the irregular case but decreases linearly to zero for $z > 0.7$, due to the local non-convexity of the surface. Again, exponential convergence is observed (see Fig. 5.8b). These results confirm that surface non-convexity does not compromise the accuracy or stability of the moving frame-based scheme for conservation laws. However, on surfaces with steep curvature variations, sufficiently high spatial resolution is required for accurate computation. For more details, see Chun (2012).

5.2 Modeling Diffusion and Reaction Processes on Surfaces

Diffusion equations involve the Laplacian operator as the primary dynamic term. When a scalar function representing chemical or biological reactions in the medium is added to the diffusion equation, the resulting form is referred to as a **diffusion–reaction equation**. These dynamics describe a wide range of biological and chemical phenomena, typically involving the following sequence of processes:

1. Propagation by excitation of neighboring units,
2. Initiation by an influx exceeding a threshold,

3. Attraction of external resources,
4. Self-excitation,
5. Further propagation to unexcited neighbors.

For instance, the propagation of electrical signals in the heart and brain is often modeled using diffusion–reaction equations. This framework also captures the broader notion of **signal propagation**, or more generally **information propagation**, drawing parallels with electromagnetic wave propagation, including light.

On a smooth curved element $\Omega_e^h \subset \Omega$, consider the diffusion equation given by:

$$\nabla \cdot \mathscr{D}\nabla u(\mathbf{x}) = f, \tag{5.25}$$

where $\mathscr{D}$ is a symmetric **diffusivity tensor** with six independent components:

$$\mathscr{D} = \begin{bmatrix} d_{xx} & d_{xy} & d_{xz} \\ d_{xy} & d_{yy} & d_{yz} \\ d_{xz} & d_{yz} & d_{zz} \end{bmatrix}. \tag{5.26}$$

We can also express Eq. (5.25) in mixed form using an auxiliary vector field $\mathbf{q}$:

$$\nabla \cdot \mathbf{q}(\mathbf{x}) = f, \tag{5.27}$$

$$\mathbf{q}(\mathbf{x}) = \mathscr{D}\nabla u(\mathbf{x}), \tag{5.28}$$

If $\mathscr{D}$ is diagonal, the medium is said to be **isotropic**, meaning the diffusion modifies only the magnitude, not the direction, of ∇u. Conversely, if $\mathscr{D}$ contains nonzero off-diagonal terms, the medium is **anisotropic**, and the direction of diffusion may change. In anisotropic media, propagation tends to align with the dominant eigenvectors of $\mathscr{D}$ when their associated eigenvalues are large.

On a two-dimensional curved domain, the (a, a)-component of $\mathscr{D}$, aligned with curved coordinate direction ξ_a, can be expressed as:

$$d^{aa} = \sum_{i,j=1}^{3} \gamma^{ij} \frac{\partial \xi_a}{\partial z^i} \frac{\partial \xi_a}{\partial z^j}, \tag{5.29}$$

where γ^{ij} are the diffusivity coefficients in Euclidean coordinates z^i. For further insight, we diagonalize $\mathscr{D}$ via eigendecomposition:

$$\mathscr{D} = Q \Lambda Q^T = [\mathbf{v}_1\ \mathbf{v}_2] \begin{bmatrix} \lambda^1 & 0 \\ 0 & \lambda^2 \end{bmatrix} [\mathbf{v}_1\ \mathbf{v}_2]^T, \tag{5.30}$$

where λ^i and $\mathbf{v}_i$ are eigenvalues and eigenvectors, respectively. Suppose the gradient is expanded as:

$$\nabla u = \mathbf{q} = q^1 \mathbf{v}_1 + q^2 \mathbf{v}_2, \tag{5.31}$$

then the action of $\mathscr{D}$ on the gradient becomes:

$$\mathscr{D}\nabla u = q^1\lambda^1\mathbf{v}_1 + q^2\lambda^2\mathbf{v}_2. \tag{5.32}$$

Equation (5.32) shows that the diffusivity tensor scales the gradient in each eigendirection. In isotropic media ($\lambda^1 = \lambda^2$), propagation is uniform in all directions. In anisotropic media ($\lambda^1 \neq \lambda^2$), propagation is directional, favoring the axis with the larger eigenvalue. In practice, the vacuum is isotropic for electromagnetic waves, whereas biological tissues like cardiac and neural tissue often exhibit strong anisotropy in electric signal propagation.

Moving frames offer a natural way to incorporate anisotropy. We align the moving frames $\mathbf{e}_i$ with the eigenvectors $\mathbf{u}_i$ of $\mathscr{D}$:

$$\mathbf{e}_i = \mathbf{u}_i, \quad 1 \le i \le 2. \tag{5.33}$$

We further define scaled tangent vectors:

$$\mathbf{d}_i = \lambda^i\mathbf{e}_i, \quad 1 \le i \le 2. \tag{5.34}$$

In isotropic media ($\lambda^1 = \lambda^2$), all moving frames have equal magnitude. In anisotropic media, their lengths reflect the eigenvalue anisotropy. See Fig. 5.9.

The diffusion problem is often accompanied by boundary conditions for $\mathbf{x} \in \partial\Omega_e^h$:

$$u(\mathbf{x}) = g_D(\mathbf{x}), \tag{5.35}$$

$$\mathbf{q}(\mathbf{x}) \cdot \mathbf{n}(\mathbf{x}) = g_N(\mathbf{x}), \tag{5.36}$$

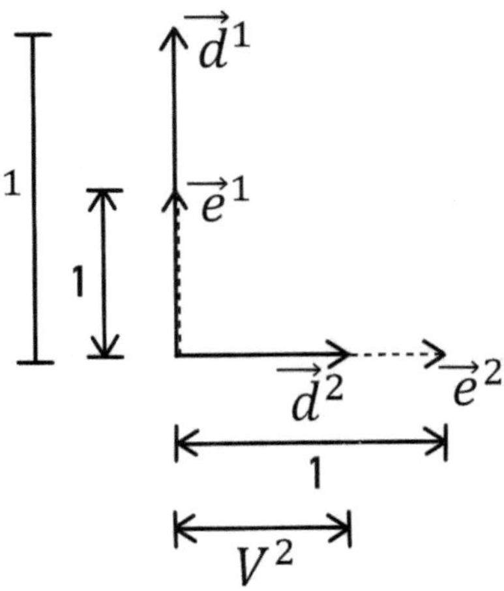

Fig. 5.9 Eigenvalues of the diffusivity tensor, denoted by V^i, and the corresponding moving frames, denoted by $\mathbf{d}^i$. The unit tangent vectors $\mathbf{e}^i$ for moving frames $\{\mathbf{e}_i\}$ are rescaled according to the magnitude of V^i, yielding the new tangent vectors $\mathbf{d}^i$, correspondingly new moving frames $\{\mathbf{d}_i\}$

where $\mathbf{n}$ is the outward unit normal, and g_D, g_N are the Dirichlet and Neumann boundary values, respectively. At the boundary, aligning $\mathbf{e}_1$ with $\mathbf{n}$, we write:

$$\mathbf{q}(\mathbf{x}) = q^1(\mathbf{x})\mathbf{e}_1(\mathbf{x}) + q^2(\mathbf{x})\mathbf{e}_2(\mathbf{x}). \tag{5.37}$$

Then Eq. (5.36) becomes for $\mathbf{x} \in \partial\Omega_e^h$:

$$q^1(\mathbf{x}) = g_N(\mathbf{x}). \tag{5.38}$$

5.2.1 Weak Formulation on Curved Domains

Multiplying Eqs. (5.27)–(5.28) by test functions φ (scalar) and $\boldsymbol{\omega}$ (vector), respectively, we obtain:

$$\int_{\Omega_e^h} (\nabla \cdot \mathbf{q})\, \varphi dx = \int_{\Omega_e^h} f\varphi dx, \tag{5.39}$$

$$\int_{\Omega_e^h} \boldsymbol{\omega} \cdot \mathbf{q} dx = \int_{\Omega_e^h} \boldsymbol{\omega} \cdot \mathscr{D}\nabla u dx. \tag{5.40}$$

Applying integration by parts yields:

$$-\int_{\Omega_e^h} \nabla\varphi \cdot \mathbf{q} dx + \int_{\partial\Omega_e^h} (\tilde{\mathbf{q}} \cdot \mathbf{n})\varphi dx = \int_{\Omega_e^h} f\varphi dx, \tag{5.41}$$

$$\int_{\Omega_e^h} \boldsymbol{\omega} \cdot \mathbf{q} dx = -\int_{\Omega_e^h} \mathscr{D}\nabla \cdot \boldsymbol{\omega} u dx + \int_{\partial\Omega_e^h} (\boldsymbol{\omega} \cdot \mathbf{n})\tilde{u} dx. \tag{5.42}$$

The tilde terms denote numerical fluxes. For isotropic media, using unit-length moving frames:

$$\mathbf{q} = q^1\mathbf{e}_1 + q^2\mathbf{e}_2, \tag{5.43}$$

$$\boldsymbol{\omega} = \varphi\mathbf{e}_m, \qquad \text{for } 1 \leq m \leq 2. \tag{5.44}$$

Substituting into Eqs. (5.41)–(5.42), we obtain:

$$\sum_{m=1}^{2} \left[\int_{\Omega_e^h} q^m (\nabla\varphi \cdot \mathbf{e}_m)\, dx + \int_{\partial\Omega_e^h} \tilde{q}^m (\mathbf{n} \cdot \mathbf{e}_m)\varphi ds \right] = \int_{\Omega_e} f\varphi dx, \tag{5.45}$$

$$\int_{\Omega_e^h} q^m \varphi dx = -\int_{\Omega_e^h} \nabla \cdot (\varphi\mathbf{e}_m) u dx + \int_{\partial\Omega_e^h} (\mathbf{n} \cdot \mathbf{e}_m)\varphi\tilde{u} ds. \tag{5.46}$$

For anisotropic media, using scaled frames:

$$\mathbf{q} = \bar{q}^1\mathbf{d}_1 + \bar{q}^2\mathbf{d}_2, \tag{5.47}$$

$$\boldsymbol{\omega} = \varphi\mathbf{d}_m, \quad \text{for } 1 \le m \le 2 \tag{5.48}$$

with the new vector $\mathbf{q}$ of which component is defined as:

$$q^i = \lambda^i\bar{q}^i, \quad 1 \le i \le 2. \tag{5.49}$$

The resulting weak formulation is:

$$\sum_{m=1}^{2}\left[\int_{\Omega_e^h} \bar{q}^m\left(\nabla\varphi \cdot \mathbf{d}_m\right)dx + \int_{\partial\Omega_e^h} \tilde{\bar{q}}^m(\mathbf{n} \cdot \mathbf{d}_m)\varphi ds\right] = \int_{\Omega_e^h} f\varphi dx, \tag{5.50}$$

$$\int_{\Omega_e^h} \|\mathbf{d}_m\|^2\bar{q}^m\varphi dx = -\int_{\Omega_e^h} \nabla\cdot(\varphi\mathbf{d}_m)u dx + \int_{\partial\Omega_e^h} (\mathbf{n}\cdot\mathbf{d}_m)\varphi\tilde{u} ds, \quad \text{for } 1 \le m \le 2. \tag{5.51}$$

Equations (5.50)–(5.51) are equivalent to Eqs. (5.45)–(5.46), with the anisotropy absorbed into the frame lengths $\|\mathbf{d}_m\|$.

5.2.2 *[NEKTAR++] Computing Diffusion Equations on Curved Surfaces*

The numerical schemes introduced in Eqs. (5.50) and (5.51) have been implemented using the spectral/hp element framework provided by `Nektar++`. These implementations follow modifications analogous to those applied in conservation law formulations, particularly tailored to handle diffusion equations on curved geometries. Central to this approach is the use of directional derivatives computed with the `IProductWRTDirectionalDerivBase` routine. This ensures that differential operations are precisely aligned with the geometry of the curved surfaces by utilizing local moving frame bases. Moreover, to maintain geometric consistency, the edge normal vector $\mathbf{n}$ is also expressed explicitly within the local moving frame.

An additional geometric correction term involving the divergence of the moving frames must also be incorporated into the numerical scheme. This compensates for the covariant characteristics intrinsic to differential operators on curved surfaces (Xu 2004). While ideally, divergence computations should be strictly covariant, in practice, an approximation using Euclidean divergence is acceptable provided that the mesh resolution is sufficiently refined. Such an approach simplifies the numerical implementation while preserving computational accuracy.

Below is the annotated code snippet detailing the practical implementation in `Nektar++`, explicitly showcasing how diffusion flux computations and geometric corrections are integrated within local moving frame formulations:

```
Array<OneD, NekDouble> MMFSystem::WeakDGMMFLDG2D(
    const int var, const Array<OneD, const NekDouble> &inarray,
    const Array<OneD, const Array<OneD, NekDouble>> &movingframes
{
    // Total number of quadrature points and coefficients
    int nq        = m_fields[var]->GetTotPoints();
    int ncoeffs   = m_fields[var]->GetNcoeffs();
    int nTracePts = m_fields[var]->GetTrace()->GetTotPoints();

    // Temporary workspaces for physical and coefficient space
    Array<OneD, NekDouble> tmp(nq);
    Array<OneD, NekDouble> tmpc(ncoeffs);

    // For storing projected derivatives
    Array<OneD, NekDouble> qfieldc(ncoeffs);

    // Storage for numerical fluxes
    Array<OneD, NekDouble> flux(nTracePts);
    Array<OneD, NekDouble> fluxFwd(nTracePts);
    Array<OneD, NekDouble> fluxBwd(nTracePts);

    // Storage for directional derivatives in moving frame basis
    Array<OneD, Array<OneD, NekDouble>> qfieldMMF(m_shapedim);
    for (int j = 0; j < m_expdim; ++j)
    {
        qfieldMMF[j] = Array<OneD, NekDouble>(nq, 0.0);
    }

    // Compute upwind numerical flux for scalar variable u
    flux = ComputeufluxMMF(var, inarray);

    // Compute dot products of edge normals with moving frame vectors (forward/backward)
    Array<OneD, Array<OneD, NekDouble>> ncdotMFFwd;
    Array<OneD, Array<OneD, NekDouble>> ncdotMFBwd;
    ComputencdotMF(movingframes, ncdotMFFwd, ncdotMFBwd, 0);

    // Compute divergence of moving frames: \nabla \cdot e^i for each direction
    Array<OneD, Array<OneD, NekDouble>> DivMF;
    ComputeDivMF(m_DerivType, movingframes, DivMF, 0);

    // Main loop over spatial directions (e.g., \xi and \eta)
    for (int j = 0; j < m_expdim; ++j)
    {
        // Step 1: Weak derivative: (\nabla_{e_i} \phi, u) = IProductWRTDirectionalDerivBase
        m_fields[0]->IProductWRTDirectionalDerivBase(movingframes[j], inarray, qfieldc);

        // Step 2: Multiply by -1 to prepare for weak DG formulation
        Vmath::Neg(ncoeffs, qfieldc, 1);

        // Step 3: Add numerical flux integrals from forward and backward states
        Vmath::Vmul(nTracePts, &flux[0], 1, &ncdotMFFwd[j][0], 1, &fluxFwd[0], 1);
        Vmath::Vmul(nTracePts, &flux[0], 1, &ncdotMFBwd[j][0], 1, &fluxBwd[0], 1);
        m_fields[var]->AddFwdBwdTraceIntegral(fluxFwd, fluxBwd, qfieldc);

        // Step 4: Add geometric correction: -((\nabla \cdot e^i) u, \phi)
```

```
        Vmath::Vmul(nq, &inarray[0], 1, &DivMF[j][0], 1, &tmp[0], 1);
        Vmath::Neg(nq, &tmp[0], 1);
        m_fields[0]->IProductWRTBase(tmp, tmpc);
        Vmath::Vadd(ncoeffs, tmpc, 1, qfieldc, 1, qfieldc, 1);

        m_fields[0]->SetPhysState(false);

        // Step 5: Multiply by inverse mass matrix to get physical values
        m_fields[0]->MultiplyByElmtInvMass(qfieldc, qfieldc);
        m_fields[0]->BwdTrans(qfieldc, qfieldMMF[j]);
    }

    // Compute numerical flux based on qfieldMMF and surface geometry
    fluxFwd = ComputeqfluxMMF(var, inarray, ncdotMFFwd, ncdotMFBwd,
         qfieldMMF);

    // Final projection: (\nabla_{e_i} \phi, qfieldMMF) for i = \xi, \eta
    qfieldc = Array<OneD, NekDouble>(ncoeffs, 0.0);
    for (int j = 0; j < m_expdim; ++j)
    {
        m_fields[0]->IProductWRTDirectionalDerivBase(movingframes[j],
                                                      qfieldMMF[j], tmpc);
        Vmath::Vadd(ncoeffs, tmpc, 1, qfieldc, 1, qfieldc, 1);
    }

    // Final assembly: apply minus sign
    Vmath::Neg(ncoeffs, qfieldc, 1);

    // Add interface flux contributions to weak form
    m_fields[var]->AddTraceIntegral(fluxFwd, qfieldc);
    m_fields[var]->SetPhysState(false);

    // Output: weak DG result in coefficient space
    return qfieldc;
}
```

Listing 5.2 Computation of diffusion equations with a spatially varying diffusivity tensor on curved surfaces using moving frames. This Nektar++ implementation utilizes weak DG formulations with directional derivatives rigorously adapted to the local geometry

The provided implementation ensures accurate representation of geometric effects, significantly enhancing numerical stability and precision when solving diffusion equations on complex curved geometries.

5.2.3 Numerical Test Problems

Consider the following linear diffusion–reaction system on the spherical shell of unit radius, originally proposed by Turing (1953) and further developed in Pudykiewicz (2006). Let $u(\mathbf{x}, t)$ and $v(\mathbf{x}, t)$ be scalar variables satisfying:

$$\frac{\partial u(\mathbf{x})}{\partial t} = \mu \nabla^2 u(\mathbf{x}) + a u(\mathbf{x}) + b v(\mathbf{x}), \qquad \mu > 0, \;\; a, \; b \in \mathbb{R}, \tag{5.52}$$

$$\frac{\partial v(\mathbf{x})}{\partial t} = \nu \nabla^2 v(\mathbf{x}) + c u(\mathbf{x}) + d v(\mathbf{x}), \qquad \nu > 0, \;\; c, \; d \in \mathbb{R}, \tag{5.53}$$

where μ and ν denote the diffusivity constants. The spherical shell is assumed to be without boundary. If, however, a boundary is introduced by removing some elements, the following Dirichlet boundary conditions are applied:

$$u(\mathbf{x}) = u_D, \quad \mathbf{x} \in \partial\Omega, \tag{5.54}$$

$$v(\mathbf{x}) = v_D, \quad \mathbf{x} \in \partial\Omega. \tag{5.55}$$

As all points $\mathbf{x}$ lie on the spherical shell, the variables u and v can be expanded in terms of spherical harmonics $Y_n^m(\theta, \varphi)$:

$$u(\theta, \varphi, t) = \sum_{n=0}^{\infty} \sum_{m=-n}^{m=n} A_n^m(t) Y_n^m(\theta, \varphi), \tag{5.56}$$

$$v(\theta, \varphi, t) = \sum_{n=0}^{\infty} \sum_{m=-n}^{m=n} B_n^m(t) Y_n^m(\theta, \varphi), \tag{5.57}$$

where $\theta \in [0, \pi]$ is the polar angle and $\varphi \in [-\pi, \pi]$ is the azimuthal angle. The coefficients $A_n^m(t)$ and $B_n^m(t)$ depend only on time. The spherical harmonics satisfy the eigenvalue equation:

$$\nabla^2 Y_n^m = -\frac{n(n+1)}{r^2} Y_n^m. \tag{5.58}$$

Using this property, the analytic solutions of Eqs. (5.52)–(5.53) take the form:

$$u(\theta, \varphi, t) = \sum_{n=0}^{\infty} \sum_{m=-n}^{m=n} e^{\gamma_n t} (\tilde{A}_n^m \cosh(\delta_n t) + \tilde{B}_n^m \sinh(\delta_n t)) Y_n^m(\theta, \varphi), \tag{5.59}$$

$$v(\theta, \varphi, t) = \sum_{n=0}^{\infty} \sum_{m=-n}^{m=n} e^{\gamma_n t} (\tilde{C}_n^m \cosh(\delta_n t) + \tilde{D}_n^m \sinh(\delta_n t)) Y_n^m(\theta, \varphi), \tag{5.60}$$

with the following parameters:

$$a = -B - 1, \;\; b = A^2, \;\; c = B, \;\; d = -A^2, \tag{5.61}$$

$$\tilde{A}_n^m = A_n^m(0), \;\; \tilde{B}_n^m = \frac{1}{\delta_n}\left[(a_n - \gamma_n) A_n^m(0) + b B_n^m(0)\right], \tag{5.62}$$

$$\tilde{C}_n^m = B_n^m(0), \;\; \tilde{D}_n^m = \frac{1}{\delta_n}\left[c A_n^m(0) + (d_n - \gamma_n) B_n^m(0)\right], \tag{5.63}$$

$$a_n = a - \mu\frac{n(n+1)}{r^2}, \quad d_n = d - \nu\frac{n(n+1)}{r^2}, \tag{5.64}$$

$$\gamma_n = \frac{1}{2}(a_n + d_n), \quad \delta_n = \frac{1}{2}\sqrt{(a_n + d_n)^2 - 4(a_n d_n - bc)}. \tag{5.65}$$

For numerical implementation, we choose $n = 5$ and $m \in [-5, 5]$, with:

$$A_5^m = [-0.5839,\ -0.8436,\ -0.4764,\ -0.6475,\ 0.1886,\ 0.8709, \tag{5.66}$$

$$-\ 0.8338,\ 0.1795\ -\ 0.7873,\ 0.8842,\ 0.2943], \tag{5.67}$$

$$B_5^m = [-0.6263,\ 0.9803,\ 0.7222,\ 0.5945,\ 0.6026,\ -0.2076, \tag{5.68}$$

$$0.4556,\ 0.6024,\ 0.9695,\ -0.4936,\ 0.1098]. \tag{5.69}$$

The parameters are set as: $A = 2$, $B = 5$, $\mu = 10^{-3}$, and $\nu = 2 \times 10^{-3}$.

5.2.3.1 Case Study: Unit Sphere Geometry

Let ℓ denote the maximum edge length of a curved triangular element. Curved elements are defined such that their edges are polynomial approximations (typically of second or third order) of the underlying domain. Table 5.2 lists $\ell_{\max}$ values and the corresponding number of curved elements used in the domain.

For the h-convergence test, we fix the polynomial degree at $p = 5$, and for the p-convergence test, we vary p between 3 and 7 while fixing $\ell_{\max} = 0.6$. Figure 5.10 demonstrates exponential convergence for both u and v on the spherical shell using local moving frames. Dashed lines illustrate the failure of convergence when the divergence of the moving frame, $\nabla \cdot \mathbf{e}^m$, is omitted—emphasizing its essential role

Table 5.2 Maximum edge length $\ell_{\max}$ vs. number of elements for the spherical shell domain (unit radius)

$\ell_{\max}$	0.4	0.5	0.6	0.7	0.8
num. elem.	480	288	192	144	96

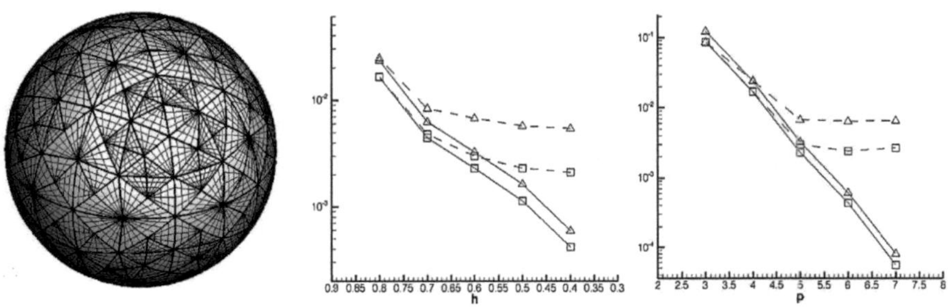

(a) Mesh (b) h-convergence (c) p-convergence

Fig. 5.10 (**a**) Mesh of a unit sphere with maximum edge length $\ell_{\max} = 0.4$. (**b**) h-convergence is displayed for u (squares) and v (triangles). (**c**) p-convergence is displayed for u (squares) and v (triangles). Dashed lines show degraded convergence without the $\nabla \cdot \mathbf{e}^m$ term in Eqs. (5.50) and (5.51). (Courtesy of Chun 2014a) in numerical accuracy. These results indicate that local moving frames are generally not **divergence-free**.

5.2.3.2 Case Study: Irregular Surface Geometry

We next consider a smooth irregular surface with slowly varying Gaussian curvature, defined by:

$$2x^2 + y^4 + y^2 + z^4 + z^2 = 2.0, \quad (x, y, z) \in \mathbf{R}^3. \tag{5.70}$$

The Gaussian curvature on this surface ranges from 0.4 to 1.4. Due to the complexity of the geometry, analytic solutions are not available. Instead, we evaluate convergence by fixing a set of sampling nodes and comparing the computed solution against that from a highly refined reference mesh. Figure 5.11a illustrates the sampling node distribution.

For numerical tests, we use a time step $\Delta t = 10^{-4}$ and final time $T = 1.0$. Table 5.3 provides the mesh configuration. For h-convergence, $p = 5$ is fixed; for p-convergence, we vary p from 3 to 7 with $\ell_{\max} = 0.6$. Figures 5.11b, c confirm exponential convergence in both h and p. Again, the results without the divergence term $\nabla \cdot \mathbf{e}^m$ (dashed lines) exhibit degraded accuracy.

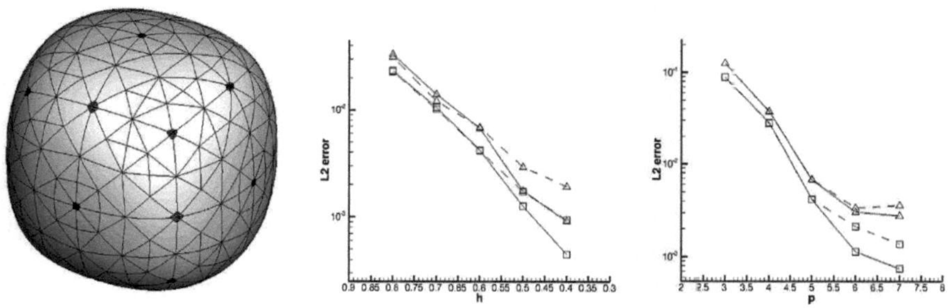

(a) Sampling points (b) h-convergence (c) p-convergence

Fig. 5.11 (**a**) Sampling points are displayed on the irregular surface of Eq. (5.70). (**b**) h- and (**c**) p-convergence for u (square) and v (triangle). Dashed lines indicate failure of convergence when omitting $\nabla \cdot \mathbf{e}^m$ in Eqs. (5.50) and (5.51). (Courtesy of Chun 2014a)

Table 5.3 Maximum edge length $\ell_{\max}$ vs. number of elements for the irregular surface

$\ell_{\max}$	0.4	0.5	0.6	0.7	0.8
num. elem.	544	336	192	144	112

5.2.3.3 Case Study: Non-Convex Surface Geometry

Lastly, we examine a smooth non-convex surface defined by:

$$2x^2 + 2(y^4 - y^2) + z^4 + z^2 = 2.0, \quad (x, y, z) \in \mathbf{R}^3. \tag{5.71}$$

This geometry yields Gaussian curvature values ranging from -1.0 to 3.0, including negative curvature. Sampling nodes are used to evaluate convergence, as shown in Fig. 5.12. Table 5.4 presents the mesh configuration.

While negative Gaussian curvature and non-convexity do not impair the accuracy of the moving frame scheme, rapid curvature variation does. Figure 5.12b illustrates steep curvature changes within a single element, which hinder exponential convergence as shown in Fig. 5.12c.

To restore exponential convergence, one can locally refine the mesh in regions of high curvature variation. Figure 5.13a shows a refined region with smoother curvature distribution. This refinement maintains the same overall geometry but reduces curvature variation within each element. Figure 5.13b confirms that exponential convergence is recovered with adaptive mesh refinement based on curvature distribution.

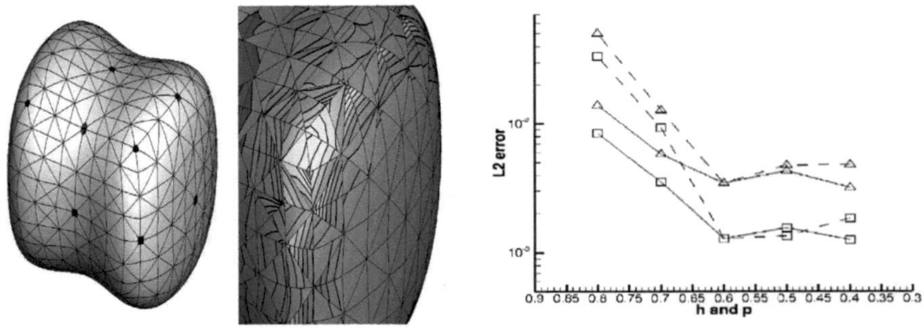

(a) Sampling points (b) Gaussian curvature (c) Convergence

Fig. 5.12 (**a**) Sampling nodes for the non-convex surface defined in Eq. (5.71). (**b**) Rapidly varying Gaussian curvature, where brighter areas correspond to higher curvature. (**c**) h- and p-convergence failure with a uniform grid without mesh refinement. (Courtesy of Chun 2014a)

Table 5.4 Maximum edge length $\ell_{\max}$ vs. number of elements for the non-convex surface

$\ell_{\max}$	0.4	0.5	0.6	0.7	0.8
num. elem.	784	496	336	256	176

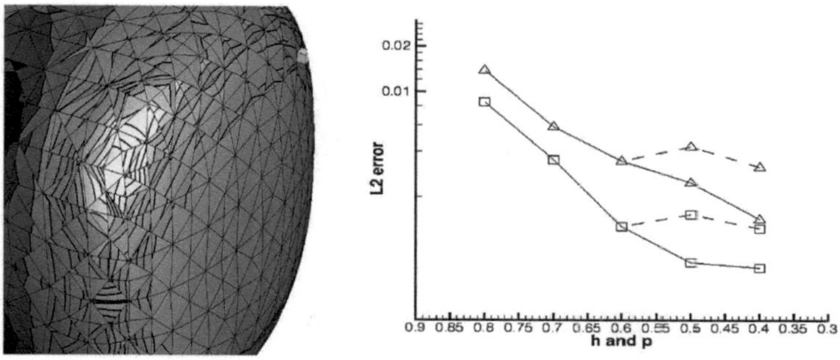

(a) Gaussian curvature (b) Convergence

Fig. 5.13 (**a**) Smaller elements in highly curved regions to reduce rapid curvature variations. (**b**) Restored exponential convergence (solid lines) for u (squares) and v (triangles), compared with the degraded results (dashed lines) without refinement in Fig. 5.12. (Courtesy of Chun 2014a)

5.3 Shallow Water Equations on the Sphere

The shallow water equations (SWEs) form a two-dimensional system of nonch5/linear, hyperbolic-type partial differential equations. These equations incorporate all horizontal operators commonly used in atmospheric and oceanic models. By assuming that the vertical scale is negligible compared to the horizontal scale, the velocity in the SWEs is treated as depth-uniform. The SWEs on the unit sphere have been widely used as a testbed for numerical weather prediction. In ocean and coastal engineering, they also serve in practical applications such as large-scale tidal, tsunami, and storm-surge forecasting. Standard benchmark problems for SWEs on the unit sphere are well-established and have been extensively tested using various numerical schemes and computational platforms. Incorporating moving frames into the numerical solution of the SWEs enables the assessment of the efficiency and accuracy of these advanced schemes relative to classical methods.

Consider the following non-conservative form of the SWEs on a smooth curved surface Ω:

$$\frac{\partial H}{\partial t} + \nabla \cdot (H\mathbf{u}) = 0\,, \tag{5.72}$$

$$\frac{\partial (H\mathbf{u})}{\partial t} + \nabla \cdot (H\mathbf{u}\mathbf{u}) + \frac{g}{2}\nabla H^2 = fH(\mathbf{u} \times \mathbf{k}) + gH\nabla H_0\,, \tag{5.73}$$

where $H = H_0 + \eta$ denotes the total water depth, composed of the free surface elevation η and the still water depth H_0. The Coriolis parameter is denoted by f, and g is the gravitational constant. The velocity vector $\mathbf{u}$ lies in the domain Ω, and the surface normal vector $\mathbf{k}$ is orthogonal to it. In a two-dimensional Cartesian plane, the velocity vector is expressed as

$$\mathbf{u} = u_x\mathbf{x} + v_y\mathbf{y}. \tag{5.74}$$

Using the Cartesian invariance of the divergence operator,

$$\nabla \cdot (H\mathbf{uu}) \cdot \mathbf{x} = \nabla \cdot (Hu_x\mathbf{u}), \tag{5.75}$$

$$\nabla \cdot (H\mathbf{uu}) \cdot \mathbf{y} = \nabla \cdot (Hv_y\mathbf{u}). \tag{5.76}$$

Eqs. (5.72) and (5.73) reduce to:

$$\frac{\partial H}{\partial t} + \frac{\partial}{\partial x}(Hu_x) + \frac{\partial}{\partial y}(Hv_y) = 0\,, \tag{5.77}$$

$$\frac{\partial\,(Hu_x)}{\partial t} + \frac{\partial}{\partial x}(Hu_x^2) + \frac{\partial}{\partial y}(Hu_xv_y) + \frac{g}{2}\frac{\partial H^2}{\partial x} = fHv_y + gH\frac{\partial H_0}{\partial x}, \tag{5.78}$$

$$\frac{\partial\,(Hv_y)}{\partial t} + \frac{\partial}{\partial x}(Hu_xv_y) + \frac{\partial}{\partial y}(Hv_y^2) + \frac{g}{2}\frac{\partial H^2}{\partial y} = -fHu_x + gH\frac{\partial H_0}{\partial y}. \tag{5.79}$$

On a curved surface, such as a spherical shell, Eqs. (5.77)–(5.79) are no longer valid. Consider two curved coordinate axes ξ and η on the domain. Let $\boldsymbol{\xi}$ and $\boldsymbol{\eta}$ denote the tangent vectors along ξ and η, respectively. The depth-averaged velocity $\mathbf{u}$ is then expressed as:

$$\mathbf{u} = u_\xi\boldsymbol{\xi} + v_\eta\boldsymbol{\eta}. \tag{5.80}$$

The first equation, Eq. (5.72), represents mass conservation and can be solved using existing numerical schemes developed for conservation laws. However, solving the second equation, Eq. (5.73), requires additional assumptions. Taking the inner product of Eq. (5.73) with $\boldsymbol{\xi}$ or $\boldsymbol{\eta}$ yields:

$$\begin{aligned}\frac{\partial(Hu_\xi\boldsymbol{\xi})}{\partial t}\cdot\boldsymbol{\xi} + \nabla\cdot(Hu_\xi\mathbf{u}) - H(\mathbf{u}\cdot\nabla\boldsymbol{\xi})\cdot\mathbf{u} + \frac{g}{2}\nabla H^2\cdot\boldsymbol{\xi}&\\ = fHv_\eta + gH\nabla H_0\cdot\boldsymbol{\xi},&\end{aligned} \tag{5.81}$$

$$\begin{aligned}\frac{\partial(Hv_\eta\boldsymbol{\eta})}{\partial t}\cdot\boldsymbol{\eta} + \nabla\cdot(Hv_\eta\mathbf{u}) - H(\mathbf{u}\cdot\nabla\boldsymbol{\eta})\cdot\mathbf{u} + \frac{g}{2}\nabla H^2\cdot\boldsymbol{\eta}&\\ = -fHu_\xi + gH\nabla H_0\cdot\boldsymbol{\eta},&\end{aligned} \tag{5.82}$$

where $\mathbf{u}\cdot\nabla\mathbf{v}$ denotes the **intrinsic derivative**. For a vector field $\mathbf{v} = v_\xi\boldsymbol{\xi} + v_\eta\boldsymbol{\eta}$, it is computed as:

$$\mathbf{u}\cdot\nabla\mathbf{v} = u_\xi\nabla v_\xi + u_\eta\nabla v_\eta. \tag{5.83}$$

Moreover, $\mathbf{u}\cdot\nabla\mathbf{v}\cdot\mathbf{u}$ represents the projection of the intrinsic derivative in the direction of $\mathbf{u}$, also known as the **tendency** (Weatherburn 1957). We assume zero tendency for the spherical coordinate axes (ϕ, θ) and their corresponding tangent vectors $(\boldsymbol{\phi}, \boldsymbol{\theta})$:

$$\mathbf{u} \cdot \nabla \boldsymbol{\phi} \cdot \mathbf{u} = 0, \tag{5.84}$$

$$\mathbf{u} \cdot \nabla \boldsymbol{\theta} \cdot \mathbf{u} = 0. \tag{5.85}$$

Although these conditions are not strictly valid on the unit sphere, they are standard assumptions in many numerical test cases for SWEs, including the popular test sets by Williamson et al. (1992). To satisfy Eqs. (5.84) and (5.85), the coordinate axes should approximate Fermi normal coordinates—i.e., be **locally Euclidean**. Spherical coordinates are far from Euclidean in this sense. The equivalent conditions for moving frames are:

$$\mathbf{u} \cdot \nabla \boldsymbol{\xi} \cdot \mathbf{u} = 0, \tag{5.86}$$

$$\mathbf{u} \cdot \nabla \boldsymbol{\eta} \cdot \mathbf{u} = 0. \tag{5.87}$$

Moving frames, as element-wise discontinuous **local** reference frames, more closely approximate Euclidean coordinates than global spherical coordinates. This can be verified by computing the divergence of the tangent vectors (smaller divergence indicates a more Euclidean-like frame). Hence, we assume Eqs. (5.86) and (5.87) to hold, leading to the following formulations:

$$\frac{\partial H}{\partial t} + \nabla \cdot (H u_\xi \boldsymbol{\xi} + H v_\eta \boldsymbol{\eta}) = 0, \tag{5.88}$$

$$\begin{aligned} &\frac{\partial H u_\xi}{\partial t} + H \left(u_\xi \frac{\partial \boldsymbol{\xi}}{\partial t} + v_\eta \frac{\partial \boldsymbol{\eta}}{\partial t} \right) \cdot \boldsymbol{\xi} + \nabla \cdot (H u_\xi \mathbf{u}) + \frac{g}{2} \nabla H^2 \cdot \boldsymbol{\xi} \\ &= f H v_\eta + g H \nabla H_0 \cdot \boldsymbol{\xi}, \end{aligned} \tag{5.89}$$

$$\begin{aligned} &\frac{\partial H v_\eta}{\partial t} + H \left(u_\xi \frac{\partial \boldsymbol{\xi}}{\partial t} + v_\eta \frac{\partial \boldsymbol{\eta}}{\partial t} \right) \cdot \boldsymbol{\eta} + \nabla \cdot (H v_\eta \mathbf{u}) + \frac{g}{2} \nabla H^2 \cdot \boldsymbol{\eta} \\ &= - f H u_\xi + g H \nabla H_0 \cdot \boldsymbol{\eta}. \end{aligned} \tag{5.90}$$

A distinctive feature in Eqs. (5.89) and (5.90) is the time derivative of the tangent vectors. This term becomes nontrivial when $\boldsymbol{\xi}$ and $\boldsymbol{\eta}$ depend on time, as in rotating or deforming domains. Applying the chain rule to the time derivatives yields:

$$\frac{\partial \boldsymbol{\xi}}{\partial t} = \frac{\partial \boldsymbol{\xi}}{\partial \xi} \frac{\partial \xi}{\partial t} + \frac{\partial \boldsymbol{\xi}}{\partial \eta} \frac{\partial \eta}{\partial t}, \tag{5.91}$$

$$\frac{\partial \boldsymbol{\eta}}{\partial t} = \frac{\partial \boldsymbol{\eta}}{\partial \xi} \frac{\partial \xi}{\partial t} + \frac{\partial \boldsymbol{\eta}}{\partial \eta} \frac{\partial \eta}{\partial t}. \tag{5.92}$$

The time derivatives of the coordinates correspond to the velocity components:

$$u_\xi = \frac{\partial \xi}{\partial t}, \tag{5.93}$$

$$v_\eta = \frac{\partial \eta}{\partial t}. \tag{5.94}$$

For a stationary domain, Eqs. (5.93) and (5.94) vanish. Substituting these into Eqs. (5.91) and (5.92), we find:

$$\frac{\partial \boldsymbol{\xi}}{\partial t} = u_\xi \frac{\partial \boldsymbol{\xi}}{\partial \xi} + v_\eta \frac{\partial \boldsymbol{\xi}}{\partial \eta}, \tag{5.95}$$

$$\frac{\partial \boldsymbol{\eta}}{\partial t} = u_\xi \frac{\partial \boldsymbol{\eta}}{\partial \xi} + v_\eta \frac{\partial \boldsymbol{\eta}}{\partial \eta}. \tag{5.96}$$

Interestingly, the time variation of the tangent vectors is expressed in terms of spatial derivatives only; time is implicitly encoded via velocity components. The earth's rotation rate, which affects the Coriolis parameter f, indirectly alters the tangent vector dynamics.

Substituting Eqs. (5.95) and (5.96) into the time-derivative terms in Eqs. (5.89) and (5.90), we obtain:

$$u_\xi \frac{\partial \boldsymbol{\xi}}{\partial t} + v_\eta \frac{\partial \boldsymbol{\eta}}{\partial t} = u_\xi^2 \frac{\partial \boldsymbol{\xi}}{\partial \xi} + u_\xi v_\eta \left(\frac{\partial \boldsymbol{\xi}}{\partial \eta} + \frac{\partial \boldsymbol{\eta}}{\partial \xi} \right) + u_\eta^2 \frac{\partial \boldsymbol{\eta}}{\partial \eta}. \tag{5.97}$$

This expression involves covariant derivatives and can be rewritten as:

$$u_\xi \frac{\partial \boldsymbol{\xi}}{\partial t} + v_\eta \frac{\partial \boldsymbol{\eta}}{\partial t} = u_\xi^2 \nabla_{\boldsymbol{\xi}} \boldsymbol{\xi} + u_\xi v_\eta \left(\nabla_{\boldsymbol{\xi}} \boldsymbol{\eta} + \nabla_{\boldsymbol{\eta}} \boldsymbol{\xi} \right) + u_\eta^2 \nabla_{\boldsymbol{\eta}} \boldsymbol{\eta}, \tag{5.98}$$

as further detailed in Chap. 4.

5.3.1 Weak Formulation Using Moving Frames

The velocity vector $\mathbf{u}$ in the SWEs is expanded in terms of the moving frames $\mathbf{e}_i$ as follows:

$$\mathbf{u} = u\mathbf{e}_1 + v\mathbf{e}_2. \tag{5.99}$$

This expression may be compared to the expansion using the tangent vectors in Eq. (5.80). If $\mathbf{e}_1$ and $\mathbf{e}_2$ are aligned with $\boldsymbol{\xi}$ and $\boldsymbol{\eta}$, respectively, then the following relations between the vector components hold:

$$u = u_\xi \, \|\boldsymbol{\xi}\| \,, \tag{5.100}$$

$$v = u_\eta \, \|\boldsymbol{\eta}\| \,. \tag{5.101}$$

These relations illustrate that the geometric information of the domain is transferred from the basis vectors to the vector coefficients. Although the vector quantity remains unchanged, this subtle reformulation can significantly improve the accuracy and efficiency of numerical schemes. This advantage stems from the fact that vector coefficients are typically time-evolved during computation, whereas the basis vectors are often incorporated only after temporal integration.

We also assume a zero tendency condition for the moving frames:

$$\mathbf{u} \cdot \nabla \mathbf{e}^i \cdot \mathbf{u} = 0. \tag{5.102}$$

This assumption is more intuitive than the zero tendency condition for spherical coordinates, since moving frames are locally flat and nearly Euclidean in a small neighborhood around each point. While there are multiple ways to implement MMF for the SWEs, the following MMF-SWE scheme is considered among the most convenient:

$$\frac{\partial H}{\partial t} + \nabla \cdot (Hu\mathbf{e}_1 + Hv\mathbf{e}_2) = 0, \tag{5.103}$$

$$\frac{\partial Hu}{\partial t} + H\left(u\frac{\partial \mathbf{e}_1}{\partial t} + v\frac{\partial \mathbf{e}_2}{\partial t}\right) \cdot \mathbf{e}_1 + \nabla \cdot (Hu\mathbf{u}) + \frac{g}{2}\nabla H^2 \cdot \mathbf{e}_1$$
$$= fHv + gH\nabla H_0 \cdot \mathbf{e}_1 \tag{5.104}$$

$$\frac{\partial Hv}{\partial t} + H\left(u\frac{\partial \mathbf{e}_1}{\partial t} + v\frac{\partial \mathbf{e}_2}{\partial t}\right) \cdot \mathbf{e}_2 + \nabla \cdot (Hv\mathbf{u}) + \frac{g}{2}\nabla H^2 \cdot \mathbf{e}_2$$
$$= -fHu + gH\nabla H_0 \cdot \mathbf{e}_2, \tag{5.105}$$

We multiply the mass equation (5.103) by a scalar test function φ, and the momentum equations (5.104) and (5.105) by a vector test function $\boldsymbol{\omega} = \varphi\mathbf{e}^i$. Integrating over a curved element Ω_e^h, we obtain the following weak formulation of the SWEs (Chun and Eskilsson 2017):

$$\int_{\Omega_e^h} \frac{\partial H}{\partial t}\varphi dx + \Delta_1 + \Delta_2 = 0\,, \tag{5.106}$$

$$\int_{\Omega_e^h} \frac{\partial\,(Hu)}{\partial t}\varphi dx + \Gamma_1 + \Gamma_2 + \Gamma_3 + \Gamma_4 = \int_{\Omega_e^h} fHv\varphi dx + \Gamma_5\,, \tag{5.107}$$

$$\int_{\Omega_e^h} \frac{\partial\,(Hv)}{\partial t}\varphi dx + \Lambda_1 + \Lambda_2 + \Lambda_3 + \Lambda_4 = -\int_{\Omega_e^h} fHu\varphi dx + \Lambda_5. \tag{5.108}$$

Here, the individual terms are defined as:

$$\Delta_1 = -\int_{\Omega_e^h} H\left\{u\nabla\varphi \cdot \mathbf{e}^1 + v\nabla\varphi \cdot \mathbf{e}^2\right\} dx, \tag{5.109}$$

$$\Delta_2 = \int_{\partial\Omega_e^h} \widetilde{H}\widetilde{\mathbf{u}} \cdot \mathbf{n}\varphi ds, \tag{5.110}$$

$$\Gamma_1 = -\int_{\Omega_e^h} \left(Hu^2 + \frac{gH^2}{2}\right) \mathbf{e}^1 \cdot \nabla\varphi dx - \int_{\Omega_e^h} Huv\mathbf{e}^2 \cdot \nabla\varphi\, dx, \tag{5.111}$$

$$\Gamma_2 = \int_{\partial\Omega_e^h} \left[\widetilde{H}\widetilde{u}^2 + \frac{g}{2}\widetilde{H}^2\right] (\mathbf{e}_1 \cdot \mathbf{n})\varphi\, ds, \tag{5.112}$$

$$\Gamma_3 = \int_{\Omega_e^h} H \left(u\frac{\partial \mathbf{e}^1}{\partial t} + v\frac{\partial \mathbf{e}^2}{\partial t}\right) \cdot \mathbf{e}^1\varphi\, dx, \tag{5.113}$$

$$\Gamma_4 = -\int_{\Omega_e^h} \frac{gH^2}{2} \left(\nabla \cdot \mathbf{e}^1\right) \varphi\, dx, \tag{5.114}$$

$$\Gamma_5 = \int_{\Omega_e^h} gH\nabla H_0 \cdot \mathbf{e}^1\varphi\, dx\,, \tag{5.115}$$

$$\Lambda_1 = -\int_{\Omega_e^h} Huv\mathbf{e}^1 \cdot \nabla\varphi dx + \int_{\Omega_e} \left(Hv^2 + \frac{gH^2}{2}\right) \mathbf{e}^2 \cdot \nabla\varphi\, dx, \tag{5.116}$$

$$\Lambda_2 = \int_{\partial\Omega_e^h} \left[\widetilde{H}\widetilde{u}^2 + \frac{g}{2}\widetilde{H}^2\right] (\mathbf{e}_2 \cdot \mathbf{n})\varphi\, ds, \tag{5.117}$$

$$\Lambda_3 = \int_{\Omega_e^h} H \left(u\frac{\partial \mathbf{e}^1}{\partial t} + v\frac{\partial \mathbf{e}^2}{\partial t}\right) \cdot \mathbf{e}^2\varphi\, dx, \tag{5.118}$$

$$\Lambda_4 = -\int_{\Omega_e^h} \frac{gH^2}{2} \left(\nabla \cdot \mathbf{e}^2\right) \varphi\, dx, \tag{5.119}$$

$$\Lambda_5 = \int_{\Omega_e^h} gH\nabla H_0 \cdot \mathbf{e}^2\varphi\, dx\,. \tag{5.120}$$

The terms Δ_1, Γ_1, and Λ_1 correspond to directional derivatives, while Δ_2, Γ_2, and Λ_2 represent numerical fluxes across boundaries. The terms Γ_3 and Λ_3 capture the effects due to rotational deformation of the domain. The covariant corrections Γ_4 and Λ_4 compensate for non-Euclidean geometry. Finally, Γ_5 and Λ_5 account for the spatial variation in still water depth H_0.

For simplicity, a local Lax–Friedrichs flux is used. Consider an interface separating a left element (L) and right element (R), as shown in Fig. 5.14. Denote the vectors in the L and R elements by superscripts L and R, respectively. The outward normal vectors are defined so that:

$$\mathbf{n} = \mathbf{n}^L = -\mathbf{n}^R.. \tag{5.121}$$

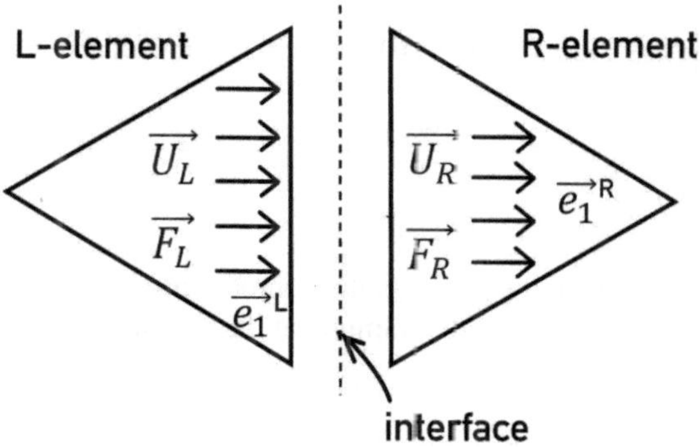

Fig. 5.14 Illustration of the Lax–Friedrichs flux between two elements, denoted as the L-element and R-element. $\mathbf{U}$ and $\mathbf{F}$ denote the velocity vector and its flux vector, respectively. Subscripts L and R indicate that the corresponding variable belongs to the L- or R-element. $\mathbf{e}_1^L$ and $\mathbf{e}_1^R$ is the first tangent vector in L and R-element, respectively

Note that vectors in each element are expanded using the local moving frames:

$$\mathbf{u}_L = u^L \mathbf{e}_1^L + v^L \mathbf{e}_2^L, \tag{5.122}$$

$$\mathbf{u}_R = u^R \mathbf{e}_1^R + v^R \mathbf{e}_2^R. \tag{5.123}$$

In general, the LOCAL moving frames $(\mathbf{e}_1^L, \mathbf{e}_2^L)$ and $(\mathbf{e}_1^R, \mathbf{e}_2^R)$ are different. For a system of hyperbolic conservation laws:

$$\frac{\partial \mathbf{U}}{\partial t} + \nabla \cdot \mathbf{F}(\mathbf{U}) = 0, \tag{5.124}$$

the Lax–Friedrichs flux is given by:

$$\mathbf{F}(\mathbf{U}) \cdot \mathbf{n} = \frac{\mathbf{F}(\mathbf{U}^L) + \mathbf{F}(\mathbf{U}^R)}{2} \cdot \mathbf{n} + |\boldsymbol{\lambda}|_{\max} \frac{\mathbf{U}^L - \mathbf{U}^R}{2}, \tag{5.125}$$

where $\boldsymbol{\lambda}$ contains the wave speeds:

$$\boldsymbol{\lambda} = \{\lambda_1, \lambda_2, \lambda_3\} = \{\mathbf{u} \cdot \mathbf{n} - \sqrt{gH},\ \mathbf{u} \cdot \mathbf{n},\ \mathbf{u} \cdot \mathbf{n} + \sqrt{gH}\}. \tag{5.126}$$

The numerical fluxes in Γ_2 and Λ_2 are then computed as:

$$\widetilde{H}\widetilde{\mathbf{u}} \cdot \mathbf{n} = \frac{H^L \mathbf{u}^L + H^R \mathbf{u}^R}{2} \cdot \mathbf{n} + \frac{|\boldsymbol{\lambda}|_{\max}}{2}(H^L - H^R), \tag{5.127}$$

$$\begin{aligned}\left[\widetilde{H}\widetilde{u}^2 + \frac{g}{2}\widetilde{H}^2\right](\mathbf{e}^1 \cdot \mathbf{n}) &= \frac{H^L u_*^L \mathbf{u}^L + H^R u_*^R \mathbf{u}^R}{2} \cdot \mathbf{n} \\ &+ \frac{g}{2}\frac{(H^L)^2 \mathbf{e}_1^L + (H^R)^2 \mathbf{e}_1^R}{2} \cdot \mathbf{n} + \frac{|\boldsymbol{\lambda}|_{\max}}{2}(H^L u_*^L - H^R u_*^R),\end{aligned} \tag{5.128}$$

$$\left[\widetilde{H}\widetilde{u}_{\perp}^{2} + \frac{g}{2}\widetilde{H}^{2}\right](\mathbf{e}^{2}\cdot\mathbf{n}) = \frac{H^{L}v_{*}^{L}\mathbf{u}^{L} + H^{R}v_{*}^{R}\mathbf{u}^{R}}{2}\cdot\mathbf{n} + \frac{g}{2}\frac{(H^{L})^{2}\mathbf{e}_{2}^{L} + H_{2}^{R}\mathbf{e}_{2}^{R}}{2}\cdot\mathbf{n} + \frac{|\boldsymbol{\lambda}|_{\max}}{2}(H^{L}v_{*}^{L} - H^{R}v_{*}^{R}), \tag{5.129}$$

where u_* and v_* refer to the components of $\mathbf{u}$ expressed in the corresponding frame (either L or R). For example, at the interface:

$$\mathbf{u}^{L} = u_{*}^{R}\mathbf{e}_{1}^{R} + v_{*}^{R}\mathbf{e}_{2}^{R}. \tag{5.130}$$

When $\mathbf{e}^1 = \mathbf{x}$ and $\mathbf{e}^2 = \mathbf{y}$, the above formulation reduces to the classical flux form. These expressions highlight the role of moving frames as local reference systems for inter-element communication.

5.3.2 [NEKTAR++] Computing SWEs on Curved Surfaces

The numerical schemes presented in Eq. (5.106), including the terms Γ_1 and Γ_2 from (5.107) and Λ_1 and Λ_2 from (5.108), are implemented within the spectral/hp element framework of `Nektar++`. Specifically designed for SWEs on curved geometries, this implementation extensively utilizes weak discontinuous Galerkin (DG) formulations. These formulations employ directional derivatives aligned explicitly with local moving frame directions using the `IProductWRTDirectionalDerivBase` function, thereby preserving geometric fidelity of derivative operations. Furthermore, the edge normal vectors $\mathbf{n}$ are expressed consistently within the local moving frame basis, ensuring geometric coherence at domain interfaces.

The first code snippet details the practical implementation of flux divergence computations using local moving frames:

```
void MMFSWE::WeakDGSWEDivergence(
    const Array<OneD, Array<OneD, NekDouble>> &InField,
    Array<OneD, Array<OneD, NekDouble>> &OutField)
{
    // Number of solution variables (e.g., height and velocity components)
    int nvariables      = m_fields.size();

    // Number of quadrature points
    int nq              = GetNpoints();

    // Number of coefficients (degrees of freedom)
    int ncoeffs         = GetNcoeffs();

    // Total number of points on the trace (element boundaries)
    int nTracePointsTot = GetTraceNpoints();

    // Initialize flux vector and physical field arrays
    Array<OneD, Array<OneD, NekDouble>> fluxvector(m_shapedim);
    Array<OneD, Array<OneD, NekDouble>> physfield(nvariables);
```

```cpp

    // Allocate memory for flux vector components
    for (int j = 0; j < m_shapedim; ++j)
    {
        fluxvector[j] = Array<OneD, NekDouble>(nq);
    }

    // Copy input primitive variables to physical fields
    for (int i = 0; i < nvariables; ++i)
    {
        physfield[i] = InField[i];
    }

    // Temporary array for intermediate results
    Array<OneD, NekDouble> tmpc(ncoeffs);

    // Loop over each solution variable to compute the flux divergence
    for (int i = 0; i < nvariables; ++i)
    {
        // Retrieve flux vector components based on shallow water equations (SWE)
        // 0: [Hu, Hv]
        // 1: [Hu^2 + 0.5gH^2, Huv ]
        // 2: [Huv, Hv^2 + 0.5 g H^2,]
        GetSWEFluxVector(i, physfield, fluxvector, 1);

        // Initialize output divergence field for the current variable
        OutField[i] = Array<OneD, NekDouble>(ncoeffs, 0.0);

        // Compute directional derivatives along moving frames
        for (int j = 0; j < m_shapedim; ++j)
        {
            // Compute directional derivative with respect to the j-th moving frame
            m_fields[i]->IProductWRTDirectionalDerivBase(
                m_movingframes[j], fluxvector[j], tmpc);

            // Add the computed directional derivative to the divergence field
            Vmath::Vadd(ncoeffs, &tmpc[0], 1, &OutField[i][0], 1, &OutField[i][0], 1);
        }
    }

    // Arrays for forward and backward numerical fluxes at element interfaces
    Array<OneD, Array<OneD, NekDouble>> numfluxFwd(nvariables);
    Array<OneD, Array<OneD, NekDouble>> numfluxBwd(nvariables);

    // Allocate numerical flux arrays
    for (int i = 0; i < nvariables; ++i)
    {
        numfluxFwd[i] = Array<OneD, NekDouble>(nTracePointsTot);
        numfluxBwd[i] = Array<OneD, NekDouble>(nTracePointsTot);
    }

    // Compute numerical fluxes based on physical fields
    NumericalSWEFlux(physfield, numfluxFwd, numfluxBwd);

    // Evaluate the integral of numerical flux across element boundaries and adjust OutField
    for (int i = 0; i < nvariables; ++i)
    {
        Vmath::Neg(ncoeffs, OutField[i], 1);
        m_fields[i]->AddFwdBwdTraceIntegral(numfluxFwd[i], numfluxBwd[i], OutField[i]);
```

```
            m_fields[i]->SetPhysState(false);  // Reset the physical state after
                modification
        }
}
```

Listing 5.3 Computation of SWEs on a curved surface using moving frames. This Nektar++ implementation employs weak DG formulations explicitly adapted to the geometry of curved surfaces, accurately computing flux divergences in the context of local frames

To account for additional dynamical complexities like the Coriolis force arising due to surface rotation, the **AddTimeVariantFrames** function computes the supplementary terms Γ_3 (in (5.107)) and Λ_3 (in (5.108)). These terms represent the influences of dynamic changes in frame orientation and translation.

The supplementary function **Compute_demdt_cdot_ek** projects time derivatives of moving frames onto their corresponding local directions, capturing the rotational and translational dynamics of the local coordinate system. This step is crucial for accurately handling dynamic geometry changes and ensuring robust numerical simulations on curved domains.

Below is the implementation excerpt detailing these computations:

```
void MMFSWE::AddTimeVariantFrames(
    const Array<OneD, const Array<OneD, NekDouble>> &inarray,
    const Array<OneD, const Array<OneD, NekDouble>> &movingframes,
    Array<OneD, Array<OneD, NekDouble>> &outarray, const int
        DivergenceRestore)
{
    // Number of spectral coefficients
    int ncoeffs = outarray[0].size();
    // Number of quadrature points
    int nq      = m_fields[0]->GetTotPoints();
    // Number of dependent variables (typically height and velocity
        components)
    int nvar    = 3;

    // Compute total water height (free surface height = fluid height +
        bathymetric depth)
    Array<OneD, NekDouble> h(nq);
    Vmath::Vadd(nq, &inarray[0][0], 1, &m_depth[0], 1, &h[0], 1);

    // Temporary array for intermediate calculations
    Array<OneD, NekDouble> tmp(nq);

    // Physical field in moving frames (velocity components, etc.)
    Array<OneD, Array<OneD, NekDouble>> physfield(nvar);
    for (int i = 0; i < nvar; ++i)
    {
        physfield[i] = Array<OneD, NekDouble>(nq, 0.0);
    }

    // Arrays for time derivatives of moving frames projected onto local
        frames
    Array<OneD, NekDouble> de0dt_cdot_e0, de0dt_cdot_e1, de1dt_cdot_e0,
        de1dt_cdot_e1;

    // Compute \partial e_m /dt \cdot e_k
    Compute_demdt_cdot_ek(0, 0, physfield, movingframes, de0dt_cdot_e0);
    Compute_demdt_cdot_ek(1, 0, physfield, movingframes, de1dt_cdot_e0);
    Compute_demdt_cdot_ek(0, 1, physfield, movingframes, de0dt_cdot_e1);
    Compute_demdt_cdot_ek(1, 1, physfield, movingframes, de1dt_cdot_e1);

    // Compute rotational terms arising from time variation of moving frames
```

```
    Array<OneD, NekDouble> Rott1(nq), Rott2(nq);

    // Project velocity components onto time-variant moving frame terms
    Vmath::Vmul(nq, inarray[1], 1, de0dt_cdot_e0, 1, Rott1, 1);
    Vmath::Vmul(nq, inarray[1], 1, de0dt_cdot_e1, 1, Rott2, 1);
    Vmath::Vvtvp(nq, inarray[2], 1, de1dt_cdot_e0, 1, Rott1, 1, Rott1, 1);
    Vmath::Vvtvp(nq, inarray[2], 1, de1dt_cdot_e1, 1, Rott2, 1, Rott2, 1);

    // Multiply the rotational terms by total height
    Vmath::Vmul(nq, &h[0], 1, &Rott1[0], 1, &Rott1[0], 1);
    Vmath::Vmul(nq, &h[0], 1, &Rott2[0], 1, &Rott2[0], 1);

    // Negate rotational terms (for correct sign convention in equations)
    Vmath::Neg(nq, Rott1, 1);
    Vmath::Neg(nq, Rott2, 1);

    // Integrate rotational terms against basis functions and update output
         arrays
    Array<OneD, NekDouble> tmpc1(ncoeffs), tmpc2(ncoeffs);
    m_fields[0]->IProductWRTBase(Rott1, tmpc1);
    m_fields[0]->IProductWRTBase(Rott2, tmpc2);

    // Add computed terms to the velocity components in the output field
    Vmath::Vadd(ncoeffs, tmpc1, 1, outarray[1], 1, outarray[1], 1);
    Vmath::Vadd(ncoeffs, tmpc2, 1, outarray[2], 1, outarray[2], 1);
}

// Compute the projection of time derivatives of moving frames onto local
     moving frame directions
template<typename ArrayType>
void MMFSWE::Compute_demdt_cdot_ek(
    const int indm, const int indk,
    const ArrayType &physarray,
    const ArrayType &movingframes,
    Array<OneD, NekDouble> &outarray)
{
    int nq = m_fields[0]->GetNpoints();  // Number of quadrature points

    Array<OneD, NekDouble> tmp(nq);

    outarray = Array<OneD, NekDouble>(nq, 0.0);
    // Loop over dimensions and spatial components to compute derivatives
    for (int j = 0; j < m_shapedim; ++j)
    {
        for (int k = 0; k < m_spacedim; ++k)
        {
            // Copy current moving frame component for differentiation
            Vmath::Vcopy(nq, &m_movingframes[indm][k * nq], 1, &tmp[0], 1);

            // Compute directional derivative of moving frame components
            MMFDirectionalDeriv(movingframes[j], tmp, tmp);

            // Weight derivatives by velocity components
            Vmath::Vmul(nq, &physarray[j + 1][0], 1, &tmp[0], 1, &tmp[0], 1)
                ;

            // Accumulate results projected onto the specified moving frame
            Vmath::Vvtvp(nq, &tmp[0], 1, &movingframes[indk][k * nq], 1,
                         &outarray[0], 1, &outarray[0], 1);
        }
    }
}
```

Listing 5.4 Computation of rotational and translational effects by projecting time derivatives of moving frames onto local frame directions. This Nektar++ code snippet explicitly uses weak DG formulations to maintain accuracy consistent with the curved geometry

These comprehensive implementations in `Nektar++` effectively integrate geometric nuances into numerical schemes for solving SWEs, significantly enhancing accuracy and stability on curved geometrical domains.

5.3.3 Benchmark Problems for SWE Validation

To evaluate the proposed numerical schemes in Eqs. (5.106)–(5.108), the weak formulation within the discontinuous Galerkin method, similar to Ref. Nair et al. (2005), is applied to five standard test problems on the sphere from the previous studies including Williamson's test suite (Williamson et al. 1992):

1. Steady zonal flow (Williamson et al. 1992; Läuter et al. 2005b; Bernard et al. 2009; Pudykiewicz 2011)
2. Unsteady zonal flow (Läuter et al. 2005a,b; Pudykiewicz 2011)
3. (Perturbed) Rossby–Haurwitz waves (Williamson et al. 1992; Giraldo et al. 2002; Giraldo and Warburton 2005; Bernard et al. 2009)
4. Zonal flow over an isolated mountain (Williamson et al. 1992; Bernard et al. 2009; Pudykiewicz 2011)
5. Perturbed unstable zonal jet (Galewsky et al. 2004; Läuter et al. 2005b; Pudykiewicz 2011)

Tests 1 and 2 have exact analytical solutions, enabling direct error computation to demonstrate exponential convergence. The error is presented in terms of the L_2 norm of the surface elevation η:

$$\|\eta\|_{L_2} = \sqrt{\int_{\Omega} (\eta_{\text{exact}} - \eta)^2 dx}. \tag{5.131}$$

For Tests 3, 4, and 5, no analytic solutions are available. Instead, results are evaluated based on mass and energy conservation, under the assumption that both should be conserved on a closed surface like the sphere. The conservation error is defined as:

$$\text{Conservation error} = \frac{\displaystyle\int_{\Omega} f(t)\, dx - \int_{\Omega} f(t=0)\, dx}{\displaystyle\int_{\Omega} f(t=0)\, dx}, \tag{5.132}$$

where the function f is defined as:

$$f = \begin{cases} H & \text{(mass)}, \\ \frac{1}{2}(H\mathbf{u}\cdot\mathbf{u} + g(H^2 + 2HH_b)) & \text{(energy)}, \end{cases} \tag{5.133}$$

with H_b denoting the bottom topography datum.

5.3.4 Case Study: Steady Zonal Flow

This test examines steady zonal flow, where both the surface elevation η and the velocity $\mathbf{u}$ remain constant over time. **Steady** implies that the right-hand sides of Eqs. (5.106)–(5.108) vanish. The still water depth H_0 is constant, with $gH_0 = 2.94 \times 10^4\,\mathrm{m}^2/\mathrm{s}^2$. The Earth's angular velocity is given by $\Omega = 7.292 \times 10^{-5}\,\mathrm{s}^{-1}$. The Coriolis parameter is defined as:

$$f = 2\Omega(-\cos\phi\cos\theta\sin\alpha + \sin\theta\cos\alpha), \tag{5.134}$$

where α is the inclination angle of the velocity vector relative to the polar axis. A steady-state current is observed at various inclination angles α when the velocity field corresponds to a solid-body rotation. In this test, steady-state zonal flow is evaluated at three different angles: $\alpha = 0$, $\pi/4$, and $\pi/2$, as illustrated in Fig. 5.15. Simulations are conducted up to $T = 5$ days using sufficiently small time steps. The results, shown in Table 5.5, demonstrate that the convergence rate is generally of order $p + 1$, although for odd polynomial degrees, the rate may degrade slightly to $p + 1/2$. Overall, the high-order DG scheme exhibits the expected exponential convergence.

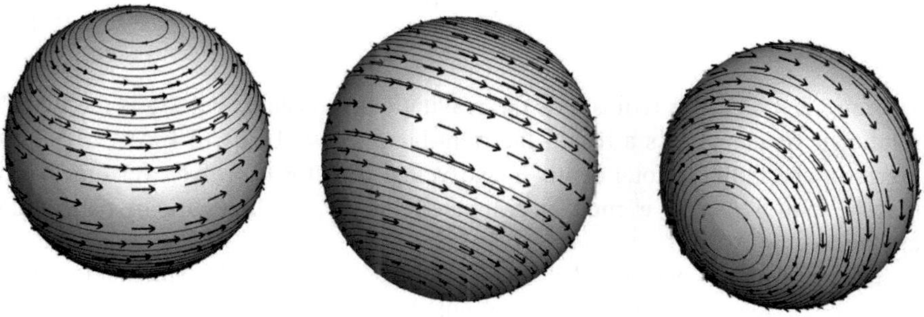

(a) $\alpha = 0$ (b) $\alpha = \pi/4$ (c) $\alpha = \pi/2$

Fig. 5.15 Steady zonal flow on the unit sphere for various angles of the velocity vector, denoted by α. Surface elevation (contours) and velocity (arrows) on a mesh with $h = 0.4$ and $p = 4$ after 5 days. (Courtesy of Chun and Eskilsson 2017). (**a**) $\alpha = 0$. (**b**) $\alpha = \pi/4$. (**c**) $\alpha = \pi/2$

Table 5.5 L_2 error and convergence order for the surface elevation η in the steady zonal flow test. (Courtesy of Chun and Eskilsson 2017)

		$h = 0.8$	$h = 0.4$		$h = 0.2$		$h = 0.1$	
α	p	Error	Error	Order	Error	Order	Error	Order
0	1	2.94E-02	5.59E-03	2.48	1.44E-03	2.03	4.83E-04	1.69
	2	3.13E-03	2.46E-04	3.80	1.72E-05	3.98	1.75E-06	3.54
	3	1.15E-03	5.76E-05	4.48	4.26E-06	3.90	4.03E-07	3.65
	4	8.01E-04	3.43E-05	4.71	1.16E-06	5.06	3.45E-08	5.44
$\pi/4$	1	2.61E-02	5.45E-03	2.35	1.61E-03	1.82	5.06E-04	1.79
	2	4.04E-03	1.66E-04	4.77	1.37E-05	3.73	1.64E-06	3.29
	3	1.16E-03	6.35E-05	4.35	4.71E-06	3.89	4.85E-07	3.52
	4	9.72E-04	3.48E-05	4.98	1.11E-06	5.15	3.14E-08	5.51
$\pi/2$	1	2.09E-02	7.60E-03	1.51	2.39E-03	1.73	6.42E-04	2.03
	2	2.75E-03	2.53E-04	3.57	1.78E-05	3.97	1.79E-06	3.56
	3	1.30E-03	9.38E-05	3.94	6.59E-06	3.97	4.84E-07	4.04
	4	7.79E-04	2.93E-05	4.91	9.03E-07	5.20	2.81E-08	5.36

5.3.5 *Case Study: Unsteady Zonal Flow*

The unsteady zonal flow test is similar to the steady case, as the velocity vector again corresponds to a solid-body rotation. However, the unsteady test involves a variable still surface elevation $H_0(\theta)$ defined by:

$$H_0(\theta) = \frac{133681}{r_a g} - \frac{10}{r_a g} - \frac{(\Omega_f \sin\theta)^2}{2g}, \tag{5.135}$$

where r_a is the Earth's radius, g is the gravitational constant, and Ω_f is the Earth's rotational frequency. As a result, the right-hand sides of Eqs. (5.106)–(5.108) are nonzero, and both the total depth H and velocity vector $\mathbf{u}$ evolve over time. Note that the term $gH\nabla H_0 \cdot \mathbf{e}^i$ must be computed directly, as integration by parts is not applicable.

The simulation is performed up to $T = 0.5$ days (12 hours). The inclination angle α may be chosen arbitrarily, as it has minimal effect on the numerical outcome. For this test, $\alpha = \pi/4$ is selected. The solution is advanced in time over 1,000 time steps. As shown in Fig. 5.16, the vector field exhibits a periodic structure with a 1-day cycle. Table 5.6 confirms a convergence rate of approximately $p + 1$, consistent with theoretical expectations.

Figure 5.17 shows that the distribution of numerical error is nearly uniform, without distinct features, even near the poles ($\theta \approx \pm\pi/2$). This robustness is attributed to the local nature of the moving frames, which mitigate geometric singularities at the pole-to-pole interface.

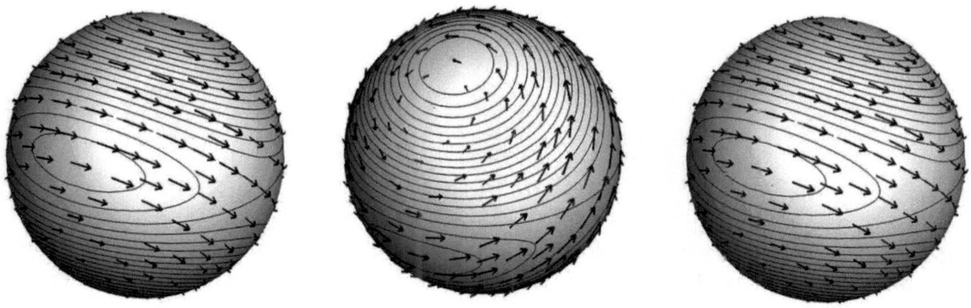

(a) initial condition (b) after 12 hrs (c) after 60 hrs

Fig. 5.16 Unsteady zonal flow on the unit sphere. (**a**) Initial velocity field, (**b**) after 12 hours, and (**c**) after 60 hours. Surface elevation (contours) and velocity (arrows) are shown on a mesh with $h = 0.4$ and $p = 4$. (Courtesy of Chun and Eskilsson 2017)

Table 5.6 Unsteady zonal flow: L_2 error and convergence order for the η component

	$h = 0.8$	$h = 0.4$		$h = 0.2$		$h = 0.1$	
p	Error	Error	Order	Error	Order	Error	Order
1	1.79E-01	4.34E-02	2.12	1.31E-02	1.79	3.79E-03	1.92
2	2.68E-02	2.81E-03	3.38	3.98E-04	2.92	5.51E-05	3.06
3	5.74E-03	3.14E-04	4.35	1.79E-05	4.28	1.12E-06	4.29
4	2.48E-03	7.35E-05	5.26	2.12E-06	5.30	6.14E-08	5.47

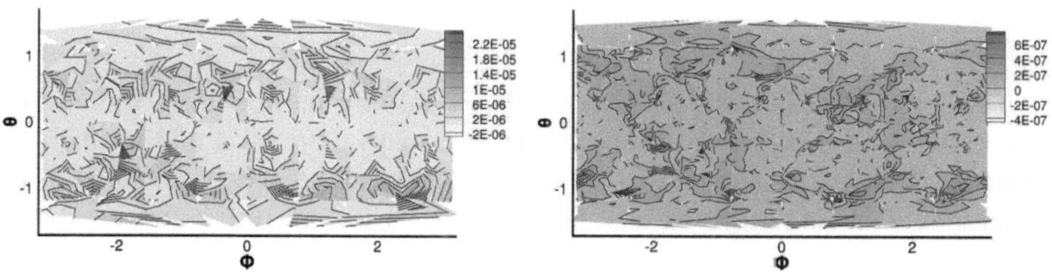

(a) $p = 3$ (b) $p = 5$

Fig. 5.17 Error distribution of η in θ–ϕ maps for the unsteady zonal flow. (**a**) Distribution of η for $p = 3$, and (**b**) for $p = 5$. The maps represent the projection of the spherical coordinate system onto a two-dimensional plane. (Courtesy of Chun and Eskilsson 2017)

5.3.6 *Case Study: Rossby–Haurwitz Wave*

The Rossby–Haurwitz wave is a steadily propagating solution to the nonlinear vorticity equation (Smith and Dritschel 2006). This wave is widely used as a benchmark test because the initial height field rotates non-dispersively on the sphere. The periodic distribution of the free surface elevation (η) and the velocity vector ($\mathbf{u}$) along the longitudinal (ϕ) direction rotates around the polar axis. As a result, the divergence of the Rossby–Haurwitz wave is zero on the sphere.

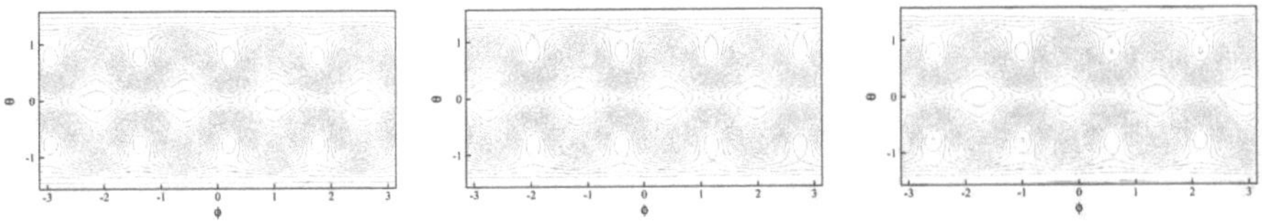

(a) 5 days (b) 10 days (c) 15 days

Fig. 5.18 Distribution of η for the Rossby–Haurwitz wave at different times in θ-ϕ map: (**a**) 5 days, (**b**) 10 days, and (**c**) 15 days from the initial state. Surface elevation contours are shown at 100 m intervals using $h = 0.4$ and $p = 4$. (Courtesy of Chun and Eskilsson 2017)

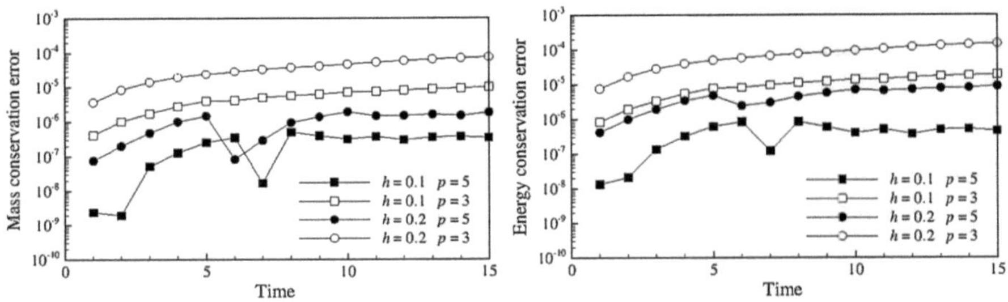

(a) Mass conservation error (b) Energy conservation error

Fig. 5.19 (**a**) Mass conservation error and (**b**) energy conservation error for the Rossby–Haurwitz wave up to 15 days. Time evolution of the conservation errors is shown using $h = 0.4$ and $p = 4$. (Courtesy of Chun and Eskilsson 2017)

The wave is numerically simulated using a uniform mesh with $h = 0.4$ on a unit-radius sphere. Figure 5.18 shows the θ–ϕ maps at $p = 4$ and a time step of 0.001 days, evaluated at 5, 10, and 15 days. The wave pattern obtained using the proposed moving frame-based scheme in Eqs. (5.106)–(5.108) is compared with known benchmark solutions. Figure 5.19 illustrates the time evolution of mass and energy conservation errors for various values of h and p. The errors decrease with finer mesh spacing (smaller h) or higher polynomial order (p).

In terms of mass conservation, the performance of the proposed scheme is slightly suboptimal compared to numerical methods that achieve machine-precision conservation. Machine precision is attainable only when the connection between grid points is analytically known and the metric tensor and Christoffel symbols are explicitly available throughout the domain. In contrast, the moving frame method assumes no **a priori** geometric information; all geometric connections must be computed numerically. As a result, mass conservation errors are typically non-trivial and significantly larger than machine precision.

The Rossby–Haurwitz wave maintains its periodicity due to the absence of any sources or sinks. When a perturbation is added to the initial height field, a new wave pattern emerges, which is also used to validate computational robustness. This perturbed wave test was conducted on two meshes with different resolutions ($h = 0.4$ and $h = 0.2$) at fixed $p = 4$. The simulation was advanced using a time

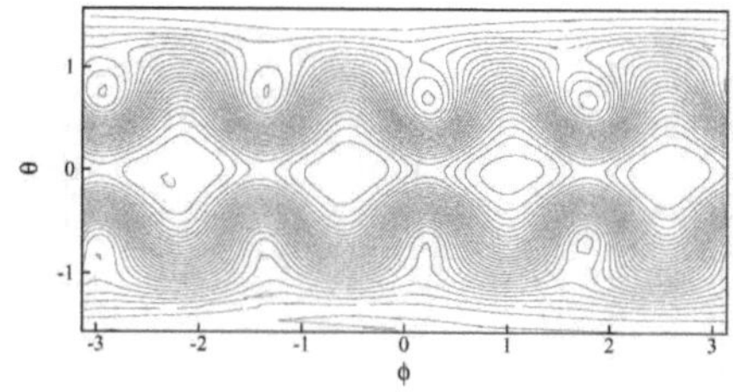

(a) $h = 0.4$ at 5 days

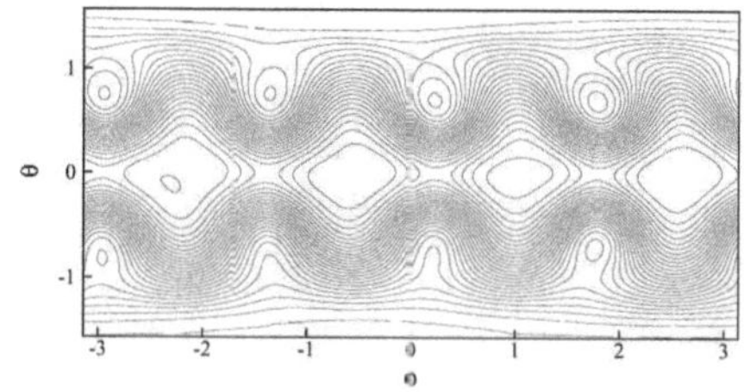

(b) $h = 0.2$ at 5 days

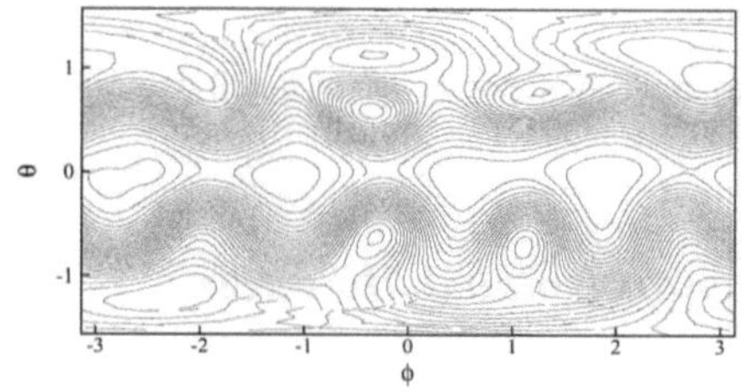

(c) $h = 0.4$ at 10 days

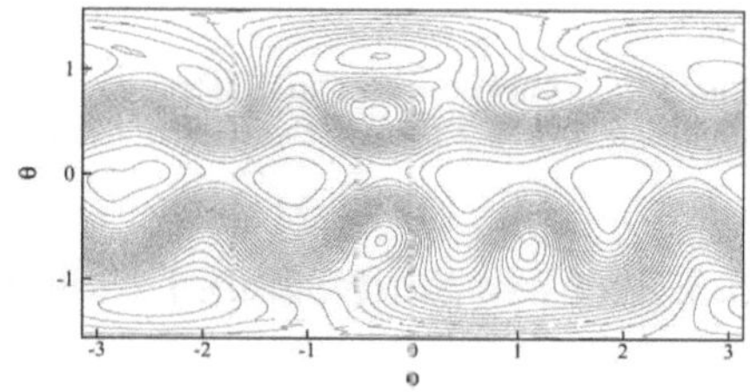

(d) $h = 0.2$ at 10 days

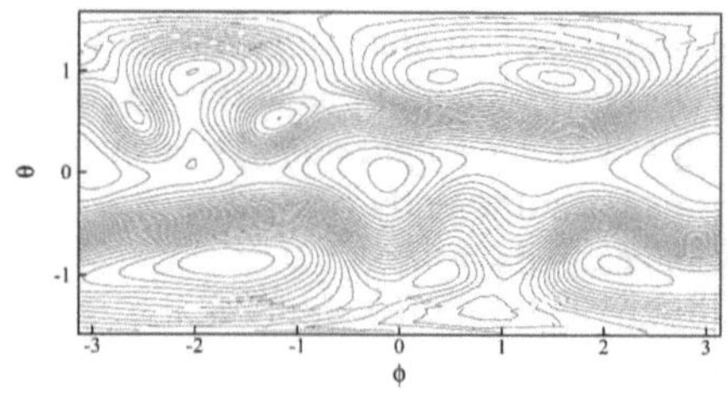

(e) $h = 0.4$ at 15 days

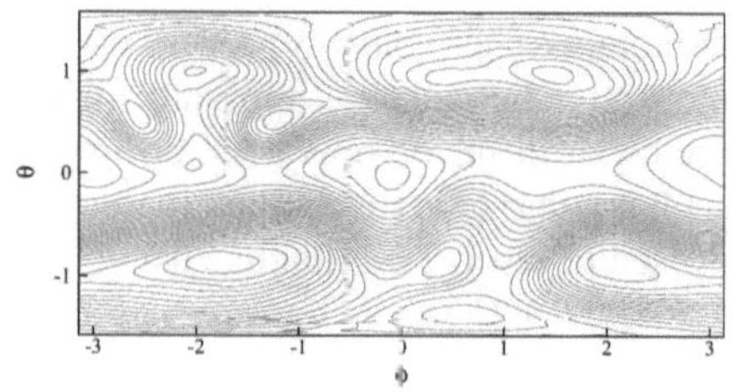

(f) $h = 0.2$ at 15 days

Fig. 5.20 Perturbed Rossby–Haurwitz wave at 5, 10, and 15 days for mesh resolutions $h = 0.4$ and $h = 0.2$ in θ-ϕ map. Surface elevation contours are shown at 100 m intervals. (**a**) $h = 0.4$ at 5 days. (**b**) $h = 0.2$ at 5 days. (**c**) $h = 0.4$ at 10 days. (**d**) $h = 0.2$ at 10 days. (**e**) $h = 0.4$ at 15 days. (**f**) $h = 0.2$ at 15 days. (Courtesy of Chun and Eskilsson 2017)

step of 5×10^{-3} days. The resulting θ–ϕ maps at 5, 10, and 15 days are shown in Fig. 5.20. Only minor differences appear between the two mesh resolutions, confirming the convergence of the scheme. Moreover, the results closely agree with previously reported solutions.

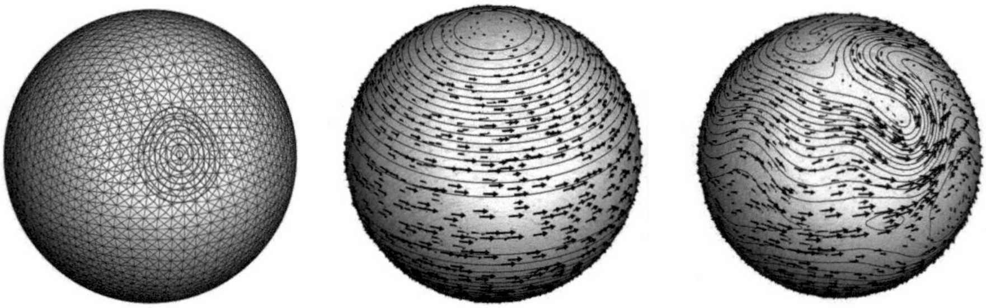

(a) Mesh (b) Initial condition (c) Final solution at 15 days

Fig. 5.21 Zonal flow over an isolated mountain: (**a**) mesh and topography, (**b**) initial state, and (**c**) final state at $T = 15$ days. (Courtesy of Chun and Eskilsson 2017)

5.3.7 Case Study: Zonal Flow Over an Isolated Mountain

This test introduces a conical mountain into the zonal flow, resulting in an unsteady and non-periodic solution (Takacs 1988). It is similar to the steady-state zonal flow case, except for the presence of an isolated conical obstacle in the domain. The mountain's sharp base introduces oscillations due to its non-smooth geometry. Common remedies include applying filtering techniques, spectral vanishing viscosity, nonlinear diffusion operators, or limiters with low-order polynomials. However, no such regularization is applied in this test. Instead, a sufficiently fine mesh ($h = 0.1$) is used to resolve oscillations using high-order methods.

Figure 5.21 shows the mesh, initial condition, and the final solution at $T = 15$ days. A polynomial degree of $p = 5$ and time step of 2×10^{-3} days were used. The final surface elevation and velocity fields are smooth near the mountain and match prior benchmark solutions. The elevation is presented as a θ–ϕ map in Fig. 5.22. Conservation properties for $p = 3, 4, 5$ are plotted in Fig. 5.23. While conservation improves with higher p, the errors plateau at approximately 10^{-9} for mass and 10^{-4} for energy. Further reductions would require smaller h.

5.3.8 Case Study: Perturbed Unstable Jet

A perturbation to the free surface can destabilize an otherwise balanced jet stream, causing it to roll up into vortices with sharp vorticity gradients. This test simulates such a scenario and is solved on two meshes with different spatial resolutions: $h = 0.15$ and $h = 0.08$, using a polynomial order of $p = 6$. The simulation is run for 8 days with a total of 32,000 time steps. No artificial diffusion terms are added.

As the simulation progresses, the vortices continue to develop finer-scale structures with increasingly sharp gradients, requiring high spatial resolution to accu-

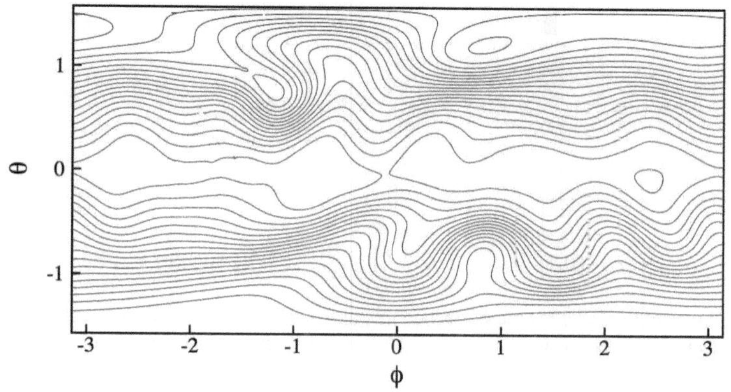

Fig. 5.22 Zonal flow over an isolated mountain in θ-ϕ map: Surface elevation contours at $T = 15$ days, with 50 m between isolines. (Courtesy of Chun and Eskilsson 2017)

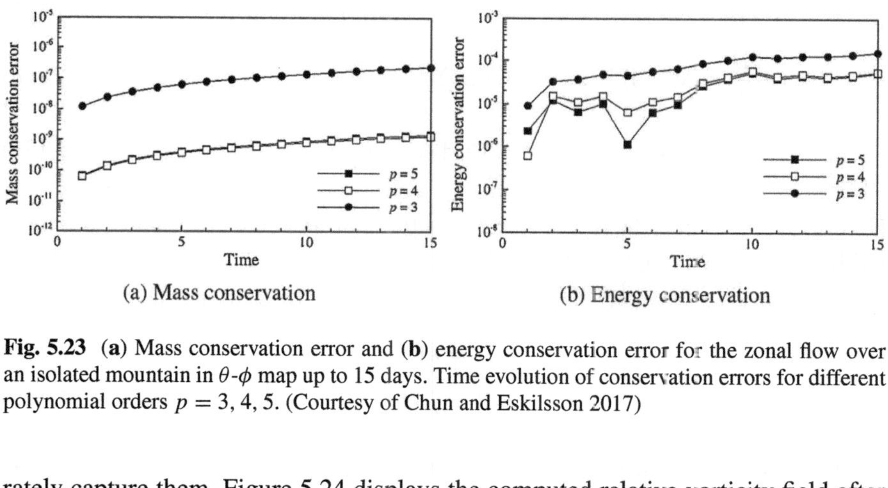

(a) Mass conservation

(b) Energy conservation

Fig. 5.23 (**a**) Mass conservation error and (**b**) energy conservation error for the zonal flow over an isolated mountain in θ-ϕ map up to 15 days. Time evolution of conservation errors for different polynomial orders $p = 3, 4, 5$. (Courtesy of Chun and Eskilsson 2017)

rately capture them. Figure 5.24 displays the computed relative vorticity field after 6 days for both mesh configurations.

The numerical solution produced by the scheme defined in Eqs. (5.106)–(5.108) shows remarkable consistency between the two meshes. Both capture the key vortical structures, although the finer mesh ($h = 0.08$) yields smoother and more sharply resolved results. The overall flow pattern at day 6 aligns well with previously reported benchmark solutions.

5.4 Maxwell's Equations in Moving Frame Formulations

The fourth class of partial differential equations considered for the application of moving frames is Maxwell's equations, which consist of the following four equations in SI units (Jackson 1999; Griffiths 2017):

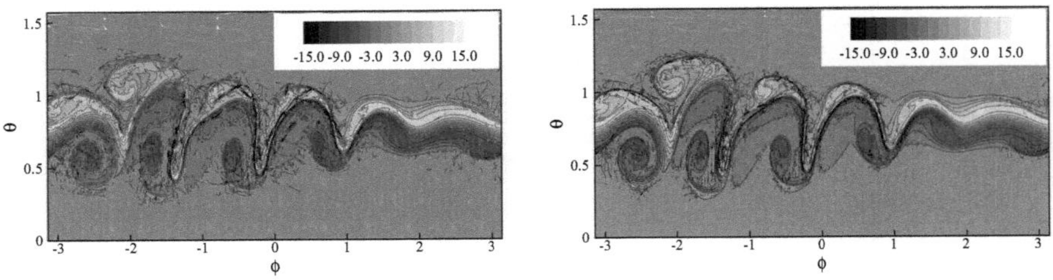

(a) $h = 0.15$ (b) $h = 0.08$

Fig. 5.24 Distribution of η for the perturbed unstable jet in θ–ϕ maps for $h = 0.15$ and $h = 0.08$. Relative vorticity at day 6 with $p = 6$. (**a**) $h = 0.15$. (**b**) $h = 0.08$. (Courtesy of Chun and Eskilsson 2017)

$$\nabla \cdot \mathbf{D} = \rho, \qquad \text{(Gauss's law)} \tag{5.136}$$

$$\nabla \cdot \mathbf{B} = 0, \qquad \text{(Gauss's law for magnetism)} \tag{5.137}$$

$$\frac{\partial \mathbf{D}}{\partial t} = \nabla \times \mathbf{H} - \mathbf{J}, \qquad \text{(Ampère's law)} \tag{5.138}$$

$$\frac{\partial \mathbf{B}}{\partial t} = -\nabla \times \mathbf{E}, \qquad \text{(Faraday's law)} \tag{5.139}$$

Here, **E** is the electric field, **B** is the magnetic field, **D** is the electric displacement field, and **H** is the magnetic field intensity. Also, **J** represents the electric current density, ρ is the charge density, and t denotes time.

$$\mathbf{D} = \hat{\varepsilon}\mathbf{E}, \tag{5.140}$$

$$\mathbf{B} = \hat{\mu}\mathbf{H}, \tag{5.141}$$

where $\hat{\varepsilon}$ and $\hat{\mu}$ are the permittivity and permeability tensors, respectively. These tensors describe the material properties of the domain with respect to electric and magnetic responses.

We assume further that the charge density ρ and current density **J** vanish, and that the material tensors are time-independent (static), possibly anisotropic and discontinuous across elements. Substituting Eqs. (5.140) and (5.141) into Eqs. (5.136)–(5.139) yields:

$$\nabla \cdot \hat{\varepsilon}\mathbf{E} = 0, \tag{5.142}$$

$$\nabla \cdot \hat{\mu}\mathbf{H} = 0, \tag{5.143}$$

$$\hat{\varepsilon}\frac{\partial \mathbf{E}}{\partial t} = \nabla \times \mathbf{H}, \tag{5.144}$$

$$\hat{\mu}\frac{\partial \mathbf{H}}{\partial t} = -\nabla \times \mathbf{E}, \tag{5.145}$$

Gauss's laws, Eqs. (5.142) and (5.143), represent the divergence constraints (or conservation laws) for the electric and magnetic fields. In most electromagnetic wave propagation problems, these equations are enforced as boundary conditions through the divergence theorem:

$$\int_{\partial\Omega} \hat{\varepsilon}\mathbf{E}\cdot\mathbf{n}ds = \int_{\Omega} \rho d\mathbf{x}, \tag{5.146}$$

$$\int_{\partial\Omega} \hat{\mu}\mathbf{H}\cdot\mathbf{n}ds = 0. \tag{5.147}$$

The remaining two equations—Ampère's and Faraday's laws—are typically solved within the domain. Let the spatial position vector $\mathbf{x}$ be represented in Cartesian coordinates by:

$$\mathbf{x} = x^1\mathbf{x} + x^2\mathbf{y} + x^3\mathbf{z}, \quad \text{or} \quad \mathbf{x} = (x^1, x^2, x^3), \tag{5.148}$$

where indices 1, 2, and 3 correspond to the x, y, and z directions, respectively. The electric and magnetic fields are expressed in component form as:

$$\mathbf{E} = \begin{bmatrix} E^1 \\ E^2 \\ E^3 \end{bmatrix}, \tag{5.149}$$

$$\mathbf{H} = \begin{bmatrix} H^1 \\ H^2 \\ H^3 \end{bmatrix}. \tag{5.150}$$

The full 3×3 permittivity and permeability tensors in Cartesian coordinates are:

$$\hat{\varepsilon} = \begin{bmatrix} \varepsilon^{11} & \varepsilon^{12} & \varepsilon^{13} \\ \varepsilon^{21} & \varepsilon^{22} & \varepsilon^{23} \\ \varepsilon^{31} & \varepsilon^{32} & \varepsilon^{33} \end{bmatrix}, \tag{5.151}$$

$$\hat{\mu} = \begin{bmatrix} \mu^{11} & \mu^{12} & \mu^{13} \\ \mu^{21} & \mu^{22} & \mu^{23} \\ \mu^{31} & \mu^{32} & \mu^{33} \end{bmatrix}. \tag{5.152}$$

The tensors are symmetric: $\varepsilon^{ij} = \varepsilon^{ji}$ and $\mu^{ij} = \mu^{ji}$ for $1 \leq i, j \leq 3$, so each contains six independent components. The Cartesian form in Eqs. (5.151) and (5.152) is convenient for multiplication with the field vectors in Eqs. (5.149) and (5.150):

$$\hat{\varepsilon}\mathbf{E} = \begin{bmatrix} \varepsilon^{11} & \varepsilon^{12} & \varepsilon^{13} \\ \varepsilon^{21} & \varepsilon^{22} & \varepsilon^{23} \\ \varepsilon^{31} & \varepsilon^{32} & \varepsilon^{33} \end{bmatrix} \begin{bmatrix} E^1 \\ E^2 \\ E^3 \end{bmatrix} = \begin{bmatrix} \varepsilon^{11}E^1 + \varepsilon^{12}E^2 + \varepsilon^{13}E^3 \\ \varepsilon^{21}E^1 + \varepsilon^{22}E^2 + \varepsilon^{23}E^3 \\ \varepsilon^{31}E^1 + \varepsilon^{32}E^2 + \varepsilon^{33}E^3 \end{bmatrix}, \tag{5.153}$$

$$\hat{\mu}\mathbf{H} = \begin{bmatrix} \mu^{11} & \mu^{12} & \mu^{13} \\ \mu^{21} & \mu^{22} & \mu^{23} \\ \mu^{31} & \mu^{32} & \mu^{33} \end{bmatrix} \begin{bmatrix} H^1 \\ H^2 \\ H^3 \end{bmatrix} = \begin{bmatrix} \mu^{11}H^1 + \mu^{12}H^2 + \mu^{13}H^3 \\ \mu^{21}H^1 + \mu^{22}H^2 + \mu^{23}H^3 \\ \mu^{31}H^1 + \mu^{32}H^2 + \mu^{33}H^3 \end{bmatrix}. \tag{5.154}$$

It is important to note that a change of coordinate system alters the component representation of vectors and tensors. For example, Ward and Pendry Ward and Pendry (1996) proposed transforming material tensors through coordinate transformations. Suppose q_i $(i = 1, 2, 3)$ are new orthogonal (but not necessarily orthonormal) coordinates. Then, the metric tensor is given by:

$$g^{ij} = \left(\frac{\partial}{\partial q_i}\right) \cdot \left(\frac{\partial}{\partial q_j}\right), \tag{5.155}$$

and the line element becomes:

$$ds^2 = dx^2 + dy^2 + dz^2 = Q_{11}dq_1^2 + Q_{22}dq_2^2 + Q_{33}dq_3^2,$$

where each Q_{ii} is computed as:

$$Q_{ii} = \frac{\partial x}{\partial q_i}\frac{\partial x}{\partial q_i} + \frac{\partial y}{\partial q_i}\frac{\partial y}{\partial q_i} + \frac{\partial z}{\partial q_i}\frac{\partial z}{\partial q_i}.$$

Under this coordinate transformation, the permittivity and permeability tensors are mapped as:

$$\hat{\mu}_q^{ij} = \mu^{ij} g^{ij} \left| \frac{\partial}{\partial q_1} \cdot \frac{\partial}{\partial q_2} \times \frac{\partial}{\partial q_3} \right| \frac{Q_1 Q_2 Q_3}{Q_i Q_j}, \tag{5.156}$$

$$\hat{\varepsilon}_q^{ij} = \varepsilon^{ij} g^{ij} \left| \frac{\partial}{\partial q_1} \cdot \frac{\partial}{\partial q_2} \times \frac{\partial}{\partial q_3} \right| \frac{Q_1 Q_2 Q_3}{Q_i Q_j}. \tag{5.157}$$

Note that the electric and magnetic fields themselves are not transformed; only their component representations and the material tensors change with the coordinate system. Accordingly, the Maxwell equations in the new coordinates become:

$$\hat{\varepsilon}_q \frac{\partial \mathbf{E}}{\partial t} = \nabla \times \mathbf{H}, \tag{5.158}$$

$$\hat{\mu}_q \frac{\partial \mathbf{H}}{\partial t} = -\nabla \times \mathbf{E}. \tag{5.159}$$

Let us consider three types of material tensors, focusing particularly on the nature of their off-diagonal components. For simplicity, we consider only the permittivity tensor $\hat{\varepsilon}$, although the formulation applies analogously to the permeability tensor $\hat{\mu}$.

5.4.1 Full Anisotropy

In the case of full anisotropy, the permittivity tensor includes nonzero off-diagonal components. Mathematically, it is expressed as:

$$\hat{\varepsilon} = \begin{bmatrix} \varepsilon^{11} & \varepsilon^{12} & \varepsilon^{13} \\ \varepsilon^{21} & \varepsilon^{22} & \varepsilon^{23} \\ \varepsilon^{31} & \varepsilon^{32} & \varepsilon^{33} \end{bmatrix}. \tag{5.160}$$

According to Maxwell's equation (5.158), the full anisotropic permittivity tensor implies that the time evolution of each electric field component may be influenced by the others. Substituting Eq. (5.160) into Eq. (5.158) yields:

$$\begin{bmatrix} \varepsilon^{11}(\partial E^1/\partial t) + \varepsilon^{12}(\partial E^2/\partial t) + \varepsilon^{13}(\partial E^3/\partial t) \\ \varepsilon^{21}(\partial E^1/\partial t) + \varepsilon^{22}(\partial E^2/\partial t) + \varepsilon^{23}(\partial E^3/\partial t) \\ \varepsilon^{31}(\partial E^1/\partial t) + \varepsilon^{32}(\partial E^2/\partial t) + \varepsilon^{33}(\partial E^3/\partial t) \end{bmatrix} = \begin{bmatrix} (\nabla \times \mathbf{H}) \cdot \mathbf{x} \\ (\nabla \times \mathbf{H}) \cdot \mathbf{y} \\ (\nabla \times \mathbf{H}) \cdot \mathbf{z} \end{bmatrix}. \tag{5.161}$$

Let $\hat{\varepsilon}^{-1}$ denote the inverse of $\hat{\varepsilon}$, satisfying

$$\hat{\varepsilon}\hat{\varepsilon}^{-1} = \hat{\varepsilon}^{-1}\hat{\varepsilon} = \hat{I}, \tag{5.162}$$

where $\hat{I}$ is the identity tensor. Since $\hat{\varepsilon}$ is a full matrix, its inverse generally also has full off-diagonal structure. Applying the inverse to Eq. (5.161) yields:

$$\begin{bmatrix} \partial E^1/\partial t \\ \partial E^2/\partial t \\ \partial E^3/\partial t \end{bmatrix} = \begin{bmatrix} \left(\varepsilon^{11}\right)^{-1} & \left(\varepsilon^{12}\right)^{-1} & \left(\varepsilon^{13}\right)^{-1} \\ \left(\varepsilon^{21}\right)^{-1} & \left(\varepsilon^{22}\right)^{-1} & \left(\varepsilon^{23}\right)^{-1} \\ \left(\varepsilon^{31}\right)^{-1} & \left(\varepsilon^{32}\right)^{-1} & \left(\varepsilon^{33}\right)^{-1} \end{bmatrix} \begin{bmatrix} (\nabla \times \mathbf{H}) \cdot \mathbf{x} \\ (\nabla \times \mathbf{H}) \cdot \mathbf{y} \\ (\nabla \times \mathbf{H}) \cdot \mathbf{z} \end{bmatrix}. \tag{5.163}$$

This expression confirms that under full anisotropy, the evolution of each electric field component is coupled to all components of the magnetic field's curl.

5.4.2 Axial Anisotropy

In the case of axial (or principal-axis) anisotropy, the permittivity tensor contains only diagonal entries:

$$\hat{\varepsilon} = \begin{bmatrix} \varepsilon^{11} & 0 & 0 \\ 0 & \varepsilon^{22} & 0 \\ 0 & 0 & \varepsilon^{33} \end{bmatrix} \tag{5.164}$$

with $\varepsilon^{ii} \neq \varepsilon^{jj}$ for $i \neq j$ in general. Substituting this form into Eq. (5.158) yields:

$$\begin{bmatrix} \varepsilon^{11}\partial E^1/\partial t \\ \varepsilon^{22}\partial E^2/\partial t \\ \varepsilon^{33}\partial E^3/\partial t \end{bmatrix} = \begin{bmatrix} (\nabla \times \mathbf{H}) \cdot \mathbf{x} \\ (\nabla \times \mathbf{H}) \cdot \mathbf{y} \\ (\nabla \times \mathbf{H}) \cdot \mathbf{z} \end{bmatrix}, \tag{5.165}$$

Multiplying each row by the inverse of the corresponding permittivity component, we obtain:

$$\begin{bmatrix} \partial E^1/\partial t \\ \partial E^2/\partial t \\ \partial E^3/\partial t \end{bmatrix} = \begin{bmatrix} (1/\varepsilon^{11})\,(\nabla \times \mathbf{H}) \cdot \mathbf{x} \\ (1/\varepsilon^{22})\,(\nabla \times \mathbf{H}) \cdot \mathbf{y} \\ (1/\varepsilon^{33})\,(\nabla \times \mathbf{H}) \cdot \mathbf{z} \end{bmatrix}. \tag{5.166}$$

This formulation indicates that the time derivative of each component of the electric field is influenced only by the corresponding component of the magnetic field's curl. In other words, axial anisotropy decouples the system across coordinate axes. The wave propagation speed along each axis is modulated by the value of ε^{ii} for $1 \leq i \leq 3$:

- Propagation slows down when $\varepsilon^{ii} > 1$,
- Speeds up when $\varepsilon^{ii} < 1$,
- Remains unchanged when $\varepsilon^{ii} = 1$.

5.4.3 Isotropy

Finally, in an isotropic material, the permittivity tensor has unit values along its diagonal and zero off-diagonal components:

$$\hat{\varepsilon} = \begin{bmatrix} 1\,0\,0 \\ 0\,1\,0 \\ 0\,0\,1 \end{bmatrix}. \tag{5.167}$$

In an isotropic medium, the propagation speed is the same in all directions; that is, no preferential direction exists. Since the identity matrix represents an isotropic permittivity tensor, it effectively disappears from Maxwell's equations, yielding:

$$\frac{\partial \mathbf{E}}{\partial t} = \nabla \times \mathbf{H}, \tag{5.168}$$

$$\frac{\partial \mathbf{H}}{\partial t} = -\nabla \times \mathbf{E}. \tag{5.169}$$

Equations (5.168) and (5.169) describe electromagnetic wave propagation, analogous in form to the axial anisotropy case. However, unlike axial anisotropy, isotropy neither amplifies nor attenuates propagation along any specific direction. In isotropy, the behavior of the wave is governed solely by the shape of the domain, not by the material's directional preferences. ch5/ These equations are the most convenient for mathematical analysis and computational simulation. While pure isotropy may seem unrealistic in complex physical and biological systems, it is often a consequence of coordinate system selection. In fact, full anisotropy, axial anisotropy, and isotropy are coordinate-dependent manifestations of the same physical property.

5.4.4 The Role of Coordinate System Choice

The type of material tensor representation—whether full anisotropy, axial anisotropy, or isotropy—might appear to be intrinsic to the material. However, this is not the case. The apparent form of the tensor depends entirely on the choice of coordinate system and is independent of the actual physical anisotropy of the domain. In other words, a full anisotropic tensor may simply result from an inconvenient or non-aligned coordinate frame.

To illustrate, consider an example of permittivity anisotropy aligned with the longitudinal axis, shown in Fig. 5.25a. In the Cartesian coordinate system ($\mathbf{x}, \mathbf{y}, \mathbf{z}$), the electric field is represented as:

$$\mathbf{E} = E_x \mathbf{x} + E_y \mathbf{y} + E_z \mathbf{z} = \begin{bmatrix} E_x \\ E_y \\ E_z \end{bmatrix}. \tag{5.170}$$

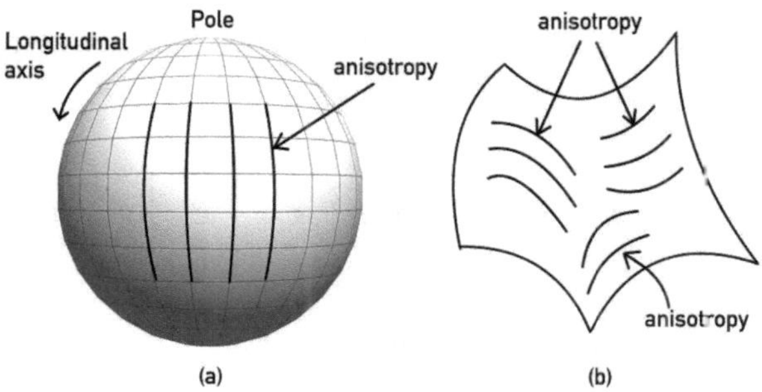

Fig. 5.25 (**a**) Anisotropy aligned with the θ-axis of the spherical coordinate system. (**b**) Anisotropy aligned with a curved axis on a general curved surface

The corresponding permittivity tensor in Cartesian coordinates becomes:

$$\hat{\varepsilon} = \mathbf{R} \begin{bmatrix} \lambda_\theta & 0 & 0 \\ 0 & \lambda_\phi & 0 \\ 0 & 0 & 1 \end{bmatrix} \mathbf{R}^T, \tag{5.171}$$

where **R** is the rotation matrix from spherical to Cartesian coordinates:

$$\mathbf{R} = \begin{bmatrix} \cos\theta\cos\varphi & -\sin\varphi & \sin\theta\cos\varphi \\ \cos\theta\sin\varphi & \cos\varphi & \sin\theta\sin\varphi \\ -\sin\theta & 0 & \cos\theta \end{bmatrix}. \tag{5.172}$$

For example, at $(\theta, \phi) = (\pi/4, \pi/4)$, corresponding to $(x, y, z) = (1/2, 1/2, \sqrt{2}/2)$ in Cartesian coordinates, and with $\lambda_\theta = 2.0$, the full permittivity tensor is:

$$\hat{\varepsilon}\left(\frac{1}{2}, \frac{1}{2}, \frac{\sqrt{2}}{2}\right) = \begin{bmatrix} \frac{5}{4} & \frac{1}{4} & -\frac{\sqrt{2}}{4} \\ \frac{1}{4} & \frac{1}{4} & -\frac{\sqrt{2}}{4} \\ -\frac{\sqrt{2}}{4} & -\frac{\sqrt{2}}{4} & \frac{3}{2} \end{bmatrix}. \tag{5.173}$$

Multiplying this tensor by **E** yields:

$$\hat{\varepsilon}\mathbf{E} = \begin{bmatrix} \frac{5}{4} & \frac{1}{4} & -\frac{\sqrt{2}}{4} \\ \frac{1}{4} & \frac{1}{4} & -\frac{\sqrt{2}}{4} \\ -\frac{\sqrt{2}}{4} & -\frac{\sqrt{2}}{4} & \frac{3}{2} \end{bmatrix} \begin{bmatrix} E_x \\ E_y \\ E_z \end{bmatrix}. \tag{5.174}$$

This equation implies that:

1. The permittivity tensor transforms **E** into its principal directions .
2. Each principal component is scaled according to the anisotropy.
3. The result is then rotated back into the original Cartesian coordinate system.

Now consider the same physical configuration but expressed in spherical coordinates (θ, ϕ, r):

$$\mathbf{E} = E_\theta \boldsymbol{\theta} + E_\phi \boldsymbol{\phi} + E_r \mathbf{r} = \begin{bmatrix} E_\theta \\ E_\phi \\ E_r \end{bmatrix}. \tag{5.175}$$

In this coordinate system, the permittivity tensor becomes:

$$\hat{\varepsilon} = \begin{bmatrix} 1 & 0 & 0 \\ 0 & 1 & 0 \\ 0 & 0 & 1 \end{bmatrix} \begin{bmatrix} \lambda_\theta & 0 & 0 \\ 0 & \lambda_\phi & 0 \\ 0 & 0 & 1 \end{bmatrix} \begin{bmatrix} 1 & 0 & 0 \\ 0 & 1 & 0 \\ 0 & 0 & 1 \end{bmatrix}^T, \tag{5.176}$$

where $\mathbf{I}$ is the identity matrix. This tensor is now axially anisotropic rather than fully anisotropic. Multiplying by $\mathbf{E}$:

$$\hat{\varepsilon}\mathbf{E} = \begin{bmatrix} \lambda_\theta & 0 & 0 \\ 0 & \lambda_\phi & 0 \\ 0 & 0 & 1 \end{bmatrix} \begin{bmatrix} E_\theta \\ E_\phi \\ E_r \end{bmatrix} = \lambda_\theta E_\theta \boldsymbol{\theta} + \lambda_\phi E_\phi \boldsymbol{\phi} + \lambda_r E_r \mathbf{r}. \tag{5.177}$$

Furthermore, let us redefine the basis vectors by incorporating the anisotropy magnitudes:

$$(\boldsymbol{\theta}', \boldsymbol{\phi}', \mathbf{r}') = (\lambda_\theta \boldsymbol{\theta}, \lambda_\phi \boldsymbol{\phi}, \lambda_r \mathbf{r}'). \tag{5.178}$$

The electric field is now represented as:

$$\mathbf{E} = E_{\theta'} \boldsymbol{\theta}' + E_{\phi'} \boldsymbol{\phi}' + E_{r'} \mathbf{r}' = \begin{bmatrix} E_{\theta'} \\ E_{\phi'} \\ E_{r'} \end{bmatrix}. \tag{5.179}$$

In this new basis, the permittivity tensor simplifies to the identity:

$$\hat{\varepsilon} = \begin{bmatrix} 1\,0\,0 \\ 0\,1\,0 \\ 0\,0\,1 \end{bmatrix} \begin{bmatrix} 1\,0\,0 \\ 0\,1\,0 \\ 0\,0\,1 \end{bmatrix} \begin{bmatrix} 1\,0\,0 \\ 0\,1\,0 \\ 0\,0\,1 \end{bmatrix}^T = \begin{bmatrix} 1\,0\,0 \\ 0\,1\,0 \\ 0\,0\,1 \end{bmatrix}, \tag{5.180}$$

which represents an isotropic medium.

This re-coordination offers significant computational benefits. Matrix–vector multiplication with a full anisotropic tensor is replaced by simple scalar multiplication in the axially anisotropic case, or eliminated entirely in the isotropic case. Since this operation must be repeated at every time step, the reduction substantially improves computational speed. It also decreases numerical round-off errors from repeated approximations. These simplifications enable easier visualization, interpretation, and modification of anisotropic effects.

However, to realize axial anisotropy or isotropy representations, local reference frames are generally required. If the anisotropy aligns with a global coordinate axis, such as in spherical coordinates, then the representation is naturally aligned (as in Eq. (5.175)). But in practice, most real-world materials exhibit anisotropy that is either discontinuous or aligned along nonuniform directions (see Fig. 5.25b). In such cases, adopting local reference frames—i.e., moving frames—is essential, just as it is for the expansion of vectors and gradients on curved domains.

5.4.5 Representing Anisotropy Using Moving Frames

Consider a case of anisotropy on a two-dimensional curved domain. Let the anisotropy be represented by a linear permittivity tensor $\hat{\varepsilon}$, aligned obliquely at an angle θ, as illustrated in Fig. 5.26. In this figure, the solid lines denote the direction of material anisotropy (i.e., preferential propagation), while the arrows indicate basis vectors.

Figure 5.26a depicts a situation in which the basis vectors are not aligned with the direction of anisotropy. In such a case, a full anisotropy tensor is required to express the material property. Figure 5.26b shows the scenario where basis vectors are aligned with the anisotropy, although their direction may vary pointwise across the domain. This configuration allows an axial anisotropy representation, assuming the magnitude of anisotropy (i.e., eigenvalues) varies with location. Figure 5.26c illustrates a further refinement: the basis vectors are both aligned and scaled according to the anisotropic strength, which results in an identity permittivity tensor.

Assume $\hat{\varepsilon}$ is a full anisotropic tensor in a two-dimensional domain. Then, the action of $\hat{\varepsilon}$ on the electric field vector $\mathbf{E}$ in the Cartesian basis is given by:

$$\hat{\varepsilon}\mathbf{E} = [\mathbf{x}\ \mathbf{y}] \begin{bmatrix} \varepsilon^{xx} & \varepsilon^{xy} \\ \varepsilon^{yx} & \varepsilon^{yy} \end{bmatrix} \begin{bmatrix} E_x \\ E_y \end{bmatrix}. \tag{5.181}$$

Evaluating the matrix–vector multiplication yields:

$$\hat{\varepsilon}\mathbf{E} = (\varepsilon^{xx} E_x + \varepsilon^{xy} E_y)\mathbf{x} + (\varepsilon^{yx} E_x + \varepsilon^{yy} E_y)\mathbf{y}. \tag{5.182}$$

Assuming the material tensor is symmetric ($\varepsilon^{xy} = \varepsilon^{yx}$), its eigenvalue decomposition is:

$$\begin{bmatrix} \varepsilon^{xx} & \varepsilon^{xy} \\ \varepsilon^{yx} & \varepsilon^{yy} \end{bmatrix} = \begin{bmatrix} e_1^x & e_2^x \\ e_1^y & e_2^y \end{bmatrix} \begin{bmatrix} \varepsilon^{11} & 0 \\ 0 & \varepsilon^{22} \end{bmatrix} \begin{bmatrix} e_1^x & e_2^x \\ e_1^y & e_2^y \end{bmatrix}^T, \tag{5.183}$$

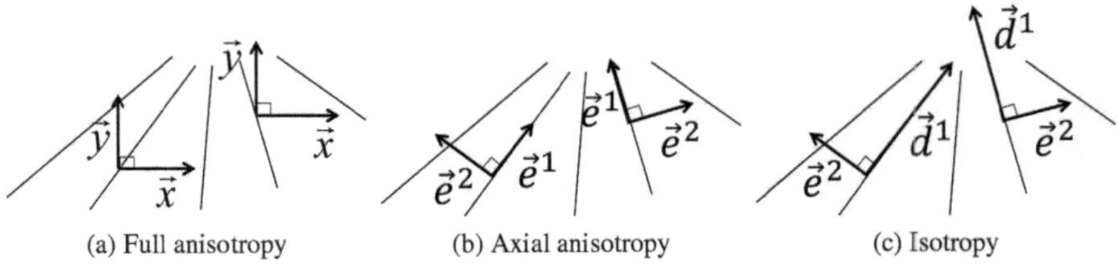

(a) Full anisotropy (b) Axial anisotropy (c) Isotropy

Fig. 5.26 Representation of (**a**) full anisotropy, (**b**) axial anisotropy, and (**c**) isotropy. (**a**) Full anisotropy: In the global coordinate system, anisotropy is expressed as a full diffusivity matrix with off-diagonal components. (**b**) Axial anisotropy: When the first tangent vector $\mathbf{e}_1$ is aligned with the anisotropic direction, the diffusivity matrix becomes diagonal without off-diagonal components. (**c**) Isotropy: When the first tangent vector is scaled according to the anisotropy magnitude, the diffusivity reduces to the identity matrix in the local moving frames where $\mathbf{e}_i = [e_i^x \; e_i^y]^T$ are normalized eigenvectors and ε^{ii} are the eigenvalues of $\hat{\varepsilon}$ for $i = 1, 2$.

Multiplying the matrices in Eq. (5.183) gives:

$$\begin{bmatrix} \varepsilon^{xx} & \varepsilon^{xy} \\ \varepsilon^{yx} & \varepsilon^{yy} \end{bmatrix} = \begin{bmatrix} \varepsilon^{11}(e_1^x)^2 + \varepsilon^{22}(e_1^x)^2 & \varepsilon^{11}e_1^x e_1^y + \varepsilon^{22}e_2^x e_2^y \\ \varepsilon^{11}e_1^x e_1^y + \varepsilon^{22}e_2^x e_2^y & \varepsilon^{11}(e_1^y)^2 + \varepsilon^{22}(e_1^y)^2 \end{bmatrix}. \tag{5.184}$$

Substituting Eq. (5.183) into Eq. (5.181) leads to the following expression for $\hat{\varepsilon}\mathbf{E}$:

$$\hat{\varepsilon}\mathbf{E} = [\mathbf{x}\,\mathbf{y}] \begin{bmatrix} e_1^x & e_2^x \\ e_1^y & e_2^y \end{bmatrix} \begin{bmatrix} \varepsilon^{11} & 0 \\ 0 & \varepsilon^{22} \end{bmatrix} \begin{bmatrix} e_1^x & e_2^x \\ e_1^y & e_2^y \end{bmatrix}^T \begin{bmatrix} E_x \\ E_y \end{bmatrix} \tag{5.185}$$

$$= [\mathbf{e}_1 \, \mathbf{e}_2] \begin{bmatrix} \varepsilon^{11} & 0 \\ 0 & \varepsilon^{22} \end{bmatrix} \begin{bmatrix} E^1 \\ E^2 \end{bmatrix} \tag{5.186}$$

$$= \varepsilon^{11}E^1\mathbf{e}_1 + \varepsilon^{22}E^2\mathbf{e}_2, \tag{5.187}$$

where E^i is the electric field component in the $\mathbf{e}_i$ direction:

$$E^1 = E^x e_1^x + E^y e_1^y, \tag{5.188}$$

$$E^2 = E^x e_2^x + E^y e_2^y. \tag{5.189}$$

The eigenvalues in the rotated frame can be expressed as:

$$\varepsilon^{11} = \varepsilon^{xx}\cos\theta + \varepsilon^{xy}\sin\theta, \tag{5.190}$$

$$\varepsilon^{22} = -\varepsilon^{yx}\sin\theta + \varepsilon^{yy}\cos\theta. \tag{5.191}$$

Now, suppose the moving frames are chosen such that their directions align with the eigenvectors $\mathbf{e}_i$ of the permittivity tensor. Then, Eqs. (5.182) and (5.187) describe the same physical behavior under different bases. Using the Cartesian coordinate system requires a full anisotropy tensor, while rotating to the eigenbasis yields an axial anisotropy.

Next, let us define new basis vectors scaled according to the eigenvalues:

$$\mathbf{d}_i = \varepsilon^{ii}\mathbf{e}_i, \quad 1 \le i \le 2. \tag{5.192}$$

Then, Eq. (5.187) simplifies further:

$$\hat{\varepsilon}\mathbf{E} = [\mathbf{d}_1 \, \mathbf{d}_2] \begin{bmatrix} 1 & 0 \\ 0 & 1 \end{bmatrix} \begin{bmatrix} E^1 \\ E^2 \end{bmatrix} = E^1\mathbf{d}_1 + E^2\mathbf{d}_2. \tag{5.193}$$

Here, $\hat{\varepsilon}$ becomes the identity matrix, but the basis vectors $\mathbf{d}_i$ are no longer orthonormal. Nevertheless, Eq. (5.193) describes the same dynamics as Eqs. (5.181) and (5.182).

This demonstrates that full anisotropy, axial anisotropy, and isotropy are not fundamentally different material behaviors but rather different representations of the same physics under different coordinate bases. By adopting moving frames that align with and scale according to the principal axes of anisotropy, one can reduce the permittivity tensor to diagonal or even identity form—yielding significant simplifications in computation and analysis.

5.4.6 Electromagnetic Field Formulation in Moving Frames

Two polarizations of an electromagnetic wave are considered for the numerical formulation of Maxwell's equations on a two-dimensional surface domain. Polarization on a two-dimensional domain specifies whether the electric field (**E**) or the magnetic field (**H**) lies on the domain, while the other field is aligned along the orthogonal direction.

The first mode is referred to as the transverse magnetic (TM) mode. In this mode, the magnetic field **H** lies on the surface, or equivalently, within the tangent plane of the surface, as illustrated in Fig. 5.27. The electric field **E** is aligned with the surface normal vector. For the moving frames $\mathbf{e}_i$ defined on the domain, the field **H** is expanded using the tangent vectors $\mathbf{e}_1$ and $\mathbf{e}_2$, while the electric field **E** is aligned with the third frame vector $\mathbf{e}_3$, as follows:

$$\mathbf{H} = H^1\mathbf{e}_1 + H^2\mathbf{e}_2, \tag{5.194}$$

$$\mathbf{E} = E^3\mathbf{e}_3. \tag{5.195}$$

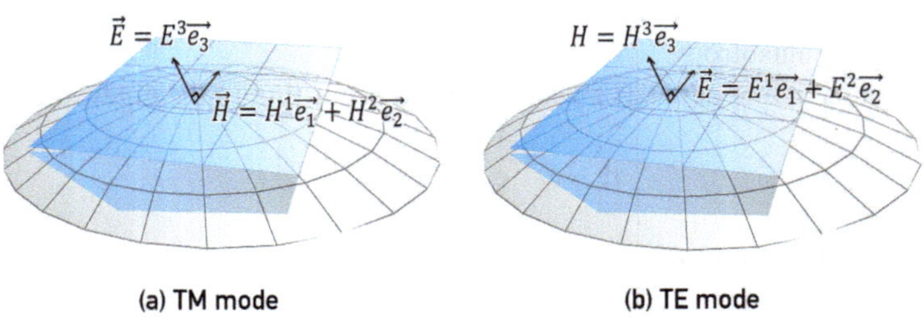

Fig. 5.27 Expansion of the electric field **E** and magnetic field **H** in moving frames $\mathbf{e}_i$ for transverse magnetic (TM) and transverse electric (TE) modes. (**a**) TM mode: **H** expands in $\mathbf{e}_1$ and $\mathbf{e}_2$, while **E** expands in $\mathbf{e}_3$. (**b**) TE mode: **E** expands in $\mathbf{e}_1$ and $\mathbf{e}_2$, while **H** expands in $\mathbf{e}_3$

The second mode is known as the transverse electric (TE) mode. In contrast to the TM mode, the electric field **E** lies in the tangent plane, while the magnetic field **H** is aligned with the surface normal vector. Accordingly, the fields are expanded using the moving frames as follows:

$$\mathbf{E} = E^1\mathbf{e}_1 + E^2\mathbf{e}_2, \tag{5.196}$$

$$\mathbf{H} = H^3\mathbf{e}_3. \tag{5.197}$$

The first tangent vector of the moving frame, $\mathbf{e}_1$, is aligned with an arbitrary direction. However, it is advantageous to align the frame vectors with the anisotropy to facilitate mathematical analysis and numerical computation. More complex applications of moving frames in electromagnetic propagation arise from the use of various materials with directional properties. The general material tensor in electromagnetic theory is expressed as a complex-valued tensor:

$$\widehat{\varepsilon} = \widehat{\varepsilon}_r + \frac{\widehat{\sigma_\varepsilon}}{i\omega} = \widehat{\varepsilon}_r - i\frac{\widehat{\sigma_\varepsilon}}{\omega}, \tag{5.198}$$

$$\widehat{\mu} = \widehat{\mu}_r + \frac{\widehat{\sigma_\mu}}{i\omega} = \widehat{\mu}_r - i\frac{\widehat{\sigma_\mu}}{\omega}, \tag{5.199}$$

where $\widehat{\varepsilon}_r$ and $\widehat{\mu}_r$ are the real permittivity and permeability tensors, respectively. The tensors $\widehat{\sigma_\varepsilon}$ and $\widehat{\sigma_\mu}$ represent the electric and magnetic conductivities for a given frequency ω. If each material tensor exhibits anisotropy, then distinct moving frames should be constructed for each tensor. This implies that four moving frames are required—one for each material tensor. For example, moving frames aligned with anisotropic directions can be expressed in complex form as:

$$\mathbf{e}_i = \mathbf{e}_i^\varepsilon + i\mathbf{e}_i^\sigma, \quad \text{for } 1 \le i \le 3, \tag{5.200}$$

where $i = \sqrt{-1}$ is the imaginary unit. The vectors $\mathbf{e}_i^\varepsilon$ and $\mathbf{e}_i^\sigma$ are aligned with the anisotropic directions of the real permittivity and electric conductivity, respectively.

For instance, consider how the term $\widehat{\varepsilon}(\partial\mathbf{E}/\partial t)$ is expressed using complex moving frames. The permittivity tensor is given in complex form as in Eq. (5.198). In many electromagnetic models, the electric field **E** in a one-dimensional domain is expressed as a Fourier series:

$$\mathbf{E} = \sum_{k=-\infty}^{\infty} E^k \exp(kx - i\omega t), \tag{5.201}$$

where k is the wave number and ω the angular frequency. Taking the time derivative yields:

$$\frac{\partial\mathbf{E}}{\partial t} = \sum_{k=-\infty}^{\infty} -i\omega E^k \exp(kx - i\omega t), \tag{5.202}$$

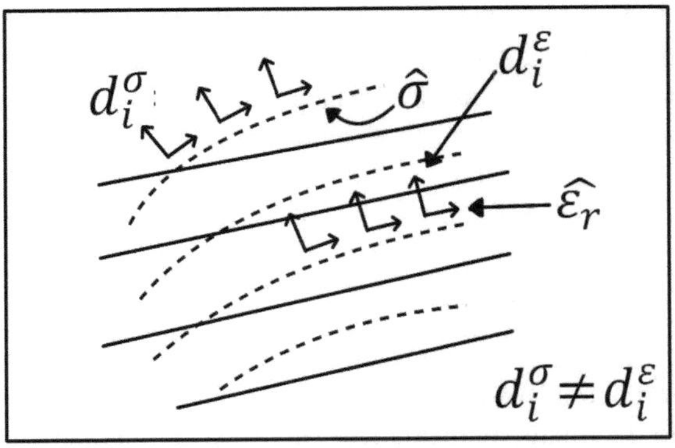

Fig. 5.28 Two moving frames in a single domain for two types of electromagnetic media: (i) a permittivity medium, denoted by $\widehat{\epsilon}_r$, and (ii) a conductivity medium, denoted by $\widehat{\sigma}$. The lengths of the moving frames vary with the magnitudes of $\widehat{\epsilon}_r$ and $\widehat{\sigma}$, represented by d_i^ϵ and d_i^σ, respectively. In general, $d_i^\epsilon \neq d_i^\sigma$

Neglecting the imaginary part, the real component of $\widehat{\varepsilon}(\partial \mathbf{E}/\partial t)$ is approximated as:

$$\widehat{\varepsilon}\frac{\partial \mathbf{E}}{\partial t} = \widehat{\varepsilon}_r \frac{\partial \mathbf{E}}{\partial t} + \widehat{\sigma}_\varepsilon \mathbf{E}_\sigma. \tag{5.203}$$

When two distinct material tensors exist within the domain, as shown in Fig. 5.28, two corresponding moving frames can be employed in Eq. (5.203). The first, denoted by $\mathbf{e}_i^\varepsilon$, is aligned with the permittivity tensor $\widehat{\varepsilon}_r$, and the second, $\mathbf{e}_i^\sigma$, with the conductivity tensor $\widehat{\sigma}_\varepsilon$. These two frame sets are generally distinct. The action of each tensor on the field $\mathbf{E}$ is then expressed as:

$$\widehat{\varepsilon}_r \mathbf{E} = \varepsilon_r^{11} E_\varepsilon^1 \mathbf{e}_1^\varepsilon + \varepsilon_r^{22} E_\varepsilon^2 \mathbf{e}_2^\varepsilon, \tag{5.204}$$

$$\widehat{\sigma}_\varepsilon \mathbf{E} = \sigma_\varepsilon^{11} E_\sigma^1 \mathbf{e}_1^\sigma + \sigma_\varepsilon^{22} E_\sigma^2 \mathbf{e}_2^\sigma.. \tag{5.205}$$

By rescaling the tangent vectors $\mathbf{e}_i^\varepsilon$ and $\mathbf{e}_i^\sigma$ using the corresponding tensor coefficients, we define the new vectors for $1 \leq i \leq 2$:

$$\mathbf{d}_i^\varepsilon = \varepsilon_r^{ii} \mathbf{e}_i^\varepsilon, \tag{5.206}$$

$$\mathbf{d}_i^\sigma = \sigma_\varepsilon^{ii} \mathbf{e}_i^\sigma \tag{5.207}$$

yielding the simplified expressions:

$$\widehat{\varepsilon}_r \mathbf{E} = E_\varepsilon^1 \mathbf{d}_1^\varepsilon + E_\varepsilon^2 \mathbf{d}_2^\varepsilon, \tag{5.208}$$

$$\widehat{\sigma}_\varepsilon \mathbf{E} = E_\sigma^1 \mathbf{d}_1^\sigma + E_\sigma^2 \mathbf{d}_2^\sigma.. \tag{5.209}$$

Substituting Eqs. (5.208) and (5.209) into Eq. (5.203) gives:

$$\widehat{\varepsilon}\frac{\partial \mathbf{E}}{\partial t} = \sum_{m=1}^{2} \left[\frac{\partial E_\varepsilon^m}{\partial t} \mathbf{d}_m^\varepsilon + E_\sigma^m \mathbf{d}_m^\sigma \right]. \tag{5.210}$$

In this formulation, the same electric field $\mathbf{E}$ is expanded differently using two distinct sets of moving frames: $\mathbf{e}_i^{\varepsilon}$ and $\mathbf{e}_i^{\sigma}$, which are aligned with $\widehat{\varepsilon}_r$ and $\widehat{\sigma}_{\varepsilon}$, respectively. By aligning each frame set with the anisotropy of its corresponding material tensor, the resulting matrix representations of these tensors become isotropic in their respective frames, as seen in Eq. (5.210). For simplicity, the following formulation assumes the presence of only one type of anisotropic material in the domain.

5.4.7 Transverse Magnetic (TM) Mode Formulation

Expand $\mathbf{E}$ and $\mathbf{H}$ as in Eqs. (5.194) and (5.195). Align the first tangent vector $\mathbf{e}_1$ with the direction of anisotropy in the permeability tensor $\widehat{\mu}$. For simplicity, we assume that the same anisotropic direction applies to both the real and imaginary parts of the permeability tensor, i.e., $\widehat{\mu}_r$ and $\widehat{\sigma}_{\mu}$. The magnetic field $\mathbf{H}$ is expanded using $\mathbf{e}_1$ and $\mathbf{e}_2$ as:

$$\mathbf{H} = H^1\mathbf{e}_1 + H^2\mathbf{e}_2. \tag{5.211}$$

Consequently, the tensor vector multiplication $\widehat{\mu}\mathbf{H}$ is expressed as:

$$\hat{\mu}\mathbf{H} = \mu^1 H^1\mathbf{e}_1 + \mu^2 H^2\mathbf{e}_2, \tag{5.212}$$

where μ^1 and μ^2 denote the eigenvalues of the anisotropic permeability tensor—either its real or imaginary part—and the corresponding eigenvectors are $\mathbf{e}_1$ and $\mathbf{e}_2$. Substituting these into Maxwell's equations (5.158) and (5.159) yields:

$$\mu_r^i\frac{\partial H^i}{\partial t} + \mathscr{R}_H^i + \sigma_\mu^i H^i + \nabla E^3\cdot(\mathbf{e}^3\times\mathbf{e}^i) + E^3\mathbf{e}^i\cdot(\nabla\times\mathbf{e}^3) = 0, \quad 1\le i\le 2, \tag{5.213}$$

$$\varepsilon_r^3\frac{\partial E^3}{\partial t} + \mathscr{R}_E + \sigma_\varepsilon^3 E^3 - \sum_{m=1}^{2}\Big[\nabla H^m\cdot(\mathbf{e}^m\times\mathbf{e}^3) - H^m\mathbf{e}^3\cdot(\nabla\times\mathbf{e}^m)\Big] = 0, \tag{5.214}$$

where the terms $\mathscr{R}_H^i$ and $\mathscr{R}_E$ represent the effects of moving domains (e.g., rotation) on the $\mathbf{H}$ and $\mathbf{E}$ fields, respectively, given by:

$$\mathscr{R}_H^i = \sum_{m=1}^{2}\mu_r^m H^m\left(\frac{\partial\mathbf{e}^m}{\partial t}\cdot\mathbf{e}^i\right), \tag{5.215}$$

$$\mathscr{R}_E = \varepsilon_r^3 E^3\left(\mathbf{e}^3\cdot\frac{\partial\mathbf{e}^3}{\partial t}\right). \tag{5.216}$$

For static domains without temporal variation, $\partial \mathbf{e}^m/\partial t = 0$, and hence Eqs. (5.215) and (5.216) vanish. Even for rotating domains, these terms may be negligible if the angular velocity is small relative to the speed of electromagnetic propagation. Therefore, we assume a stationary domain Ω such that:

$$\frac{\partial \mathbf{e}^m}{\partial t} \cdot \mathbf{e}^i = 0, \quad \text{for} \quad 1 \leq i, m \leq 3. \tag{5.217}$$

Let φ be a scalar polynomial test function. The weak forms of Eqs. (5.213) and (5.214) are given by:

$$\int \mu^i \frac{\partial H^i}{\partial t} \varphi \, dx + \int \sigma_\mu^i H^i \varphi \, dx - \int E^3 \nabla \varphi \cdot (\mathbf{e}_3 \times \mathbf{e}_i) \, dx + \int E^3 \mathbf{e}^3 \cdot (\nabla \times \mathbf{e}_i) \varphi \, dx + \int_{\partial \Omega} \mathbf{n} \cdot (\mathbf{E}^* \times \mathbf{e}_i) \varphi \, ds = 0, \tag{5.218}$$

$$\int \varepsilon^3 \frac{\partial E^3}{\partial t} \varphi \, dx - \sum_{m=1}^{2} \int H^m \nabla \varphi \cdot (\mathbf{e}_3 \times \mathbf{e}_m) \, dx - \sum_{m=1}^{2} \int H^m \mathbf{e}^m \cdot (\nabla \times \mathbf{e}_3) \varphi \, dx - \int_{\partial \Omega} \mathbf{n} \cdot (\mathbf{H}^* \times \mathbf{e}_3) \varphi \, ds = 0. \tag{5.219}$$

Here, $\mathbf{n}$ is the outward unit normal vector on the boundary. The most critical challenge in deriving a numerical scheme for Maxwell's equations under full anisotropy lies in the final term:

$$\mathbf{n} \cdot \mathbf{E}^* \times \mathbf{e}_i, \quad \mathbf{n} \cdot \mathbf{H}^* \times \mathbf{e}_3, \tag{5.220}$$

known as the **numerical flux**. In conservative form, this flux must satisfy the Rankine–Hugoniot condition to ensure the conservation property of the governing PDEs (Toro 2009). While flux derivation under isotropy is straightforward and widely used, it remains a major difficulty for full anisotropy—especially in discontinuous Galerkin methods.

To address this, one can construct moving frames aligned with the anisotropic directions, allowing the numerical flux to resemble that of a locally isotropic medium, as illustrated in Fig. 5.26. The conservative form of Eqs. (5.213) and (5.214) becomes:

$$\frac{\partial}{\partial t}\begin{bmatrix} H^1 \\ H^2 \\ E^3 \end{bmatrix} + \begin{bmatrix} 0 & 0 & 0 \\ 0 & 0 & -1/\mu^2 \\ 0 & -1/\varepsilon^3 & 0 \end{bmatrix} \nabla \begin{bmatrix} H^1 \\ H^2 \\ E^3 \end{bmatrix} \cdot \mathbf{e}_1$$

$$+ \begin{bmatrix} 0 & 0 & 1/\mu^1 \\ 0 & 0 & 0 \\ 1/\varepsilon^3 & 0 & 0 \end{bmatrix} \nabla \begin{bmatrix} H^1 \\ H^2 \\ E^3 \end{bmatrix} \cdot \mathbf{e}_2 = \begin{bmatrix} U^1 \\ U^2 \\ U^3 \end{bmatrix}, \tag{5.221}$$

where U^i denotes source terms in Eqs. (5.213)–(5.214). In the case of full anisotropy, the coefficient matrix has six eigenvalues. However, when the moving frames align with anisotropy, it yields three distinct eigenvalues $(-c, 0, c)$, independent of the anisotropic tensor's orientation. The Rankine–Hugoniot condition then gives, for $i = 1, 2$:

$$Z_i^-(\mathbf{H}^* - \mathbf{H}^-) = -\mathbf{e}_i \times (\mathbf{E}^* - \mathbf{E}^-), \tag{5.222}$$

$$Z_i^+(\mathbf{H}^{**} - \mathbf{H}^+) = \mathbf{e}_i \times (\mathbf{E}^{**} - \mathbf{E}^+), \tag{5.223}$$

$$\mathbf{e}_i \times (\mathbf{E}^{**} - \mathbf{E}^*) = 0, \tag{5.224}$$

$$\mathbf{e}_i \times (\mathbf{H}^{**} - \mathbf{H}^*) = 0, \tag{5.225}$$

with $Z_i^{\pm}$ defined as:

$$Z_i^{\pm} = (Y_i^{\pm})^{-1} \equiv \sqrt{\frac{\mu^{3-i}}{\varepsilon^3}}, \tag{5.226}$$

where the superscripts $(-)$ and $(+)$ indicate fields in a given cell and its neighboring cell, while $(*)$ and $(**)$ denote numerical flux approximations for those cells (Mohammadian et al. 1991). See Fig. 5.29 for illustration. Solving these relations yields the following upwind fluxes for the TM mode ($i = 1, 2$):

$$-\mathbf{e}_i \cdot (\mathbf{n} \times \mathbf{E}^*) = \mathbf{e}_i \cdot \frac{(-\mathbf{n} \times \mathbf{e}_3)\,\{\{Y_i E^3\}\} + 0.5\alpha \mathbf{n} \times (\mathbf{n} \times [\mathbf{H}])}{\{\{Y_i\}\}}, \tag{5.227}$$

$$\mathbf{e}_3 \cdot (\mathbf{n} \times \mathbf{H}^*) = \mathbf{e}_3 \cdot \frac{\sum_{m=1}^{2} n^m \mathbf{e}_m \times \{\{Z_m \mathbf{H}\}\} - 0.5\alpha [E^3]}{\sum_{m=1}^{2} \{\{Z_m\}\}}, \tag{5.228}$$

where we assume $\mathbf{e}^{3-} = \mathbf{e}^{3+} = \mathbf{e}^3$ and $\mathbf{n} = \sum_{i=1}^{2} n^i \mathbf{e}_i$. The averaging and jump operators are defined as:

$$\{\{A\}\} \equiv \frac{1}{2}(A^+ + A^-), \tag{5.229}$$

$$[A] \equiv A^- - A^+. \tag{5.230}$$

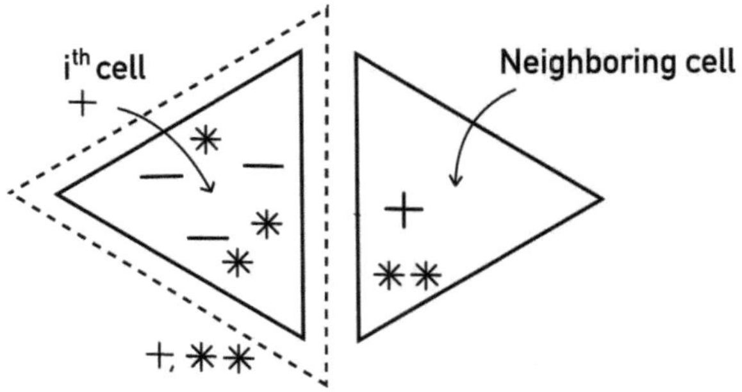

Fig. 5.29 Illustration of the $(-, +)$ and $(*, **)$ notations used in numerical flux construction. Here, $(-)$ and $(+)$ indicate the field values inside a reference cell and its neighboring cell, respectively. The notations $(*)$ and $(**)$ denote the numerical flux approximations evaluated for those cells

In Eq. (5.228), the parameter α is less than, or equal to 1.0. When $\alpha = 0$ and $Z_i = Y_i = 1$, the flux reduces to the central flux, which uses the average of $(-)$ and $(+)$ values.

5.4.8 *Transverse Electric (TE) Mode Formulation*

Similarly, expand $\mathbf{E}$ and $\mathbf{H}$ as in Eqs. (5.196) and (5.197). Align the first tangent vector $\mathbf{e}_1$ with the direction of anisotropy in the permittivity tensor. We assume that the same anisotropic direction applies to both the real and imaginary parts of the permittivity tensor, i.e., $\widehat{\varepsilon}_r$ and $\widehat{\sigma}_\varepsilon$. The electric field $\mathbf{E}$ is expanded as:

$$\mathbf{E} = E^1\mathbf{e}_1 + E^2\mathbf{e}_2. \tag{5.231}$$

Consequently, the tensor-vector product $\widehat{\varepsilon}\mathbf{E}$ becomes:

$$\hat{\varepsilon}\mathbf{E} = \varepsilon^1 E^1\mathbf{e}_1 + \varepsilon^2 E^2\mathbf{e}_2, \tag{5.232}$$

where ε^1 and ε^2 are the eigenvalues of either the real or imaginary part of the anisotropic permittivity tensor. Ignoring the effects of domain motion, we can derive the weak formulation of the TE mode equations for $1 \leq i \leq 2$ as follows:

$$\begin{aligned}\int \varepsilon^i \frac{\partial E^i}{\partial t}\varphi\, dx + \int \sigma_\varepsilon^i E^i\varphi\, dx + \int H^3\nabla\varphi\cdot(\mathbf{e}_3\times\mathbf{e}_i)\, dx\\ - \int H^3\mathbf{e}_3\cdot(\nabla\times\mathbf{e}_i)\varphi\, dx - \int_{\partial\Omega}\mathbf{n}\cdot(\mathbf{H}^*\times\mathbf{e}_i)\varphi\, ds = 0,\end{aligned} \tag{5.233}$$

$$\int \mu^3 \frac{\partial H^3}{\partial t} \varphi \, dx + \sum_{m=1}^{2} \int E^m \nabla \varphi \cdot (\mathbf{e}_3 \times \mathbf{e}_m) \, dx$$

$$+ \sum_{m=1}^{2} \int E^m \mathbf{e}_m \cdot (\nabla \times \mathbf{e}_3) \varphi \, dx + \int_{\partial \Omega} \mathbf{n} \cdot (\mathbf{E}^* \times \mathbf{e}_3) \varphi \, ds = 0. \tag{5.234}$$

The numerical fluxes for the TE mode are given by:

$$\mathbf{e}_i \cdot (\mathbf{n} \times \mathbf{H}^*) = \mathbf{e}_i \cdot \frac{(\mathbf{n} \times \mathbf{e}_3)\{Z_i H^3\} + 0.5\alpha \mathbf{n} \times (\mathbf{n} \times [\mathbf{E}])}{\{Z_i\}}, \tag{5.235}$$

$$- \mathbf{e}_3 \cdot (\mathbf{n} \times \mathbf{E}^*) = -\mathbf{e}_3 \cdot \frac{\sum_{m=1}^{2} n^m \mathbf{e}_m \times \{Y_m \mathbf{E}\} + 0.5\alpha [H^3] \mathbf{e}_3}{\sum_{m=1}^{2} \{Y_m\}}, \tag{5.236}$$

where the upwind parameters are defined as:

$$Z_i^{\pm} = (Y_i^{\pm})^{-1} \equiv \sqrt{\frac{\mu^3}{\varepsilon^{3-i}}}, \quad 1 \leq i \leq 2. \tag{5.237}$$

where the superscripts $(-)$ and $(+)$ denote field values from the left and right sides of the interface, while $*$ represents the numerical flux approximation. The notation $\{\{\cdot\}\}$ indicates the average, and $[\cdot]$ denotes the jump across the interface.

5.4.9 *[NEKTAR++] Computing Maxwell's Equations on Curved Surfaces*

The numerical schemes described in Eqs. (5.218) and (5.219) or alternatively in Eqs. (5.233) and (5.234) are implemented in the `Nektar++` spectral/hp element framework. This implementation is carefully structured to handle the intricate geometries of curved surfaces by employing the concept of moving frames. Utilizing weak Discontinuous Galerkin (DG) formulations, the computational approach aligns the differential operators with the local geometric context of the curved domain. Such alignment is particularly crucial in electromagnetics, where accurate representation of curls and fluxes significantly affects the solution accuracy.

The file `WeakDGMaxwellDirDeriv.cpp` details the computation of directional derivatives required for Maxwell's equations within a weak DG framework. Its main functionalities include the followings.

The directional derivative function first computes the physical flux vector, titled as `fluxvector`, derived from Maxwell's equations in physical space. This is accomplished through the function `GetMaxwellFluxVector`, which translates the physical fields (`physfield`) into directional flux vectors specific to Maxwell's equations. It employs the function `IProductWRTDirectionalDerivBase`, crucially aligning the derivatives with the local coordinate directions defined by moving frames. This ensures the geometric coherence of the computational domain and accurately captures the curvature effects on Maxwell's field evolution.

Numerical flux integration calculates numerical fluxes, titled as `numfluxFwd` and `numfluxBwd`, at domain interfaces using the method `NumericalMaxwellFlux`. These fluxes are vital to the DG method, ensuring continuity across element boundaries in a weak sense and allowing for effective capturing of discontinuities or sharp gradients common in electromagnetic solutions.

```
/**
 * @brief Compute weak Discontinuous Galerkin (DG) advection terms for
     Maxwell equations
 * in the form \f$ \langle \phi, \hat{F}\cdot n\rangle - (\nabla \phi \cdot
     F) \f$.
 *
 * @param   InField         Input field variables in physical space.
 * @param   OutField        Output field variables after computing DG
     advection.
 * @param   time            Simulation time, used for time-dependent
     numerical flux computation.
 */
void MMFMaxwell::WeakDGMaxwellDirDeriv(
    const Array<OneD, const Array<OneD, NekDouble>> &InField,
    Array<OneD, Array<OneD, NekDouble>> &OutField, const NekDouble time)
{
    // Number of quadrature points
    int nq                = GetNpoints();
    // Number of spectral coefficients (degrees of freedom)
    int ncoeffs           = GetNcoeffs();
    // Total number of points on the trace (element interfaces)
    int nTracePointsTot = GetTraceNpoints();
     // Number of variables (e.g., electric and magnetic fields)
    int nvar              = 3;

    // Flux vector components in local moving frames
    Array<OneD, Array<OneD, NekDouble>> fluxvector(m_shapedim);
    for (int i = 0; i < m_shapedim; ++i)
    {
        fluxvector[i] = Array<OneD, NekDouble>(nq);
    }

    // Physical field values for each variable
    Array<OneD, Array<OneD, NekDouble>> physfield(nvar);
    for (int i = 0; i < nvar; ++i)
    {
        physfield[i] = InField[i];
    }

    Array<OneD, NekDouble> tmpc(ncoeffs); // Temporary array for
         intermediate computations

    // Compute flux divergence using directional derivatives
    for (int i = 0; i < nvar; ++i)
    {
        // Compute Maxwell flux vector components for each variable
        GetMaxwellFluxVector(i, physfield, fluxvector);

        OutField[i] = Array<OneD, NekDouble>(ncoeffs, 0.0); // Initialize
             output

        for (int j = 0; j < m_shapedim; ++j)
        {
```

```
            // Compute directional derivative with respect to the j-th moving frame
            m_fields[i]->IProductWRTDirectionalDerivBase(m_CrossProductMF[j], fluxvector[j], tmpc);

            // Accumulate directional derivatives to compute divergence
            Vmath::Vadd(ncoeffs, &tmpc[0], 1, &OutField[i][0], 1, &OutField[i][0], 1);
        }
    }

    // Numerical flux arrays for forward and backward evaluations at element interfaces
    Array<OneD, Array<OneD, NekDouble>> numfluxFwd(nvar);
    Array<OneD, Array<OneD, NekDouble>> numfluxBwd(nvar);
    for (int i = 0; i < nvar; ++i)
    {
        numfluxFwd[i] = Array<OneD, NekDouble>(nTracePointsTot, 0.0);
        numfluxBwd[i] = Array<OneD, NekDouble>(nTracePointsTot, 0.0);
    }

    // Compute numerical fluxes at element boundaries, incorporating interface conditions
    NumericalMaxwellFlux(physfield, numfluxFwd, numfluxBwd, time);

    // Evaluate and adjust the computed flux divergences with numerical fluxes and corrections
    Array<OneD, NekDouble> GreenCompensation(ncoeffs);
    for (int i = 0; i < nvar; ++i)
    {
        // Negate computed divergence terms to prepare for integral addition
        Vmath::Neg(ncoeffs, OutField[i], 1);

        // Add numerical flux integrals across element boundaries
        m_fields[i]->AddFwdBwdTraceIntegral(numfluxFwd[i], numfluxBwd[i], OutField[i]);

        // Compute compensation terms based on Green's derivative theorem
        GreenCompensation = GreenDerivCompensate(i, physfield);

        // Add the Green's theorem compensation to ensure numerical accuracy and stability
        Vmath::Vadd(ncoeffs, &GreenCompensation[0], 1, &OutField[i][0], 1, &OutField[i][0], 1);

        // Reset the physical state to false after updates to spectral coefficients
        m_fields[i]->SetPhysState(false);
    }
}
```

Listing 5.5 Computation of the Maxwell equations on a curved surface using moving frames. This Nektar++ implementation employs weak DG formulations with directional derivatives specifically adapted to the local surface geometry

The file `GreenDerivCompensate.cpp` specifically addresses the geometric complexities encountered when evaluating the curl operations on curved surfaces, providing critical correction terms required in the numerical schemes.

Curl of moving frames explicitly calculates the geometric curl of the local moving frames (`m_CurlMF`). These curls represent how local tangent vectors rotate or twist along the curved surface, capturing essential geometric properties that directly impact Maxwell's equations. Depending on the electromagnetic field component being computed (e.g., TM or TE mode), this function multiplies the relevant flux components by pre-computed curls of the local moving frames. Specifically, it handles that the product of flux vectors and frame curls for each Maxwell field component, ensuring that the final numerical scheme correctly respects the non-flat geometry. Moreover, Negative signs are systematically applied, aligning with the physical interpretation of flux and curl directions. Finally, it integrates these corrected geometric quantities over the domain elements using `IProductWRTBase`, converting these geometrically-aware computations back into modal coefficients utilized by Nektar++ for further numerical solution processes.

```
Array<OneD, NekDouble> MMFMaxwell::GreenDerivCompensate(
    const int var,
    const Array<OneD, const Array<OneD, NekDouble>> &physarray)
{
    int nq      = GetNpoints();  // Number of quadrature points
    int ncoeffs = GetNcoeffs();  // Number of coefficients (degrees of freedom)

    Array<OneD, NekDouble> tmp(nq);        // Temporary array for intermediate calculations
    Array<OneD, NekDouble> outarray(ncoeffs);  // Output array to store final compensated values

    // Flux vector components in the local moving frames
    Array<OneD, Array<OneD, NekDouble>> fluxvector(m_shapedim);
    for (int j = 0; j < m_shapedim; ++j)
    {
        fluxvector[j] = Array<OneD, NekDouble>(nq);
    }

    // Compute the Maxwell flux vector for the current variable
    // Depending on TM or TE mode, flux vector components differ
    // TM: (-E3,0), (0,E3), (-H1,H2)
    // TE: (H3,0), (0,-H3), (E1,-E2)
    GetMaxwellFluxVector(var, physarray, fluxvector);

    // Curl components of moving frames
    // m_CurlMF[0] = e^3 \cdot (\nabla \times e^1)
    // m_CurlMF[1] = e^3 \cdot (\nabla \times e^2)
    // m_CurlMF[2] = e^1 \cdot (\nabla \times e^3)
    // m_CurlMF[3] = e^2 \cdot (\nabla \times e^3)

    // Perform specific computations based on the variable index (var)
    switch (var)
    {
        case 0:
            // Compensate for derivative using curl of moving frame for var = 0
            Vmath::Vmul(nq, &fluxvector[0][0], 1, &m_CurlMF[0][0], 1, &tmp[0], 1);
            Vmath::Neg(nq, tmp, 1);  // Negate the result for correct sign
            break;

        case 1:
            // Compensate for derivative using curl of moving frame for var = 1
            Vmath::Vmul(nq, &fluxvector[1][0], 1, &m_CurlMF[1][0], 1, &tmp[0], 1);
            Vmath::Neg(nq, tmp, 1);  // Negate the result
            break;

        case 2:
            // Compensate derivative with two curl components for var = 2
```

```
            Vmath::Vmul(nq, &fluxvector[0][0], 1, &m_CurlMF[2][0], 1, &tmp
                [0], 1);
            Vmath::Neg(nq, tmp, 1);
            Vmath::Vvtvm(nq, &fluxvector[1][0], 1, &m_CurlMF[3][0], 1, &tmp
                [0], 1,
                         &tmp[0], 1); // Combine curl components for final
                             compensation
            break;

        default:
            break;
    }

    // Project the compensated values onto the polynomial basis
    m_fields[0]->IProductWRTBase(tmp, outarray);

    return outarray; // Return the compensated output array
}
```

Listing 5.6 Computation of geometric compensation terms via the curl of moving frames. This `Nektar++` implementation uses weak DG formulations, explicitly incorporating directional derivatives aligned with the local geometry of curved surfaces

5.4.10 Numerical Simulation of Two-Dimensional Electromagnetic Cloaks

To demonstrate a direct application of the numerical scheme with moving frames for anisotropic Maxwell's equations, we consider the computational simulation of metamaterial-based electromagnetic cloaks (Li and Pendry 2008; Schurig et al. 2006). While various designs for invisible cloaks exist, we focus on a model with the following mechanism. The term **invisibility** implies that an electromagnetic wave directed toward the object is neither dissipated nor reflected. Instead, the wave is guided around the object as if it were absent. The object is covered with a **coating** made of metamaterial that induces this invisibility for arbitrarily shaped objects, potentially composed of diverse materials. Here, **metamaterial** refers to an engineered material not found in nature.

For an arbitrarily shaped object with such a coating, we decompose the material tensor into two principal directions: the tangential direction (parallel to the surface) and the radial (surface-normal) direction, which is orthogonal to the tangential direction. Assuming that the electric field propagates, the material tensor becomes the permittivity tensor. The eigenvalues of the material tensor corresponding to the tangential and radial directions are denoted by ε_t and ε_r, respectively. One strategy for achieving invisibility is to slow down wave propagation in the radial direction while allowing faster propagation in the tangential direction. Since dissipation and reflection result primarily from the wave interacting with the surface, eliminating the radial component of the wave ensures that tangential propagation preserves its frequency and energy, thereby rendering the object invisible. See Fig. 5.30 for illustration.

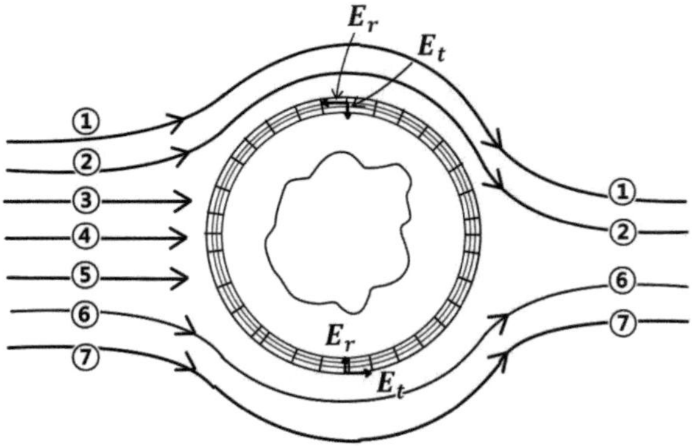

Fig. 5.30 Illustration of the mechanism of an invisible cloak. Electromagnetic rays are represented by streamlines numbered 1–7. If only streamlines 1, 2, 6, and 7 are present, the object becomes *invisible*. The quantities E_r and E_t correspond to the electric field components in the radial and tangential directions of the cloak, respectively

The underlying mechanism of the metamaterial cloak involves annihilating the field component normal to the surface—a concept often referred to as **hiding under the carpet** (Li and Pendry 2008). Figure 5.31 shows the prominent shadows generated by a perfectly electric conductor (PEC) and a perfectly magnetic conductor (PMC). Geometrically, this is equivalent to significantly reducing the **Riemannian distance** of the cloak in the radial direction, thereby compressing the effective radial size of the scattering object to a scale negligible compared to the wavelength. Consequently, the tangential component of the electric field guides the energy flux with minimal loss, reducing shadows behind the object.

Consider a perfectly electric conductor (PEC) or perfectly magnetic conductor (PMC) cylinder defined by $x^2 + y^2 = a^2$ (Cai et al. 2007). The cloak is formed from $r = a$ to $r = b$ using a metamaterial. Let $\widehat{t}$ and $\widehat{r}$ be unit vectors along the tangential and radial directions of the circle, respectively. For TE-polarized waves, we assume unit permeability, i.e., $\widehat{\mu} = \widehat{I}$, and define the permittivity tensor as:

$$\widehat{\varepsilon} = \varepsilon_t \widehat{t} + \varepsilon_r \widehat{r}, \tag{5.238}$$

with the magnitude of each component specified by:

$$\varepsilon_t = \left(\frac{b}{b-a}\right)^2, \tag{5.239}$$

$$\varepsilon_r = \left(\frac{b}{b-a}\right)^2 \left(\frac{r-a}{r}\right)^2. \tag{5.240}$$

The value of ε_t is constant across the cloak, while ε_r varies continuously from $a \leq r \leq b$. As $r \to a$, $\varepsilon_r \to 0$. Assuming vacuum permittivity equals one, a permittivity value less than one is nonphysical in natural materials. The permittivity ε_r and permeability μ_r in the radial direction determine the phase velocity v:

$$v = \sqrt{\frac{1}{\varepsilon_r \mu_r}} = \frac{c}{n}. \tag{5.241}$$

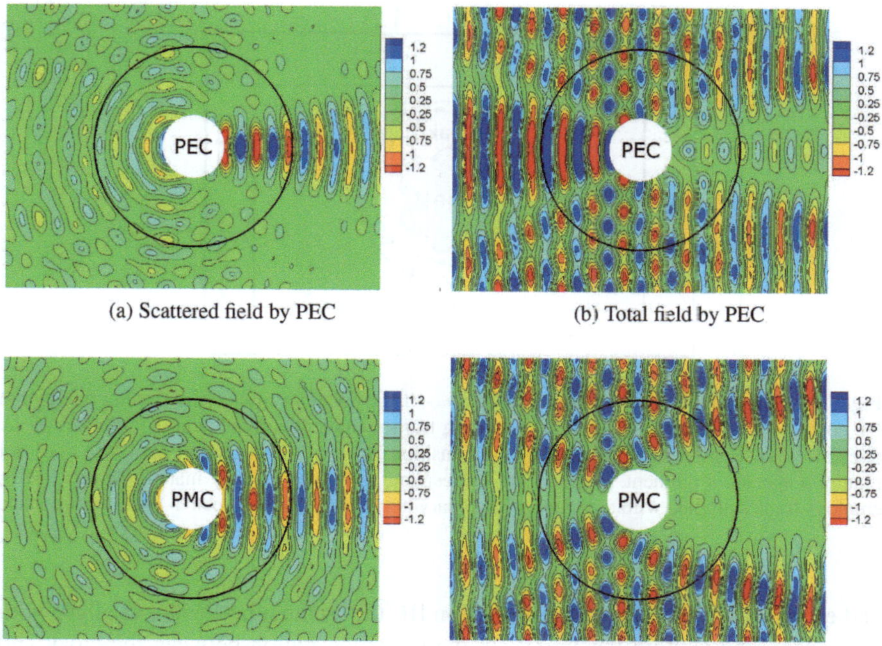

(a) Scattered field by PEC (b) Total field by PEC

(c) Scattered field by PMC (d) Total field by by PMC

Fig. 5.31 Scattering of H_z by perfect electric conductor (PEC) and perfect magnetic conductor (PMC) objects. The scattered field is obtained by subtracting the incident field from the total field. Neither case exhibits the characteristics of an invisible cloak. The object radius is $a = 1$ (632.8 nm). (**a**) Scattered field by PEC. (**b**) Total field by PEC. (**c**) Scattered field by PMC. (**d**) Total field by by PMC. (Modified from Chun 2017)

Given that the wave speed cannot exceed the speed of light in a vacuum, the vanishing ε_r implies an extremely large phase velocity due to a near-zero refractive index. This reduced index helps suppress absorption and reflection in the radial direction.

In the numerical modeling of optical cloaks, the phase velocity becomes infinite at $r = a$ because $\varepsilon_r = 0$. In time-domain simulations, this results in a material singularity. Even near $r = a$, the CFL condition demands extremely small time steps. To overcome this, two common approaches are used: (1) the lossless Drude model (Li and Xiao 2010; Zhao et al. 2008), and (2) the piecewise-constant layered cloak without dissipation (König et al. 2010). Each will be described in the following subsections.

In both models, moving frames are constructed as follows. The domain is divided into three regions: (I) the cloak region $a \leq r \leq b$, (II) the transition zone between the cloak and the perfectly matched layer (PML), and (III) the rectangular PML. The PML is used to absorb outgoing waves and reduce artificial reflection. It also enables easier comparison of the wave pattern behind the object. In regions I and II, $\mathbf{e}_1$ is aligned along the tangential direction of the circular geometry ($x^2 + y^2 = r^2$),

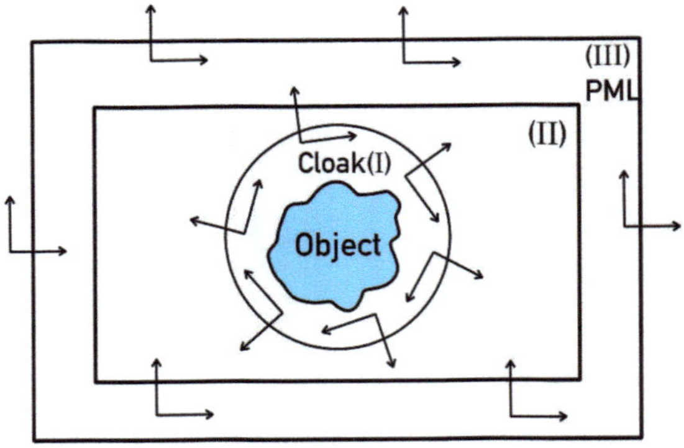

Fig. 5.32 Alignment of moving frames in the two-dimensional invisible cloak simulation. Region (I), corresponding to the cloak, employs moving frames aligned with the polar coordinates of the circular cloak. Region (II), representing the transition between the cloak and the PML, does not require a specific alignment. Region (III), corresponding to the perfectly matched layer (PML), employs frames aligned with the outward boundary normals to ensure effective PML performance

and $\mathbf{e}_2$ along the radial direction. In region III (the rectangular PML), $\mathbf{e}_1$ aligns with the x-axis, and $\mathbf{e}_2$ with the y-axis. In the transition region between the cloak and PML, which is isotropic, conventional local moving frames are used. See Fig. 5.32 for the configuration of moving frames.

5.4.11 Drude Model-Based Cloaking Simulation

The lossless Drude model represents ε_r using the following dispersive material formulation (Li and Xiao 2010; Zhao et al. 2008):

$$\varepsilon_r = 1 - \frac{\omega_p^2}{\omega^2 - i\omega\gamma}, \tag{5.242}$$

where ω and ω_p denote the angular and plasma frequencies, respectively. For a lossless material, the collision frequency γ is set to zero. The moving frames are aligned as shown in Fig. 5.32:

$$\mathbf{e}_1 = \mathbf{e}_t = \widehat{t}, \tag{5.243}$$

$$\mathbf{e}_2 = \mathbf{e}_r = \widehat{r}. \tag{5.244}$$

To facilitate simulations of the invisible cloak, we employ the total and scattered field formulation for the electromagnetic field. The total electric field is expressed as the sum of the incident and scattered fields:

$$\mathbf{E}_{total} = \mathbf{E}_{inc} + \mathbf{E}_{scat}, \tag{5.245}$$

where $\mathbf{E}_{\rm inc}$ represents the incident field (typically a plane wave), and $\mathbf{E}_{\rm scat}$ is the field resulting from scattering due to objects in the domain. These fields are expanded in the moving frame basis as:

$$\mathbf{E}_{inc} = E^t_{inc}\mathbf{e}_t + E^r_{inc}\mathbf{e}_r, \tag{5.246}$$

$$\mathbf{E}_{scat} = E^t_{scat}\mathbf{e}_t + E^r_{scat}\mathbf{e}_r. \tag{5.247}$$

Applying this formulation to the TE mode of Maxwell's equations yields:

$$\varepsilon_t \frac{\partial E^t_{scat}}{\partial t} = \nabla H_{scat} \cdot \mathbf{e}_r - (\varepsilon_t - 1)\frac{\partial E^t_{inc}}{\partial t}, \tag{5.248}$$

$$\frac{\partial E^r_{scat}}{\partial t} = -\nabla H_{scat} \cdot \mathbf{e}_t - \omega_p^2 \int_0^t E^r_{inc} dt' - Q, \tag{5.249}$$

$$\frac{\partial H_{scat}}{\partial t} = \nabla E^t_{scat} \cdot \mathbf{e}_r - \nabla E^r_{scat} \cdot \mathbf{e}_t, \tag{5.250}$$

$$\frac{\partial Q}{\partial t} = \omega_p^2 E^r_{scat}, \tag{5.251}$$

where $\partial E^t_{\rm inc}/\partial t$ and $\int_0^t E^r_{\rm inc}\, dt'$ are evaluated analytically. The computational domain is $[-6, 6] \times [-4.5, 4.5]$, with a PML layer of thickness 0.5 applied at the outer boundary. The cloak is divided into 16 radial layers, and within each layer, ε_r and ω_p vary continuously with radius r. Figure 5.33 shows the results of the simulation. The scattered fields beyond the cloak are significantly attenuated for both PEC and PMC objects, and the total fields beyond the cloak remain nearly undisturbed. Notably, the magnitude of the scattered field decreases with increased spatial resolution.

5.4.12 *Layered Cloak Simulation with Piecewise-Constant Parameters*

Another model for wave propagation in invisible cloaks employs multiple concentric layers. In this model, the material parameters are constant within each layer (König et al. 2010). The cloak is divided into N layers, as shown in Fig. 5.34a. The radius r of the i-th layer is given by:

$$r = 1 + \frac{b-a}{N(i-0.5)}, \quad \text{if} \quad i \le \frac{Nr}{b-a} \le i+1. \tag{5.252}$$

As a result, ε_r becomes piecewise constant across layers (Fig. 5.34b). As N increases, the model approaches the theoretical cloaking limit. In our simulations, we set $N = 16$. The same mesh is used as in the Drude model. However, the time

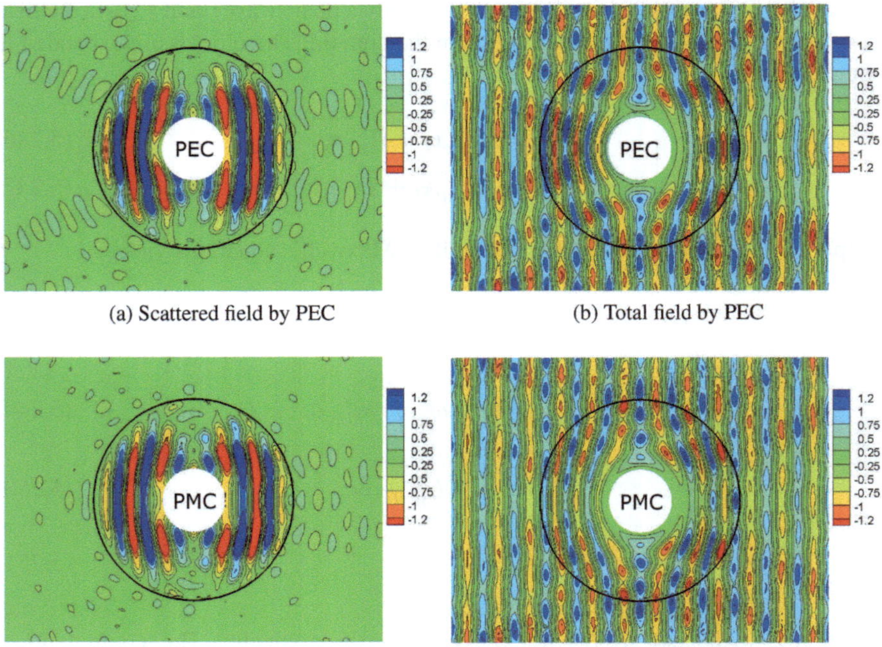

(a) Scattered field by PEC

(b) Total field by PEC

(c) Scattered field by PMC

(d) Total field by by PMC

Fig. 5.33 Scattering of H_z by objects modeled as a perfect electric conductor (PEC) and a perfect magnetic conductor (PMC), computed with the lossless Drude model to demonstrate the behavior of an invisible cloak. The scattered field is defined as the difference between the total and incident fields. The object radius is $a = 1$ (632.8 nm), and the cloak radius is $b = 1/0.314$ (2015.3 nm). (**a**) Scattered field by PEC. (**b**) Total field by PEC. (**c**) Scattered field by PMC. (**d**) Total field by by PMC. (Adapted from Chun 2017)

step Δt depends not only on the magnitude of ε_r but also inversely on the grid resolution:

$$\Delta t \propto \frac{\varepsilon_r}{h^{p+1}}, \tag{5.253}$$

where h is the element size and p is the polynomial order of the numerical solution. Compared to the Drude model, this method is advantageous as it does not require auxiliary variables such as Q. Furthermore, the electromagnetic fields reach a steady state immediately, unlike the Drude model which exhibits a delayed steady state due to accumulated dissipation effects (Li and Xiao 2010; Zhao et al. 2008).

Figure 5.35 illustrates the simulation results for PEC and PMC objects cloaked with the layered metamaterial. The distribution of the magnetic field H_z closely resembles that of the Drude model, with reduced dielectric scattering. The observation time is $T = 12$ (0.0253 ps in physical units), which is half the duration used in the Drude model.

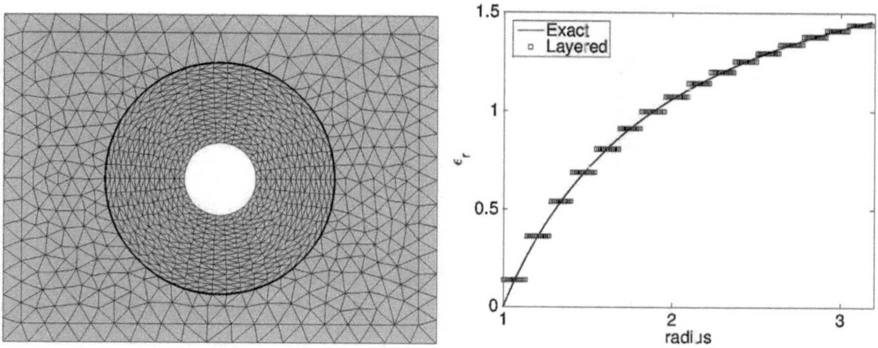

(a) Layered mesh

(b) Piecewise-constant ε_r

Fig. 5.34 Layered mesh for the simulation of a circular invisible cloak and corresponding material parameter ε_r distribution of a piecewise-constant layered cloak. (**a**) Layered mesh. (**b**) Piecewise-constant ε_r. (Modified from Chun 2017)

5.4.13 ELF Wave Propagation on General Curved Geometries

The numerical scheme using moving frames is also applied to simulate extremely low-frequency (ELF) electromagnetic wave propagation on curved surfaces. ELF waves typically have frequencies between 3 Hz and 30 kHz and are observed in the Earth-ionosphere cavity as guided waves (Cummer 2000). These wavelengths can range from 10,000 km to 100,000 km. In nature, ELF waves are generated by phenomena such as lightning or perturbations in the Earth's magnetic field, and they propagate within a layer of the atmosphere (Simpson and Taflove 2002). The Coriolis force terms are not negligible and may serve as a sufficient source for novel electromagnetic propagation in objects undergoing extremely rapid angular rotation. This phenomenon, though debated, is known as the *Sagnac effect* (Post 1967). Nevertheless, the Sagnac effect is disregarded in the present simulation. Consequently, ELF waves are frequently modeled as propagating along the surface of the Earth.

The governing equations used for curved surfaces follow the same weak formulations as the TE mode (Eqs. (5.233) and (5.234)) or TM mode (Eqs. (5.213) and (5.214)).

Figure 5.36 illustrates the simulation results of ELF propagation on three types of surfaces: a unit sphere, an irregular surface, and a non-convex surface. The irregular and non-convex geometries are defined by:

$$\text{Irregular surface} : x^2 + y^4 + y + z^4 + z^2 = 1, \tag{5.254}$$

$$\text{Non-convex surface} : 2x^2 + 2(y^4 - y^2) + z^4 + z^2 = 2.0. \tag{5.255}$$

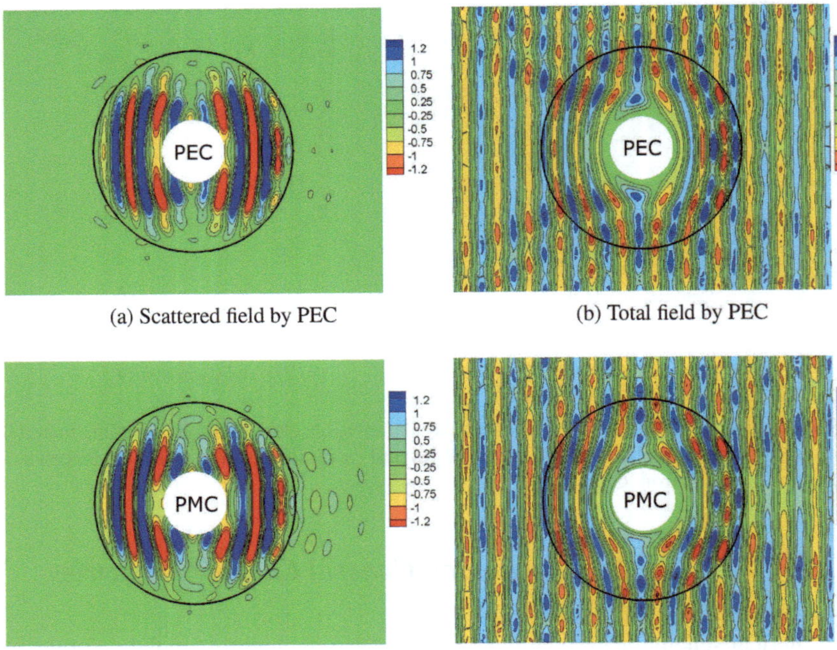

(a) Scattered field by PEC (b) Total field by PEC

(c) Scattered field by PMC (d) Total field by by PMC

Fig. 5.35 Scattering of H_z by objects modeled as a perfect electric conductor (PEC) and a perfect magnetic conductor (PMC), computed with the piecewise-parameterized layered cloak to demonstrate the behavior of an invisible cloak. The scattered field is defined as the difference between the total and incident fields. The object radius is $a = 1$ (632.8 nm), and the cloak radius is $b = 1/0.314$ (2015.3 nm). (**a**) Scattered field by PEC. (**b**) Total field by PEC. (**c**) Scattered field by PMC. (**d**) Total field by by PMC. (Adapted from Chun 2017)

Electromagnetic field distributions are shown at $T = 0.2$ (3.2 ms) and $T = 1.0$ (16 ms). These simulations validate the robustness and versatility of the moving frame numerical scheme for general curved domains.

Figure 5.37 shows a 180-degree phase reversal of the electric field at the antipodal point. The solid lines correspond to the static surface case, while dashed lines represent rotating surfaces. Although the differences between the two are minor, they become more apparent on non-convex surfaces after long-term integration. The second row in Fig. 5.37 demonstrates energy conservation properties of the numerical scheme under various mesh sizes h and polynomial orders p.

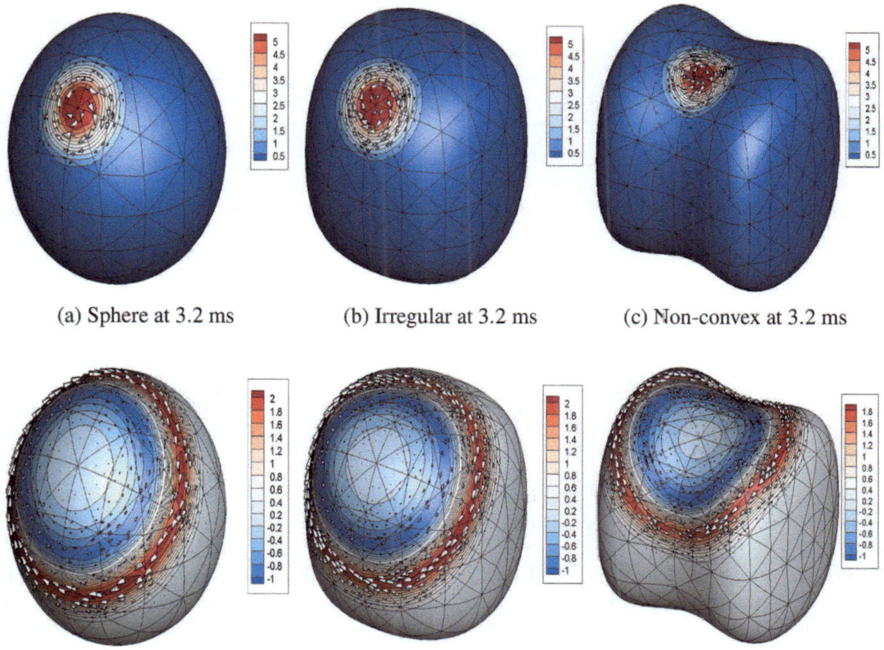

(a) Sphere at 3.2 ms (b) Irregular at 3.2 ms (c) Non-convex at 3.2 ms

(d) Sphere at 16.0 ms (e) Irregular at 16.0 ms (f) Non-convex at 16.0 ms

Fig. 5.36 Extremely low-frequency (ELF) wave propagation across various curved surfaces. A Gaussian pulse with a radius of 0.2, corresponding to 1.27×10^3 km in real scale, is initiated at a single point in each curved domain. The propagation patterns are illustrated at $t = 3.2$ ms and $t = 16.0$ ms. (**a**) Sphere at 3.2 ms. (**b**) Irregular at 3.2 ms. (**c**) Non-convex at 3.2 ms. (**d**) Sphere at 16.0 ms. (**e**) Irregular at 16.0 ms. (**f**) Non-convex at 16.0 ms (Adapted from Chun 2017)

5.5 Summary

In this chapter, we developed and demonstrated a numerical framework that incorporates **moving frames** into the solution of partial differential equations (PDEs) on complex, curved domains. The approach is especially effective in environments exhibiting **anisotropy** or geometric complexity, where standard formulations struggle with coordinate singularities or explicit dependence on the metric tensor.

We began by formulating **conservation laws** on general curved surfaces and introduced rescaled moving frames to simplify the weak form, eliminating explicit geometric terms while maintaining accuracy and conservation properties. Numerical tests on spherical, irregular, and non-convex surfaces confirmed the scheme's **exponential convergence** and robustness across diverse geometries.

Next, we addressed **diffusion-reaction equations** by aligning moving frames with the principal directions of anisotropic diffusivity tensors. This alignment allows for efficient incorporation of anisotropic effects without relying on off-diagonal

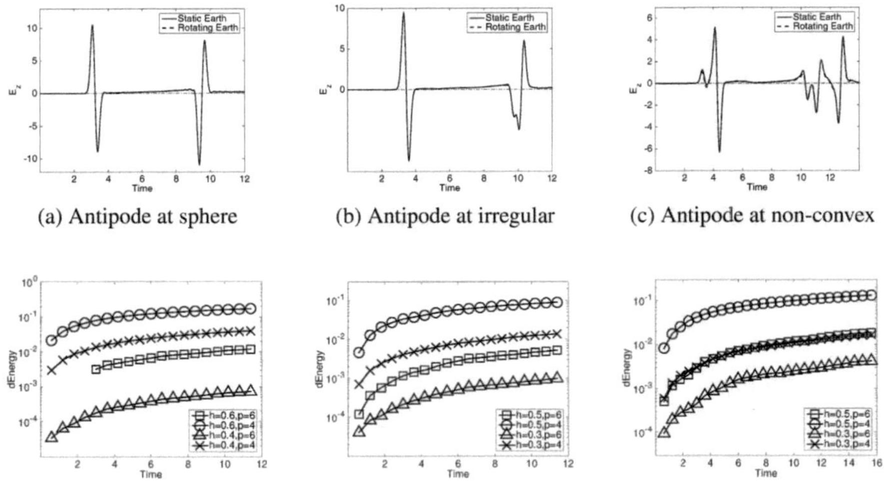

(a) Antipode at sphere (b) Antipode at irregular (c) Antipode at non-convex

(d) Energy conservation at sphere (e) Energy conservation at irregular (f) Energy conservation at non-convex

Fig. 5.37 Antipodal field values and the corresponding energy conservation demonstrated in extremely low-frequency (ELF) propagation simulations across different curved surfaces. (**a**) Antipode at sphere. (**b**) Antipode at irregular. (**c**) Antipode at non-convex. (**d**) Energy conservation at sphere. (**e**) Energy conservation at irregular. (**f**) Energy conservation at non-convex. (Adapted from Chun 2017)

terms in the diffusivity matrix. Simulations on spherical and irregular domains validated the approach, emphasizing the importance of retaining divergence terms in the frame fields.

The chapter then introduced a reformulation of the **shallow water equations (SWEs)** on the sphere using moving frames, with a particular focus on weak formulation and numerical stability. Through benchmark problems—including steady and unsteady zonal flows, Rossby–Haurwitz waves, and perturbed jet streams—we demonstrated the high accuracy, stability, and conservation properties of the moving frame-based SWE solver, even in challenging scenarios such as flows over isolated mountains.

We extended this formulation to **Maxwell's equations**, handling **full anisotropy** via frame-aligned transformations of permittivity and permeability tensors. By reformulating TM and TE polarization modes in moving frames, we enabled efficient numerical treatment of anisotropic electromagnetic propagation. This culminated in the simulation of **two-dimensional cloaking devices** using both Drude and layered metamaterial models, where frame alignment simplified the implementation of cloaking conditions.

Finally, we applied the same moving frame framework to simulate **extremely low-frequency (ELF) electromagnetic waves** on a variety of curved surfaces. The method preserved both phase and energy properties across spherical, irregular, and non-convex domains, validating the geometric adaptability and numerical strength of the moving frame approach.

Overall, this chapter establishes moving frames as a versatile and powerful tool for solving a broad class of PDEs on non-Euclidean geometries, offering geometric clarity, computational efficiency, and numerical accuracy across physical systems.

5.6 Exercises and Conceptual Applications

1. Derive the weak formulation of the scalar conservation law

$$\frac{\partial u}{\partial t} + \nabla \cdot (uv) = 0 \tag{5.256}$$

on a curved surface using a test function ϕ and moving frames $\{\mathbf{e}_1, \mathbf{e}_2\}$. Show that the spatial flux term becomes a directional derivative along scaled moving frames.

   a. Implement the cosine bell advection test on a spherical domain as described in Eqs. (5.14)–(5.19). Analyze spectral convergence for increasing polynomial degree p and decreasing mesh size h.
   b. Explain why divergence operators on curved domains are nontrivial even for constant vector fields. Provide an example using spherical coordinates.

2. Given a diffusivity tensor D with non-zero off-diagonal entries, perform eigendecomposition and explain how anisotropy affects wavefront shape. Illustrate with a 2D example.

   a. Re-derive the weak formulation for anisotropic diffusion on a curved surface (Eqs. (5.49)–(5.50)) using scaled frames $d_i = \lambda_i e_i$. Discuss the role of $\nabla \cdot \mathbf{e}_i$.
   b. Simulate Turing patterns from Eqs. (5.51)–(5.52) on a spherical domain using spectral elements. Validate your solution using spherical harmonics expansion.

3. Explain the geometric motivation for using moving frames in the SWE formulation. Why does the zero-tendency assumption (Eq. (5.81)) matter?

   a. Reproduce the steady zonal flow test ($\alpha = \pi/4$) using the DG formulation in Eqs. (5.98)–(5.100). Compute the L_2 error in elevation η and compare with Table 5.5.
   b. Show how moving frames aligned with local flow direction can reduce artificial diffusion across element boundaries in SWE solvers.

4. Given a full anisotropic permittivity tensor $\hat{\varepsilon}$, show how a change of basis aligned with its eigenvectors reduces it to a diagonal matrix. Verify using the example in the text.

c. Simulate TM mode propagation over a curved surface using Eqs. (5.208)–(5.209). Analyze the impact of curvature by comparing results on spherical and non-convex surfaces (Eq. (5.250)).
d. Discuss how the choice of moving frames affects numerical flux construction (Eq. (5.215)) under full anisotropy. Propose a strategy to minimize flux inconsistency at element boundaries.

5. Propose a new numerical benchmark problem involving conservation or diffusion-reaction on a genus-1 (toroidal) surface. Describe the setup, equations, and goals. Analyze why divergence terms of moving frames are essential for convergence. Reproduce convergence plots from Fig. 5.11 or Fig. 5.12, with and without the $\nabla \cdot \mathbf{e}_i$ term.

Chapter 6
Relative Acceleration and the Riemann Curvature Tensor

> *To see a world in a grain of sand,*
> *And a heaven in a wild flower,*
> *Hold Infinity in the palm of your hand,*
> *And eternity in an hour"*
>
> —*William Blake, "Auguries of Innocence"*

Some partial differential equations (PDEs) describe time-dependent wave propagation in multidimensional space. A well-known example is the Navier–Stokes equation, which governs time-dependent fluid flow in multidimensional domains. For instance, incompressible flow can be described by the following equation:

$$\frac{\partial \mathbf{u}}{\partial t} + (\mathbf{u} \cdot \nabla)\mathbf{u} - \nu \nabla^2 \mathbf{u} = -\frac{1}{\rho}\nabla p + \mathbf{g}, \tag{6.1}$$

where $\mathbf{u}$ is the flow velocity, ρ is the fluid density, p is the pressure, ν is the dynamic viscosity, and $\mathbf{g}$ is the gravitational acceleration. The Navier–Stokes equation characterizes how the flow velocity evolves over time. The path traveled by each fluid particle is determined by the flow velocity, which serves as the tangent vector to the trajectory (see Fig. 6.1a). This trajectory is obtained by integrating the flow velocity. In idealized settings (e.g., incompressible laminar flow), wavefronts can be approximated as contour lines orthogonal to the local flow vector.

Maxwell's equations also describe the propagation of electromagnetic waves through electric and magnetic fields, denoted by $\mathbf{E}$ and $\mathbf{B}$, respectively. In vacuum and without source terms, Maxwell's equations are given by:

$$\frac{\partial \mathbf{E}}{\partial t} = \nabla \times \mathbf{H}, \tag{6.2}$$

$$\frac{\partial \mathbf{B}}{\partial t} = -\nabla \times \mathbf{E}. \tag{6.3}$$

S. Chun, *Moving Frames for the Numerical Solution of Partial Differential Equations in Complex Domains*, Springer Asia Pacific Mathematics Series 7,
https://doi.org/10.1007/978-981-95-4495-0_6

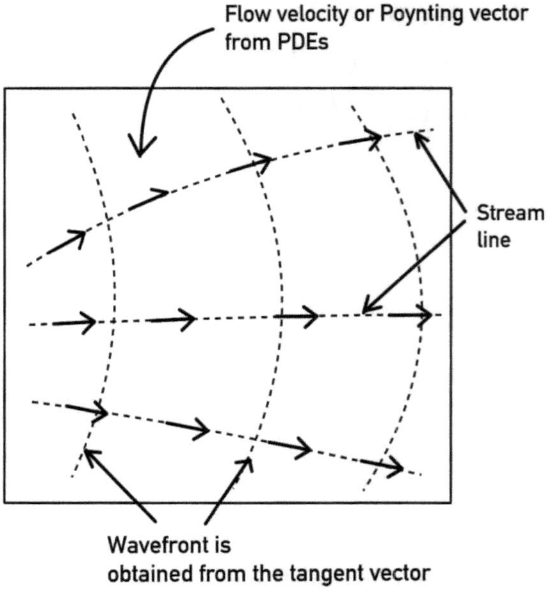

Fig. 6.1 Illustration of streamlines (dashed), their tangent vectors (arrows), and associated wavefronts (dashed lines orthogonal to the streamlines), all induced by the flow vector field in the partial differential equations

An electrically neutral particle such as a photon which mediates the electromagnetic force travels in the direction orthogonal to both **E** and **B**, along **E** × **B**—a direction known as the **Poynting vector**. Depending on whether **E** or **B** lies in the domain plane or along the surface normal, the electromagnetic wave is classified as either a transverse electric (TE) or transverse magnetic (TM) mode. In homogeneous, isotropic media, the direction of wave propagation aligns with the Poynting vector, which indicates energy flux., as illustrated in Fig. 6.1. Similar to the case of fluid flow, wavefronts in electromagnetic propagation can be constructed as contour lines orthogonal to the Poynting vector. In this context, the wavefront has a more concrete physical interpretation—it separates the region where the wave has arrived from the region it has yet to reach.

6.1 Wave Patterns in Diffusion—Reaction Equations

Another important class of equations in modern scientific computing is the diffusion—reaction equation, which can be written in a simplified form as:

$$\frac{\partial u}{\partial t} = \nabla \mathscr{D} \cdot \nabla u + \mathscr{R}(u, \cdots,), \tag{6.4}$$

where u is a scalar variable that propagates via diffusion, $\mathscr{D}$ is the diffusion tensor, and $\mathscr{R}(u, \cdots)$ is a reaction function. Most diffusion—reaction equations are multivariate and serve as governing equations for numerous chemical and

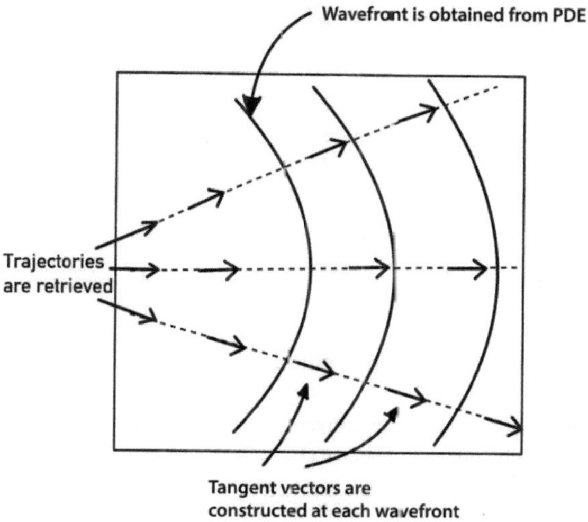

Fig. 6.2 Illustration of streamlines (dashed), their tangent vectors (arrows), and the associated wavefronts (dashed lines orthogonal to the streamlines)

biological phenomena, including pattern formation, epidemics, tumor growth, and the propagation of biological electrical signals in the heart and brain.

The first term in Eq. (6.4) represents **diffusion**, or more specifically, **anisotropic diffusion** when the medium exhibits directional dependence. This term governs the spread of u between regions. The second term represents the **reaction**, which accounts for self-excitation initiated by incoming diffusion from neighboring regions. When a neighboring region accumulates sufficient u to surpass a threshold, the reaction term recharges the region to a fully excited state. Subsequently, this excited region diffuses u further to its neighbors.

In diffusion—reaction equations, the time-dependent solution is the spatial distribution of the scalar variable u. The solution often exhibits sharp contrasts between a resting state (where u remains near the initial value) and an excited state (where u exceeds a threshold). This contrast defines the wavefront—a boundary between regions that are already excited and those about to be. Constructing trajectories in this case is the reverse of how wavefronts were derived from trajectories in the Navier–Stokes and Maxwell equations (see Fig. 6.2): the tangent vector of the trajectory is taken to be orthogonal to the wavefront at each point. The trajectory can thus be obtained as a continuous path whose tangent aligns with the local normal vector of the wavefront.

6.2 Wave Patterns and the Velocity Vector

PDEs describe wave propagation phenomena, and analyzing the resulting wave patterns is essential when they correlate with specific physical processes. Of particular interest are converging and diverging wave patterns, as illustrated in

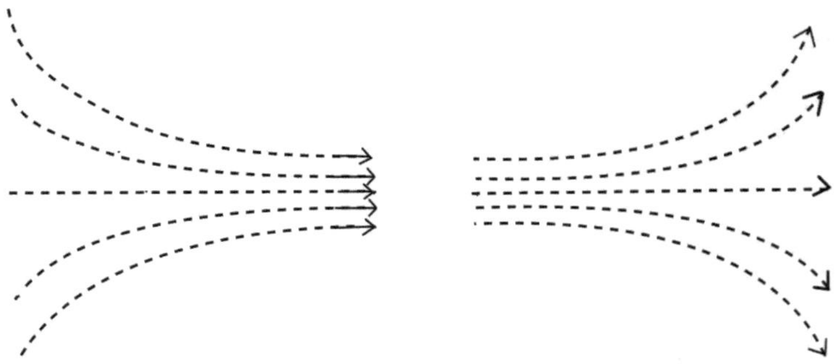

Fig. 6.3 Illustration of converging and diverging wave patterns along trajectories (dashed lines). In convergence, the distance between trajectories decreases, whereas in divergence it increases

Fig. 6.3. In a converging wave, trajectories move closer together as the wave propagates. In a diverging wave, they spread apart. These patterns reflect different physical states depending on what is being transported.

In the Navier–Stokes context, where fluid particles follow the trajectory, a converging wave corresponds to an increase in particle density, resulting in more compact trajectories. Conversely, in a diverging wave, the density decreases, and the trajectories become more dispersed. These changes in trajectory density also relate to pressure variations within the fluid.

In Maxwell's equations, the convergence and divergence of electromagnetic waves are of particular interest in the presence of massive objects such as black holes. The theory of curved spacetime, extensively studied by Einstein and others, suggests that the presence of mass warps the surrounding space. In Euclidean geometry, the shortest path between two points is a straight line. In the geometric optics approximation, electromagnetic waves follow light rays, which correspond to geodesics in spacetime under general relativity (see Fig. 6.4a). However, in curved spacetime, these paths bend due to gravitational effects, and straight lines may no longer represent the shortest route (see Fig. 6.4b).

Electromagnetic wave propagation on a macroscopic scale can be viewed as an emergent effect of microscopic processes—namely, repeated absorption and emission of photons. A photon is emitted by a source atom and travels through space to excite another atom. If enough photons arrive, the receiving atom re-emits photons to neighboring atoms. We use the term **geometry** here metaphorically, to suggest that the structure of excitation influences how signals propagate, resembling directional constraints imposed by a geometry. When many excited atoms send photons to a few excitable ones, propagation is likely to succeed and may accelerate. This corresponds macroscopically to a **converging wave**. Conversely, if a few excited atoms try to excite many targets, propagation may fail due to insufficient photon density, leading to a **diverging wave** (Fig. 6.5).

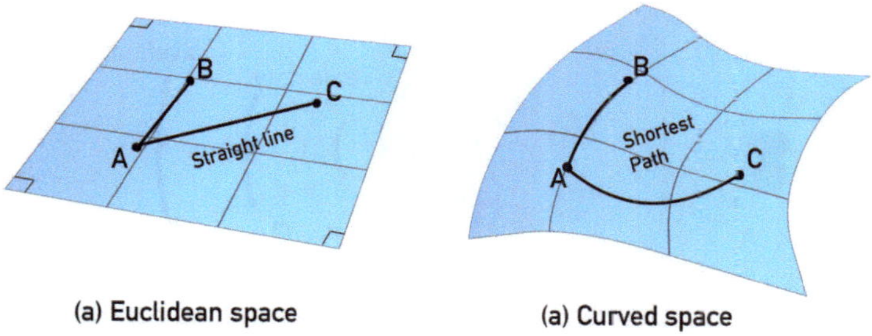

Fig. 6.4 (**a**) In a Euclidean plane, the shortest paths between A and B or A and C are straight lines. (**b**) In a curved space, the shortest paths are geodesics, which are generally curved rather than straight

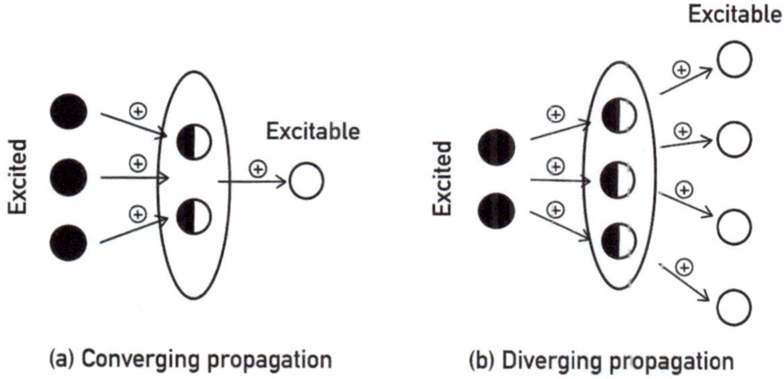

Fig. 6.5 Converging and diverging propagation at the atomic scale. (**a**) In convergence, excited atoms deliver photons to a smaller number of excitable cells. (**b**) In divergence, excited atoms deliver photons to a larger number of excitable cells

6.3 Wave Patterns and the Wavefront

A similar diffusion—reaction mechanism underlies biological signal propagation, such as in the Hodgkin–Huxley model (Hodgkin and Huxley 1952) and the FitzHugh–Nagumo equation (FitzHugh 1961). The primary signal is the membrane potential ϕ_m, defined as the difference between the intracellular potential ϕ_i and the extracellular potential ϕ_e:

$$\phi_m = \phi_i - \phi_e. \tag{6.5}$$

Changes in ϕ_m can result from ion movement either within cells or in the extracellular space. When ions move across membranes or within surrounding tissues, they alter ϕ_m and trigger signal propagation. As illustrated in Fig. 6.6a, excited cells (dark), excitable cells (white), and currently exciting cells (grey) form layered structures in tissues such as the heart or brain. In Fig. 6.6b, two excited

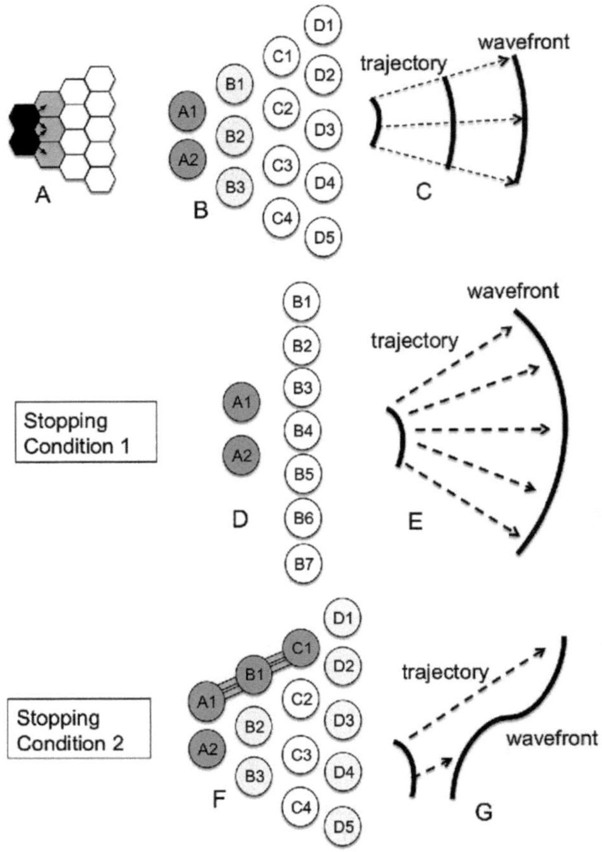

Fig. 6.6 Converging and diverging electric signal propagation in biological tissue. (**a**) Excited cells deliver charged ions to nearby excitable cells. (**b**) Cells or tissues can be treated as propagation units, leading to a unit-based description. (**c**) The increase or decrease in excited cells is represented by the length of the wavefront at each propagation step. (**d**) Stopping condition 1: when the number of excitable cells greatly exceeds the number of excited cells, the ions released by excited cells are insufficient to excite all or any excitable cells. (**e**) This condition is expressed as a significant reduction in wavefront length. (**f**) Stopping condition 2: in strongly anisotropic tissue, the number of excitable cells can again exceed that of excited cells, leading to conduction failure. (**g**) This case is represented as a sudden expansion of the wavefront. (Adapted from Chun 2014b)

cells ($A1$, $A2$) propagate the signal to $B1$–$B3$, which subsequently excite $C1$–$C4$, forming a diverging wavefront (Fig. 6.6c).

If the ratio of excited to excitable cells falls below a threshold, propagation fails. Figure 6.6d shows that $A1$ and $A2$ must excite seven neighboring cells ($B1$–$B7$), requiring each to stimulate 3.5 cells on average—an unrealistic burden due to limited ion resources and the all-or-nothing nature of excitation. Propagation stalls when this ratio becomes too large. In this stylized illustration, failure occurs when the local excitation load exceeds the resource capacity of upstream cells. The actual threshold varies with ionic dynamics and coupling strength. Figure 6.6e illustrates this as a change in the shape and curvature of the wavefront.

Anisotropy, often arising from domain geometry, can significantly alter the ratio between excited and excitable cells. In Fig. 6.6f, strong anisotropic links (solid lines) constrain excitation from $A1$ to $B1$, and from $B1$ to $C1$, bypassing other possible paths. Thus, the anisotropy controls the direction and extent of signal propagation. Figure 6.6g shows the resulting deformation in the wavefront and its trajectory. In this context, **geometry** refers to the ratio of excited to excitable cells, which depends on domain shape, initiation location, wave speed, excitation threshold, and interference—but above all, anisotropy plays the most decisive role in shaping propagation. Here, we use **geometry** as a generalized term to include structural and anisotropic features influencing signal propagation.

6.3.1 Wave Patterns by Wavefront Analysis

In cardiac excitation propagation, one of the most critical analytical objectives is to understand **conduction failure**, which is essential for analyzing the initiation and termination of many cardiac pathologies. An established analytical method for identifying conduction failure in nonlinear dynamics involves studying the shape of the wavefront. This wavefront-based analysis, known as **kinematic analysis**, expresses the approximate relationship between the propagation velocity (V). The following expression arises from kinematic wave theory and holds approximately under homogeneous, isotropic conditions. In general, full PDE simulations are required to capture nonlinear behavior (Davydov et al. 1991; Zykov and Winfree 1992; Mikhailov 1991):

$$V \approx V_0 - D\kappa, \tag{6.6}$$

where V_0 is the intrinsic conduction velocity determined by the cardiac tissue, D is the diffusivity constant, and κ is the curvature of the wavefront. In a homogeneously isotropic medium, wavefront curvature plays a key role in modulating propagation speed. When the curvature exceeds a certain threshold, the conduction velocity decreases substantially and may even reach zero.

High wavefront curvature has been shown to correlate with conduction block due to increased excitation demands at convex regions—a phenomenon sometimes termed **sink–source** or **impedance mismatch** (Rohr et al. 1997; Xie et al. 2010). A higher curvature implies a greater ratio between the number of **excited** cells (source) and **excitable** cells (sink). This ratio is constrained by a physiological upper bound, the **threshold ratio**, due to limited resources in the excited cells. When the wavefront curvature becomes too large, the sink—source ratio exceeds this threshold, leading to conduction failure.

Consider a wave propagating from a point source in an isotropic medium. The resulting wavefront is a circular arc with time-dependent radius. Let ξ_1 and ξ_2 denote

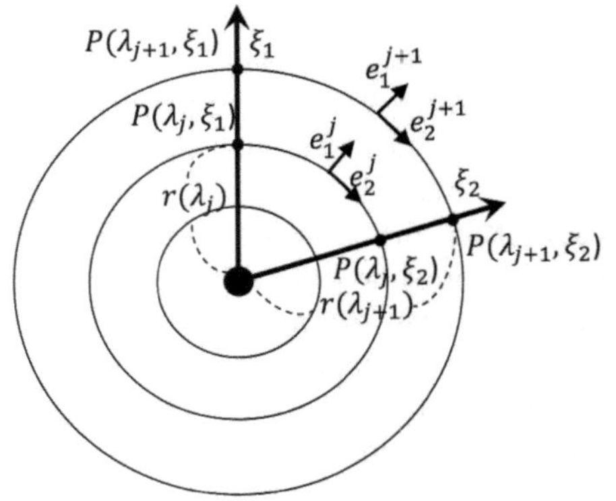

Fig. 6.7 Parametrization of a linearly expanding wavefront with parameters λ_j (wavefront) and ξ_i (trajectory). The radial function $r(\lambda_j)$ gives the distance of an excited point from the initial excitation point. Moving frames $\mathbf{e}_i^j$ are oriented so that $\mathbf{e}_1$ follows the trajectory and $\mathbf{e}_2$ lies along the wavefront

two trajectories initiated from the excitation source (see Fig. 6.7). Let λ represent the **proper time** parametrizing the wavefront. Then, the positions of points P and Q on the wavefront can be expressed as:

$$P = P(\lambda, \xi_1), \quad Q = Q(\lambda, \xi_2). \tag{6.7}$$

The arc $\overline{PQ}$ represents the wavefront at time λ, with arc length $\ell(\lambda)$. The number of excited cells is proportional to $\ell(\lambda)$. The sink–source ratio, interpreted as the change in arc length, is given by:

$$\text{Sink–source ratio } (\Delta\ell) = -\frac{\ell(\lambda_1) - \ell(\lambda_2)}{\ell(\lambda_1)} = -\frac{r(\lambda_1) - r(\lambda_2)}{r(\lambda_1)}, \tag{6.8}$$

where we used the relation $\ell = r\theta$ for a fixed angle θ between ξ_1 and ξ_2. Defining $\Delta r = r(\lambda_1) - r(\lambda_2)$, we obtain the wavefront curvature:

$$\kappa = \left.\frac{d\ell}{dr}\right|_{\lambda=\lambda_1} \approx \frac{\Delta\ell}{\Delta r} = -\frac{1}{r(\lambda_1)}, \tag{6.9}$$

where κ denotes the rate of change of arc length with respect to radius. In this simplified setting, curvature is used as a geometric proxy for the sink–source ratio. If $\ell(\lambda_2) = \ell(\lambda_1)$, then $\kappa = 0$, implying a one-to-one correspondence between source and sink. A negative κ implies more sources than sinks (acceleration), while a positive κ implies more sinks than sources (deceleration). A large positive κ indicates insufficient excitation resources, leading to propagation failure. This interpretation assumes a direct link between curvature and excitation spread, but physical accuracy depends on underlying biophysical thresholds and anisotropic properties. Differentiating Eq. (6.9) with respect to r gives:

$$\frac{\partial\kappa}{\partial r} = \left.\frac{d^2\ell}{dr^2}\right|_{\lambda_1} = \frac{1}{r^2(\lambda_1)}, \tag{6.10}$$

which describes the **acceleration or deceleration** of the sink–source ratio. Equation (6.10) is equivalent to the wavefront curvature and represents the growth rate of the sink–source mismatch. Computing such curvature becomes significantly more challenging in multidimensional anisotropic domains.

6.4 Alignment Along the Trajectory

As demonstrated above, analyzing wave patterns governed by PDEs is crucial for understanding propagation dynamics and predicting emergent behavior. Moving frames—orthonormal bases introduced by Élie Cartan—have long been employed in theoretical physics, especially general relativity (Misner et al. 2017; Dray 2015). However, this book adopts a distinct context by applying moving frames to the numerical analysis of PDEs to interpret wave propagation through **relative acceleration** and the **Riemann curvature tensor**.

To analyze wave patterns, the first step is to align the first basis vector $\mathbf{e}_1$ with the unit tangent vector of the trajectory $\hat{\mathbf{T}}$:

$$\mathbf{e}_1 = \hat{\mathbf{T}} = \frac{\mathbf{T}}{\|\mathbf{T}\|}, \tag{6.11}$$

where $\mathbf{T}$ is the trajectory's tangent vector. This construction defines the **first-order moving frame** (Cartan 1937; Akivis and Rosenfeld 2011). Higher-order frames are obtained by defining:

$$\mathbf{e}_2 = \hat{\mathbf{N}} = \frac{\hat{\mathbf{T}}'}{\|\hat{\mathbf{T}}'\|} = \frac{\mathbf{e}_1'}{\|\mathbf{e}_1'\|}, \tag{6.12}$$

$$\mathbf{e}_3 = \hat{\mathbf{B}} = \hat{\mathbf{T}} \times \hat{\mathbf{N}} = \mathbf{e}_1 \times \mathbf{e}_2. \tag{6.13}$$

In a general 2D curved domain, if the propagation velocity is derived directly from PDEs, then $\mathbf{e}_1$ is aligned accordingly. For example, in the Navier–Stokes equations:

$$\mathbf{e}_1 = \frac{\mathbf{u}}{\|\mathbf{u}\|}, \tag{6.14}$$

and in Maxwell's equations:

$$\mathbf{e}_1 = \frac{\mathbf{E} \times \mathbf{B}}{\|\mathbf{E} \times \mathbf{B}\|}. \tag{6.15}$$

For diffusion—reaction systems where the trajectory is inferred from the wavefront, gradient-based methods can be used. One example is shown in Fig. 6.8.

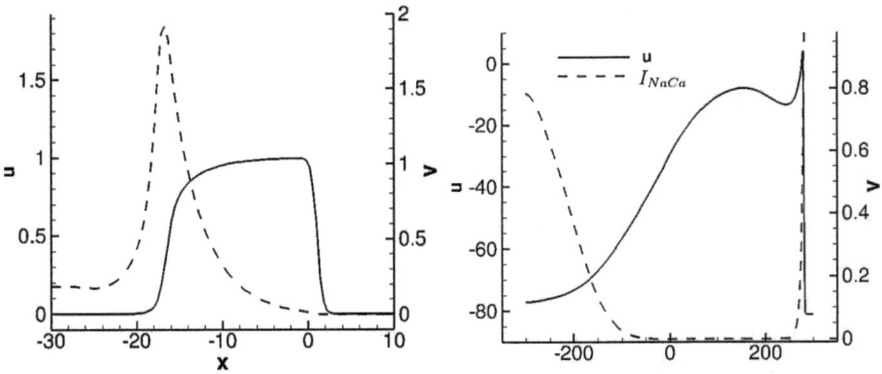

(a) Aliev–Panfilov (AP) model (b) Courtemanche–Ramirez–Nattel (CRN) model

Fig. 6.8 Membrane potential (solid line) for action potential (AP) and CRN models during wave propagation from left to right. Dashed lines denote (**a**) the total ion channel openness in the AP model and (**b**) the I_{NaCa} channel activity in the CRN model

In biological systems, wave trajectories can be approximated using the gradient of the membrane potential $u(\mathbf{x}, t)$. The normalized tangent vector $\mathbf{e}_1$ is derived from $\mathbf{T}$:

$$\mathbf{e}_1 = \frac{\mathbf{T}}{\|\mathbf{T}\|}. \tag{6.16}$$

6.4.1 Gradient Evaluation

The simplest scheme selects the gradient at the time of maximum magnitude:

$$\mathbf{t}_1(\mathbf{x}) = \nabla u(\mathbf{x}, t^*), \quad \|\nabla u(\mathbf{x}, t^*)\| \geq \|\nabla u(\mathbf{x}, t)\| \text{ for all } t. \tag{6.17}$$

While concise, this method may lack smoothness. An alternative is temporal integration of the normalized gradient. Assume that $\|\nabla u(\mathbf{x}, t)\| \geq \varepsilon > 0$ throughout the interval $[t_a, t_b]$:

$$\overline{\nabla} u = \frac{\nabla u}{\|\nabla u\|}, \qquad \mathbf{t}(\mathbf{x}) = \frac{1}{t_b - t_a} \int_{t_a}^{t_b} \overline{\nabla} u \, dt \quad . \tag{6.18}$$

To weight contributions by gradient magnitude, a more accurate scheme is:

$$\mathbf{t}(\mathbf{x}) = \frac{1}{A(t_b - t_a)} \int_{t_a}^{t_b} \|\nabla u\| \cdot \overline{\nabla} u \, dt, \quad A = \int_{t_a}^{t_b} \|\nabla u(\mathbf{x}, t)\| \, dt. \tag{6.19}$$

This weighted scheme better reflects physiological observations, as it aligns the direction of propagation with the strength of membrane potential gradients.

6.4.2 *Gradient Location*

When constructing a trajectory, it is important to consider that the action potential typically exhibits two steep slopes: the **wavefront** and the **waveback**, as illustrated in Fig. 6.8. The **wavefront** corresponds to the region where the membrane potential depolarizes—that is, where u increases rapidly. Conversely, the **waveback** corresponds to the region of repolarization, where u decreases sharply. Both the Aliev–Panfilov (AP) and Courtemanche–Ramirez–Nattel (CRN) models display steep wavefronts and wavebacks, though the waveback is generally smoother than the wavefront.

It is often assumed that the gradient directions of the wavefront and waveback are approximately aligned. The direction of the gradient ∇u can be determined from either the wavefront or the waveback, depending on the sign of the time derivative of u:

1. `WF`: Compute ∇u at the **wavefront**, where $du/dt > 0$.
2. `WB`: Compute ∇u at the **waveback**, where $du/dt < 0$.

6.5 Derivation of Relative Acceleration

Let the trajectory be parametrized by proper time λ, and expressed as $\xi(\lambda)$. The wavefront is parametrized by s and denoted by $\ell(s)$. Refer to Fig. 6.9 for a schematic overview. The tangent vector at a point along the trajectory is $\mathbf{t} = \mathbf{t}(\lambda)$. The tangent vector along the wavefront, referred to as the **separation vector**, is denoted by $\mathbf{n} = \mathbf{n}(s)$.

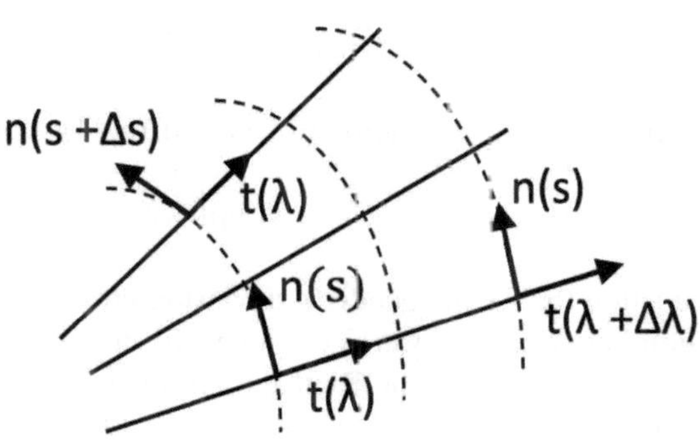

Fig. 6.9 Illustration of the tangent vector **t** and separation vector **n** defined along the trajectory (solid line) and the wavefront (dashed line). The trajectory and wavefront are parametrized by λ and s, respectively

The **relative acceleration** is derived as follows (Misner et al. 2017; Needham 2021). Along a geodesic trajectory, the covariant derivative of the tangent vector $\mathbf{t}$ along itself vanishes:

$$\nabla_{\mathbf{t}}\mathbf{t} = \mathbf{0}. \tag{6.20}$$

Taking the covariant derivative of Eq. (6.20) along the separation vector $\mathbf{n}$ yields:

$$\nabla_{\mathbf{n}}\nabla_{\mathbf{t}}\mathbf{t} = \mathbf{0}. \tag{6.21}$$

Introduce the commutator (Lie bracket) of the two vector fields:

$$[\mathbf{n}, \mathbf{t}] \equiv \nabla_{\mathbf{n}}\mathbf{t} - \nabla_{\mathbf{t}}\mathbf{n}. \tag{6.22}$$

Substituting Eq. (6.22) into Eq. (6.21), we obtain:

$$\nabla_{\mathbf{t}}\nabla_{\mathbf{n}}\mathbf{t} + [\nabla_{\mathbf{n}}, \nabla_{\mathbf{t}}]\mathbf{t} = 0. \tag{6.23}$$

Assuming the trajectory and wavefront form a local coordinate system, the symmetry of the covariant derivative implies:

$$\nabla_{\mathbf{t}}\mathbf{n} - \nabla_{\mathbf{n}}\mathbf{t} = 0. \tag{6.24}$$

Substituting Eq. (6.24) into Eq. (6.23) yields the **equation of geodesic deviation**:

$$\nabla_{\mathbf{t}}\nabla_{\mathbf{t}}\mathbf{n} + [\nabla_{\mathbf{n}}, \nabla_{\mathbf{t}}]\mathbf{t} = 0, \tag{6.25}$$

where we assume the Levi-Civita connection, which is torsion-free and metric-compatible, so the commutator of covariant derivatives corresponds to curvature action.

The first term, $\nabla_{\mathbf{t}}\nabla_{\mathbf{t}}\mathbf{n}$, is the **relative acceleration**, while the second term represents a curvature-induced force (or **tidal force**). The Riemann curvature tensor $\mathscr{R}$ can be expressed via the Lie bracket:

$$\mathscr{R}(\mathbf{n}, \mathbf{t}) = [\nabla_{\mathbf{n}}, \nabla_{\mathbf{t}}]\mathbf{t} - \nabla_{[\mathbf{n},\mathbf{t}]}, \tag{6.26}$$

where the final term vanishes due to Eq. (6.24). Therefore, Eq. (6.25) simplifies to:

$$\nabla_{\mathbf{t}}\nabla_{\mathbf{t}}\mathbf{n} + \mathscr{R}(\mathbf{n}, \mathbf{t}) = 0. \tag{6.27}$$

Equation (6.27), the **geodesic deviation equation**, shows that the relative acceleration between nearby trajectories is governed by the curvature of the underlying space—captured by the Riemann tensor. In the context of general relativity, this expresses how curvature manifests as tidal forces.

Consider again the circular wave propagation from a point source in Fig. 6.7. The trajectories are straight lines radiating outward at constant angular intervals between ξ_1 and ξ_2. These trajectories and the wavefront together form a polar coordinate system (r, θ) with corresponding tangent vectors $(\mathbf{r}, \boldsymbol{\theta})$. Assume the trajectory aligns with the r-direction and the wavefront aligns with the θ-direction. Then the Lie bracket vanishes:

$$[\mathbf{r}, \boldsymbol{\theta}] = \mathbf{0}. \tag{6.28}$$

Additionally, the mixed covariant derivatives vanish:

$$\nabla_{\mathbf{r}}\boldsymbol{\theta} = \nabla_{\boldsymbol{\theta}}\mathbf{r} = \mathbf{0}. \tag{6.29}$$

Thus, the second covariant derivative also vanishes:

$$\nabla_{\mathbf{r}}\nabla_{\mathbf{r}}\boldsymbol{\theta} = \mathbf{0}. \tag{6.30}$$

Equation (6.30) implies that the circular propagation occurs in a flat (zero-curvature) space. The absence of relative acceleration indicates that no net force or energy is acting to deviate the trajectories from their geodesic paths (Misner et al. 2017).

6.6 Relative Acceleration of Orthonormal Basis Vectors

Alternatively, relative acceleration can be evaluated using orthonormal basis vectors aligned with the trajectory and wavefront. In this context, the relative acceleration is defined as the second covariant derivative of $\mathbf{e}_2$ along the direction of $\mathbf{e}_1$:

$$\textbf{Relative acceleration of orthonormal basis vectors}: \ \nabla_{\mathbf{e}_1}\nabla_{\mathbf{e}_1}\mathbf{e}_2, \tag{6.31}$$

Let $\mathbf{e}_i(\lambda, s)$ denote the moving frames parameterized by λ (proper time along the trajectory) and s (along the wavefront). Here, $\mathbf{e}_1$ is aligned with the trajectory, and $\mathbf{e}_2$ is aligned with the wavefront. In the case of point-source propagation, as shown in Fig. 6.7, $\mathbf{e}_1$ is aligned with the radial direction, and $\mathbf{e}_2$ with the circular (azimuthal) direction:

$$\mathbf{e}_r = \mathbf{r}, \tag{6.32}$$

$$\mathbf{e}_\theta = \frac{\boldsymbol{\theta}}{\|\boldsymbol{\theta}\|}. \tag{6.33}$$

The covariant derivative of $\mathbf{e}_\theta$ along $\mathbf{e}_r$ is computed as:

$$\nabla_{\mathbf{e}_r}\mathbf{e}_\theta = -\frac{1}{r}\mathbf{e}_\theta. \tag{6.34}$$

In flat 2D Euclidean space using polar coordinates, this identity holds. In general curved geometries, this expression must be replaced by a frame-dependent connection coefficient. This curvature is geometrically equivalent to the source–sink ratio described in Eq. (6.9). Taking a further covariant derivative of Eq. (6.34) yields:

$$
\begin{aligned}
\nabla_{\mathbf{e}_r}\nabla_{\mathbf{e}_r}\mathbf{e}_\theta &= -\nabla_{\mathbf{e}_r}\left(\frac{1}{r}\mathbf{e}_\theta\right), \\
&= -\left(\nabla_{\mathbf{e}_r}\frac{1}{r}\right)\mathbf{e}_\theta - \frac{1}{r}\nabla_{\mathbf{e}_r}\mathbf{e}_\theta, \\
&= -\frac{2}{r^2}\mathbf{e}_\theta.
\end{aligned}
\tag{6.35}
$$

In contrast to the zero relative acceleration of geodesic motion, Eq. (6.35) is non-trivial. It quantifies the volumetric deformation rate, consistent with Eq. (6.10).

Although Fig. 6.7 illustrates linear divergence of trajectories, more realistic models include exponentially diverging trajectories and deviation from geodesic coordinates. In such cases, relative acceleration and the Riemann curvature tensor are non-zero. This is the classical **geodesic deviation** phenomenon (Misner et al. 2017), illustrated in Fig. 6.10. The covariant derivative of the separation vector $\mathbf{n}$ along the propagation vector $\mathbf{t}$ is:

$$(I)\ \ \nabla_{\mathbf{t}}\mathbf{n} < 0 : \text{trajectory diverges,} \tag{6.36}$$

$$(II)\ \ \nabla_{\mathbf{t}}\mathbf{n} > 0 : \text{trajectory converges.} \tag{6.37}$$

Since $\nabla_{\mathbf{n}}\mathbf{t} = \nabla_{\mathbf{t}}\mathbf{n}$ under coordinate symmetry, the sign also applies to $\nabla_{\mathbf{n}}\mathbf{t}$. It is essential to assess whether the covariant derivative of orthonormal basis vectors aligns in sign with that of the tangent vectors. The following corollary addresses this question:

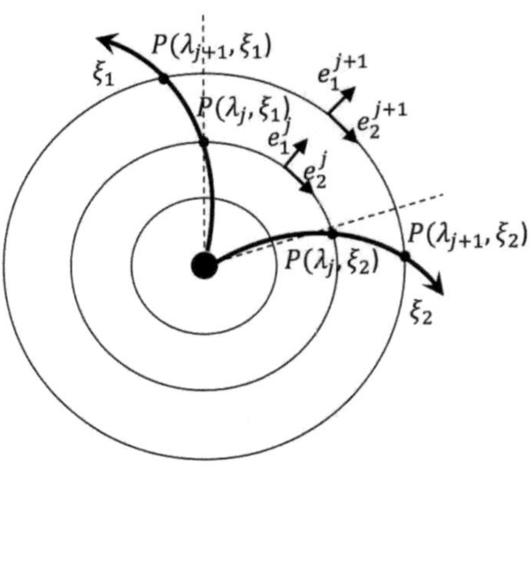

Fig. 6.10 Parametrization of a rapidly diverging wave propagation with parameters λ_j (wavefront) and ξ_i (trajectory). Moving frames $\mathbf{e}_i^j$ are oriented so that $\mathbf{e}_1$ follows the trajectory and $\mathbf{e}_2$ lies along the wavefront

Corollary 6.1 *Let* $\mathbf{e}_1$ *and* $\mathbf{e}_2$ *denote the unit tangent and separation vectors, respectively. Suppose the magnitude of the tangent vector remains approximately constant along the wavefront. Then, the covariant derivative* $\nabla_{\mathbf{n}}\mathbf{t}$ *has the same sign as* $\nabla_{\mathbf{e}_2}\mathbf{e}_1$.

Proof Let $\mathbf{t} = u\mathbf{e}_1$ and $\mathbf{n} = n\mathbf{e}_2$ where $u, n > 0$ are their magnitudes. By linearity of the covariant derivative,

$$\nabla_{\mathbf{n}}\mathbf{u} = n\nabla_{\mathbf{e}_2}u\mathbf{e}_1 = n\left[(\nabla u \cdot \mathbf{e}_2)\mathbf{e}_1 + u\nabla_{\mathbf{e}_2}\mathbf{e}_1\right]. \tag{6.38}$$

If $\nabla u \cdot \mathbf{e}_2 = 0$ (i.e., u constant along the wavefront), then

$$\nabla_{\mathbf{e}_2}\mathbf{e}_1 = \frac{1}{u}\left[\frac{1}{n}\nabla_{\mathbf{n}}\mathbf{u} - (\nabla u \cdot \mathbf{e}_2)\mathbf{e}_1\right] = \frac{1}{un}\nabla_{\mathbf{n}}\mathbf{u}, \tag{6.39}$$

Thus, their signs agree. □

Additionally, covariant derivatives of orthonormal frames are related to the connection 1-forms:

$$\nabla_{\mathbf{e}_i}\mathbf{e}_j = \sum_{k=1}^{2} \omega_{ij}(\mathbf{e}_k)\mathbf{e}_k. \tag{6.40}$$

Applying the corollary, Eqs. (6.36) and (6.37) yield:

$$\text{(I)} \quad \omega_{21}(\mathbf{e}_2) < 0 : \text{trajectory diverges}, \tag{6.41}$$

$$\text{(II)} \quad \omega_{21}(\mathbf{e}_2) > 0 : \text{trajectory converges}. \tag{6.42}$$

To connect the sign of the connection 1-form to the dynamics of wave propagation, we state the following proposition:

Proposition 6.1 *Consider a smooth two-dimensional curved domain. Let the first-order moving frames* $\mathbf{e}_i$ *be aligned such that* $\mathbf{e}_1$ *follows the gradient of the action potential. If the connection 1-form satisfies:*

$$\omega_{21}(\mathbf{e}_2) < 0 \quad \textit{and} \quad \nabla_{\mathbf{e}_1}\nabla_{\mathbf{e}_1}\mathbf{e}_2 \cdot \mathbf{e}_2 > 0, \tag{6.43}$$

then propagation fails due to a sink–source mismatch, and the region is likely to develop a conduction block.

Proof By orthonormality,

$$\nabla_{\mathbf{e}_1}\mathbf{e}_2 + \nabla_{\mathbf{e}_2}\mathbf{e}_1 = 0. \tag{6.44}$$

Differentiating Eq. (6.44) along $\mathbf{e}_1$ gives:

$$\nabla_{\mathbf{e}_1}\nabla_{\mathbf{e}_1}\mathbf{e}_2 = -\nabla_{\mathbf{e}_1}\nabla_{\mathbf{e}_2}\mathbf{e}_1. \tag{6.45}$$

If $\nabla_{\mathbf{e}_2}\mathbf{e}_1 < 0$, then the sink–source ratio is high. A further negative acceleration in this quantity implies propagation failure. Hence, the relative acceleration becomes positive, reflecting divergence. □

6.7 Numerical Scheme for Relative Acceleration

Let **u** be an arbitrary vector in the domain Ω, expanded in the orthonormal basis:

$$\mathbf{u} = u^1\mathbf{e}_1 + u^2\mathbf{e}_2. \tag{6.46}$$

Its covariant derivative along $\mathbf{e}_j$ is given by Chun (2023a):

$$\nabla_{\mathbf{e}_j}\mathbf{u} = \sum_{i=1}^{2}\left[\nabla u^i \cdot \mathbf{e}_j + \sum_{k=1}^{2}\omega_{ik}(\mathbf{e}_j)u^k\right]\mathbf{e}_i.. \tag{6.47}$$

The connection 1-forms are computed as:

$$\omega_{ij}(\mathbf{e}_k) = [e_j^x\ e_j^y\ e_j^z]\begin{bmatrix}\nabla e_i^x \cdot \mathbf{e}_k \\ \nabla e_i^y \cdot \mathbf{e}_k \\ \nabla e_i^z \cdot \mathbf{e}_k\end{bmatrix}. \tag{6.48}$$

We denote this more compactly as:

$$\omega_{ijk} \equiv \omega_{ij}(\mathbf{e}_k). \tag{6.49}$$

For scalars, directional derivatives satisfy:

$$\nabla_{\mathbf{e}_i} u = \nabla u \cdot \mathbf{e}_i. \tag{6.50}$$

Corollary 6.2 *In a smooth two-dimensional curved domain Ω with $\mathbf{e}_1$ aligned along the gradient of action potential, the relative acceleration of orthonormal basis vectors is:*

$$\nabla_{\mathbf{e}_1}\nabla_{\mathbf{e}_1}\mathbf{e}_2 = \left(\omega_{211}\omega_{212} - \nabla_{\mathbf{e}_1}\omega_{211}\right)\mathbf{e}_1 + \left(\omega_{212}^2 - \nabla_{\mathbf{e}_1}\omega_{212}\right)\mathbf{e}_2. \tag{6.51}$$

Proof From Eq. (6.47), we have:

$$\nabla_{\mathbf{e}_1}\mathbf{e}_2 = -\omega_{211}\mathbf{e}_1 - \omega_{212}\mathbf{e}_2. \tag{6.52}$$

Substitute into the second derivative:

$$\begin{aligned}\nabla_{\mathbf{e}_1}\nabla_{\mathbf{e}_1}\mathbf{e}_2 &= -\nabla_{\mathbf{e}_1}\omega_{211}\mathbf{e}_1 - \nabla_{\mathbf{e}_1}\omega_{212}\mathbf{e}_2 - \omega_{211}\nabla_{\mathbf{e}_1}\mathbf{e}_1 - \omega_{212}\nabla_{\mathbf{e}_1}\mathbf{e}_2\\ &= -\nabla_{\mathbf{e}_1}\omega_{211}\mathbf{e}_1 - \nabla_{\mathbf{e}_1}\omega_{212}\mathbf{e}_2 + \omega_{212}(\omega_{211}\mathbf{e}_1 + \omega_{212}\mathbf{e}_2).\end{aligned} \tag{6.53}$$

Grouping terms yields Eq. (6.51). □

Combining Proposition 6.1 and Corollary 6.2, we derive a practical condition for conduction failure:

Proposition 6.2 *In a smooth two-dimensional curved domain Ω, suppose the moving frames are constructed such that $\mathbf{e}_1$ follows the gradient of the action potential. If*

$$\omega_{212} < 0 \quad \textit{and} \quad \omega_{212}^2 - \nabla\omega_{212}\cdot\mathbf{e}_1 > 0, \tag{6.54}$$

then conduction fails due to sink–source mismatch, and the region is likely to develop a conduction block.

6.8 [MATLAB] Compute Relative Acceleration

The following MATLAB script, **ch4MATLABa.m**, computes the relative acceleration of the moving frames on two deformed surfaces: a **deformed sphere** and a **deformed pseudosphere**. These surfaces are parameterized as follows.

For the deformed sphere:

$$x = r\sin\theta\cos\phi, \tag{6.55}$$

$$y = 2r\sin\theta\sin\phi, \tag{6.56}$$

$$z = r\cos\theta, \tag{6.57}$$

where $\theta \in (0, \pi)$ is the polar angle, and $\phi \in [0, 2\pi)$ is the azimuthal angle. Compared to a standard sphere of radius $r = 1$, the y-component has been stretched by a factor of 2, introducing anisotropy into the geometry.

For the deformed pseudosphere:

$$x = r\,\mathrm{sech}\,\theta\cos\phi, \tag{6.58}$$

$$y = 2r\,\mathrm{sech}\,\theta\sin\phi, \tag{6.59}$$

$$z = r(\theta - \tanh\theta), \tag{6.60}$$

where $\theta > 0$ and $\phi \in [0, 2\pi)$. This surface extends the standard pseudosphere in the y-direction by the same factor of 2.

In both cases, the deformation breaks the original symmetry of the geometry, which results in nonzero relative acceleration of the moving frames, particularly along the $\mathbf{e}_\phi$ and $\mathbf{e}_\theta$ directions. For standard spherical and pseudospherical coordinate systems, such relative accelerations are zero due to the orthogonality and constancy of the local basis vectors. However, in the deformed cases, the spatial variation of the basis vectors leads to a nontrivial acceleration of the frame itself.

The MATLAB script below, **ch6MATLAB.m**, calculates and visualizes these relative accelerations using moving frames adapted to the deformed geometries.

```
function ch6MATLAB(Nphi, Ntheta, shape)
% ch6MATLAB.m
%
% This function computes and validates the relative acceoeration on a deformed spherical
% or pseudospherical coordinate grid using spectral element methods (SEM).
%
% Inputs:
%   Nphi   - Number of discretization points in the azimuthal (longitude) direction
%   Ntheta - Number of discretization points in the polar (latitude) direction
%   shape  - deformedsphere or deformedpseudosphere.
%
% Outputs:
%   Visualization of selected components of the connection 1-form on the sphere
%
% Usage example:
%   ch3MATLABb(20, 20, "deformedsphere");

%% --- Parameter and Grid Setup ---
SetupParameter;  % Ensure global parameters are consistent (radius, grid dimensions)

%% Create grids and Moving Frames
[grids,movingframes] = CreateGridTangentVec(Nphi, Ntheta, 1, shape);

%% --- Compute Connection 1-Form ---
% Compute the connection 1-form from the numerical derivatives of moving frames
wij_k = Computeconnection(Nphi, Ntheta, shape, movingframes);

%% --- Compute Relative Acceleration ---
RelAcc = Computerelacc(wij_k, movingframes, shape);

%% --- Visualization of Selected Components of the Relative acceleration ---
figure;
surf(grids(:,:,1), grids(:,:,2), grids(:,:,3), RelAcc(:,:,1), 'EdgeColor', 'none');
colorbar;
title('Relative␣acceleration␣along␣$\mathbf{e}_{\phi}$', 'Interpreter', 'latex');
xlabel('x'); ylabel('y'); zlabel('z');
axis equal; view(-45,45);
lighting gouraud; camlight;  % Lighting effects for better 3D shading
colormap(parula(256));
colorbar;

figure;
surf(grids(:,:,1), grids(:,:,2), grids(:,:,3), RelAcc(:,:,2), 'EdgeColor', 'none');
colorbar;
title('Relative␣acceleration␣along␣$\mathbf{e}_{\theta}$', 'Interpreter', 'latex');
```

```
xlabel('x'); ylabel('y'); zlabel('z');
axis equal; view(-45,45);
lighting gouraud; camlight;  % Lighting effects for better 3D shading
colormap(parula(256));
colorbar;

end
```

Listing 6.1 Compute the relative accelerations and visualize them (MATLAB script: ch6MATLAB.m)

To achieve higher accuracy, the script **Computerelacc.m** computes the same quantities using spectral element methods. This approach provides a more refined approximation of derivatives and better resolution for visualizing the acceleration fields (Fig. 6.11).

```
function RelAcc = Computerelacc(wij_k, movingframes, shape)
% Computerelacc.m
% ------------------------------------------------------------------------
% Computes the relative acceleration terms along a specific direction
% in a moving frame setup on a curved surface.
%
% Inputs:
%   wij_k          - a 5D array containing connection coefficients w_{ij}^k
%                    with dimensions (Nphi, Ntheta, 2, 1, 2)
%                    The two slices refer to w_{211} and w_{212}.
%   movingframes   - a 4D array containing orthonormal vector fields
%                    (e1 and e2) at each grid point. Dimensions: (Nphi,
    Ntheta, 2, 3)
%   shape          - a structure or parameter set used for computing
%                    directional derivatives (possibly mesh geometry info)
%
% Output:
%   RelAcc         - a (Nphi x Ntheta x 2) array of relative acceleration
    terms
%                    along the first moving frame vector (e1 direction)
%
% ------------------------------------------------------------------------

%% --- Grid and Parameter Setup ---
[Nphip, Nthetap, ~] = size(movingframes);  % Extract grid size

% Initialize output array for relative accelerations
RelAcc = zeros(Nphip, Nthetap, 2);

% Extract connection coefficients:
% w_{211} corresponds to connection coefficient component in direction 1 of
    e2
% w_{212} corresponds to component in direction 2 of e2
w211 = squeeze(wij_k(:,:,2,1,1));
w212 = squeeze(wij_k(:,:,2,1,2));

%% --- Compute Relative Acceleration Components along e1 ---

% Component 1: w_{211} * w_{212} - \nabla_{e1} w_{211}
% First compute the product term
tmp = w211 .* w212;

% Compute directional derivative of w211 along e1
dw211d1 = Computedirderiv(squeeze(movingframes(:,:,1,:)), w211, shape);

% Store the result: relative acceleration in component 1
RelAcc(:,:,1) = tmp - dw211d1;

% Component 2: w_{212} * w_{212} - \nabla_{e1} w_{212}
```

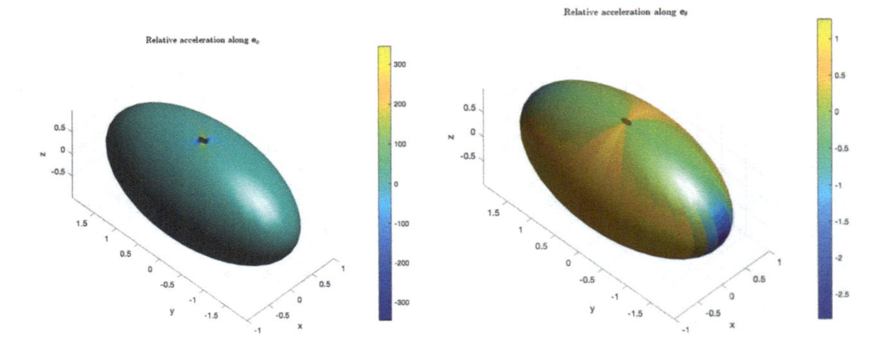

(a) (b)

Fig. 6.11 Relative acceleration on the deformed sphere in Eqs. (6.55)–(6.57) along $\mathbf{e}_\phi$ (left) and $\mathbf{e}_\theta$ (right) directions. (**a**) Relative acceleration along $\mathbf{e}_\phi$ direction. (**b**) Relative acceleration along $\mathbf{e}_\theta$ direction

```
% Square of w212
tmp = w212 .* w212;

% Directional derivative of w212 along e1
dw212d1 = Computedirderiv(squeeze(movingframes(:,:,1,:)), w212, shape);

% Store the result: relative acceleration in component 2
RelAcc(:,:,2) = tmp - dw212d1;

end
```

Listing 6.2 Compute the relative acceleration using moving frames and spectral element methods (MATLAB script: Computerelacc.m)

The results are visualized below (Figs. 6.11 and 6.12). The color-coded vector fields indicate the direction and magnitude of the relative acceleration along each coordinate direction for both deformed surfaces.

6.9 Examples of Large Relative Acceleration

We present several illustrative cases in which large relative acceleration arises, potentially leading to conduction failure.

6.9.1 Propagation Through a Narrow Gap

The first example involves conduction through a small gap. Consider a 2D rectangular domain with a narrow opening in the middle, as shown in Fig. 6.13. The Aliev–Panfilov model (Aliev and Panfilov 1996) is used to simulate diffusion—

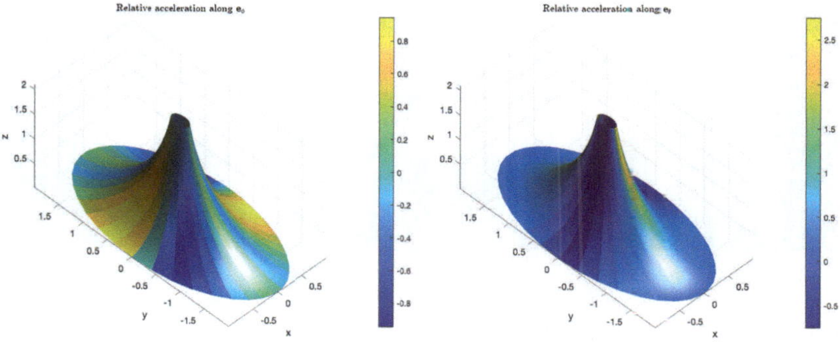

(a) (b)

Fig. 6.12 Relative acceleration on the deformed pseudosphere in Eqs. (6.58)–(6.60) along $\mathbf{e}_\phi$ (left) and $\mathbf{e}_\theta$ (right) directions. (**a**) Relative acceleration along $\mathbf{e}_\phi$ direction. (**b**) Relative acceleration along $\mathbf{e}_\theta$ direction

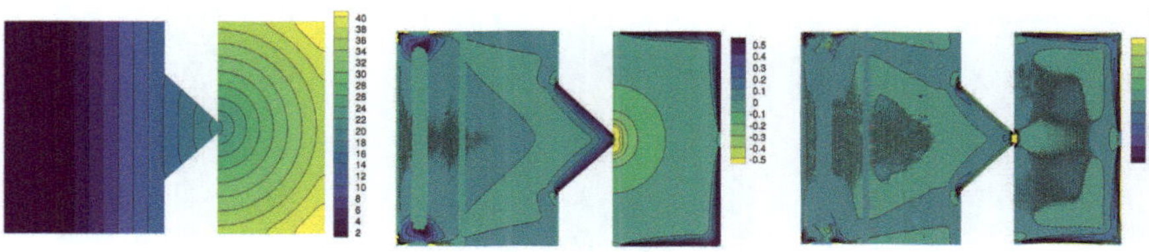

(a) (b) (c)

Fig. 6.13 Wave propagation initiated from the left wall through a narrow gap, demonstrating relative acceleration of the propagation. The gap has width 2, and the domain height is 40. (**a**) Time map. (**b**) ω_{212}. (**c**) Relative acceleration

reaction wave propagation in the domain $[-40, 20] \times [-20, 20]$. The left hole is 20 units wide, and the right hole is 10 units wide. A planar wave is initiated from the left boundary and propagates toward the right.

Figure 6.13a shows the arrival time map. Figure 6.13b displays the connection 1-form component ω_{212}, which is negative just beyond the small gap, indicating divergent propagation. A large positive relative acceleration emerges at the gap, as seen in Fig. 6.13c. When the gap is smaller, a higher sink–source ratio results, leading to even greater relative acceleration. This indicates that a threshold gap size exists for successful conduction, which depends on the properties of the medium.

6.9.2 *Scar Tissue and Fibrosis*

Large relative acceleration also appears behind structural obstructions such as holes or regions with reduced conductivity (e.g., scar tissue). A circular hole of radius 10 is placed at the center of the domain $[-40, 40] \times [-40, 40]$ with Neumann boundary

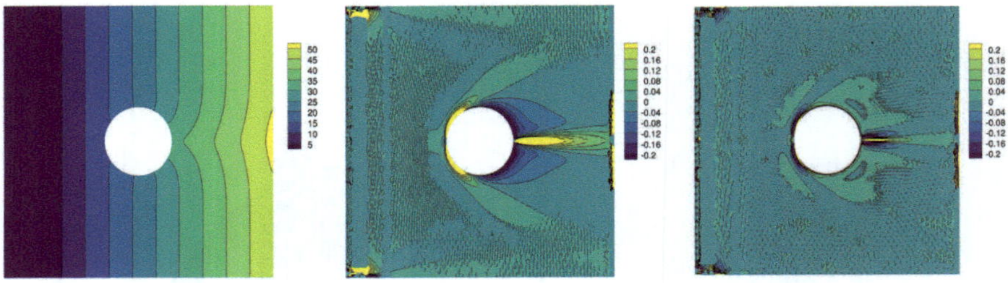

(a) (b) (c)

Fig. 6.14 Large relative acceleration occurs behind a circular hole located at the center of the domain. The wave is initiated from the left boundary and propagates toward the right wall, with the strongest relative acceleration appearing in the region behind the hole opposite to the propagation direction. (**a**) Time map. (**b**) ω_{212}. (**c**) Relative acceleration

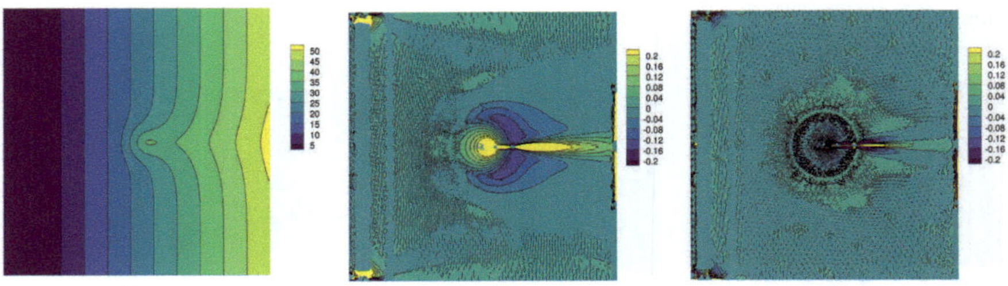

(a) (b) (c)

Fig. 6.15 Large relative acceleration occurs behind a circular region with reduced conductivity tissue located at the center of the domain. The wave is initiated from the left boundary and propagates toward the right wall, with the strongest relative acceleration appearing in the region behind the hole opposite to the propagation direction. (**a**) Time map. (**b**) ω_{212}. (**c**) Relative acceleration

conditions. Figure 6.14 shows the emergence of a large relative acceleration in the shadow of the hole.

A similar effect is observed with scar tissue. Figure 6.15 shows the case where a region of reduced conductivity (0.2) mimics a fibrotic area. A conduction block appears at its boundary due to a steep increase in the sink–source ratio.

6.10 Application to a 2D Atrium Model

We now apply the proposed relative acceleration framework to simulate electrical propagation on a realistic atrial surface. The human atrium is a thin-walled structure, with thickness ranging from 0.5 to 3.5 mm and an average of 1.89 mm (Beinart et al. 2011; Hocini et al. 1998), significantly thinner than the ventricle (mean thickness: 15.0 mm (Anderson and Cook 2007)). Therefore, the atrium can be reasonably modeled as a 2D curved surface (Chun 2014b).

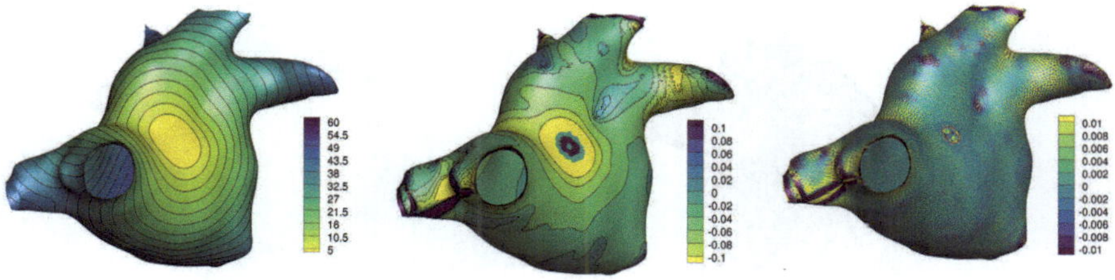

(a) (b) (c)

Fig. 6.16 Action potential propagation in an isotropic atrium. The excitation is initiated pointwise near the right pulmonary vein and propagates toward the veins. Elevated relative acceleration is observed around the veins, though its magnitude remains modest. (**a**) Time map. (**b**) ω_{212}. (**c**) Relative acceleration. (Courtesy of Chun 2023b)

6.10.1 Isotropic Atrium

Assuming isotropy (no fiber orientation), a point excitation is applied near the right pulmonary vein. Figure 6.16a shows the arrival time map. The connection 1-form ω_{212} (Fig. 6.16b) indicates regions of divergent flow. Figure 6.16c reveals that conduction blocks are likely to form near the pulmonary veins—consistent with clinical findings of atrial reentry and fibrillation (Chun 2014b).

6.10.2 Anisotropic Atrium

Next, cardiac fibers are introduced as shown in Fig. 6.17a. The fibers, manually traced from public datasets (Cantwell et al. 2014), have higher conductivity along their orientation (3:1 ratio). Propagation is dominated by the fiber direction (Fig. 6.17b), resulting in a highly directional time map.

Connection 1-form ω_{212} (Fig. 6.17c) and corresponding relative acceleration (Fig. 6.17d) show that conduction blocks align with fiber discontinuities. However, the accuracy of these findings depends on the validity of the manually drawn fiber map.

6.10.3 Atrium with Heterogeneous Conductivity

Finally, we simulate an isotropic atrium with spatially varying conductivity (Fig. 6.18a). A value of 1.0 represents normal tissue, while lower values model scar tissue or fibrosis. The arrival time map in Fig. 6.18b shows complex wavefronts due to non-uniform speed.

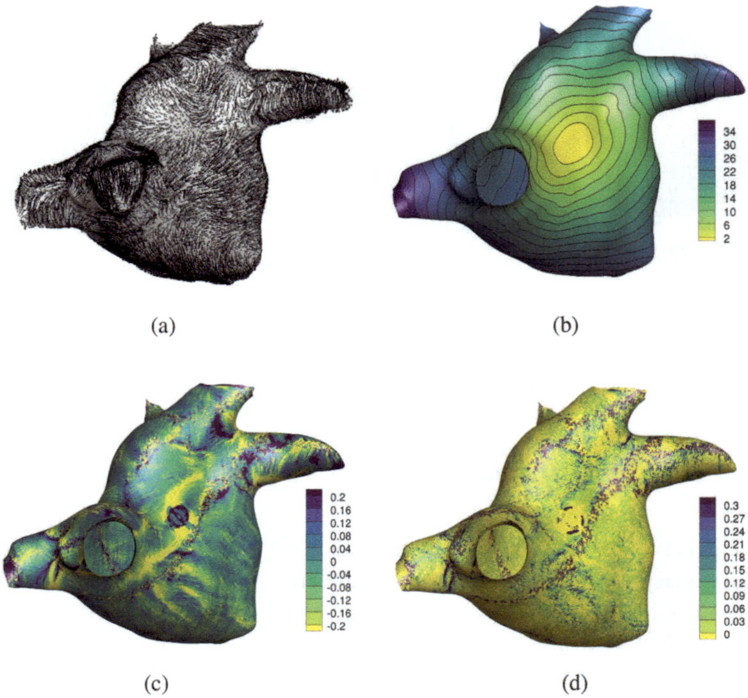

Fig. 6.17 Action potential propagation in an anisotropic atrium with cardiac fibers modeled using a 3:1 conductivity ratio, as shown in (**a**). The excitation is initiated pointwise near the right pulmonary vein and propagates toward the veins. In this case, the relative acceleration around the veins is no longer present. (**a**) Cardiac fiber. (**b**) Time map. (**c**) ω_{212}. (**d**) Relative acceleration. (Courtesy of Chun 2023b)

Figure 6.18c shows a highly variable ω_{212}, and Fig. 6.18d indicates that conduction blocks emerge across the domain, creating potential substrates for reentry and spiral wave formation—hallmarks of atrial fibrillation.

6.11 Summary

This chapter presents a geometric and dynamical framework for analyzing wave propagation in curved domains, with a particular focus on **relative acceleration** and the **Riemann curvature tensor**. It integrates classical PDE-based models with differential geometric tools to identify conduction blocks and characterize wavefront behavior.

Partial differential equations such as the Navier–Stokes and Maxwell equations describe time-dependent wave propagation. Diffusion—reaction equations, including the Hodgkin–Huxley and FitzHugh–Nagumo models, further illustrate

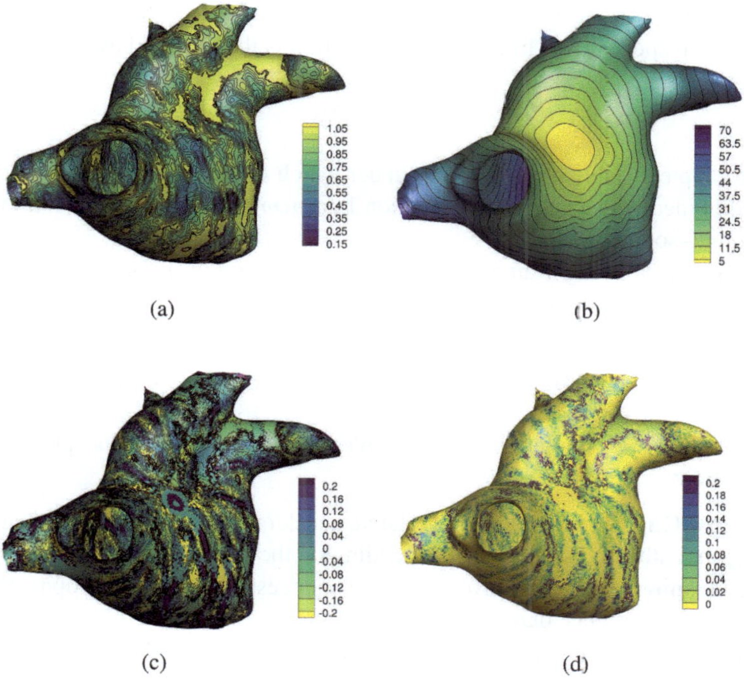

Fig. 6.18 Action potential propagation in an anisotropic atrium with cardiac fibers modeled using a heterogeneous conductivity ratio, as shown in (**a**). The excitation is initiated pointwise near the right pulmonary vein and propagates toward the veins. In this case, the relative acceleration around the veins is no longer present. (**a**) Conductivity map. (**b**) Time map. (**c**) ω_{212}. (**d**) Relative acceleration. (Courtesy of Chun 2023b)

biological signal propagation, in which wavefronts and trajectories can be naturally interpreted through geometric constructs. The wavefront is defined as a level set of a scalar field, while the trajectory is constructed as the direction orthogonal to the wavefront gradient. Diverging and converging wave patterns reflect distinct physical regimes (e.g., signal density, energy dispersion), forming the basis for relative acceleration analysis.

Using kinematic analysis, an approximate relationship between the propagation velocity V and the wavefront curvature κ is established as:

$$V \approx V_0 - D\kappa. \tag{6.61}$$

High curvature corresponds to a sink–source mismatch, potentially leading to **conduction block**. The chapter derives relative acceleration as the second covariant derivative of the wavefront's tangent vector along the trajectory, highlighting its equivalence to curvature-induced forces.

The moving frame vectors $\mathbf{e}_1$ and $\mathbf{e}_2$ are aligned along the trajectory and the wavefront, respectively. The relative acceleration is then defined as:

$$\nabla_{\mathbf{e}_1}\nabla_{\mathbf{e}_1}\mathbf{e}_2. \tag{6.62}$$

The chapter establishes conditions under which a large positive relative acceleration, combined with a negative connection 1-form ω_{212}, indicates conduction failure due to sink–source mismatch.

Within the moving frame formalism, the connection 1-form $\omega_{ij}(\mathbf{e}_k)$ and its derivatives are used to compute relative acceleration numerically. A key result is:

$$\nabla_{\mathbf{e}_1}\nabla_{\mathbf{e}_1}\mathbf{e}_2 = \left(\omega_{211}\omega_{212} - \nabla_{\mathbf{e}_1}\omega_{211}\right)\mathbf{e}_1 + \left(\omega_{212}^2 - \nabla_{\mathbf{e}_1}\omega_{212}\right)\mathbf{e}_2. \tag{6.63}$$

The following simulations are provided to validate the concept of relative acceleration:

- **Narrow Gap Simulation:** High relative acceleration emerges immediately after the gap, with conduction block depending on the gap size.
- **Pseudosphere:** The negative curvature induces non-uniform propagation and localized divergence due to geometric effects.
- **Scar Tissue and Fibrosis:** Both a structural hole and a low-conductivity region generate large positive acceleration behind them, mimicking anatomical obstructions.

Additionally, atrial models with different media are tested to evaluate the effect of tissue characteristics on propagation:

- **Isotropic Atrium:** The simulation indicates conduction blocks near the pulmonary veins, consistent with known initiation zones of atrial fibrillation.
- **Anisotropic Atrium:** Directional fiber conductivity significantly alters wave propagation and block distribution. Discontinuities in fiber orientation are correlated with block formation.
- **Heterogeneous Conductivity:** A spatially varying conductivity map produces complex wavefronts and block patterns, suggesting conditions conducive to spiral wave initiation and arrhythmogenesis.

Overall, this chapter demonstrates that relative acceleration and connection 1-forms provide powerful tools for identifying conduction failure and understanding wave propagation dynamics in curved, heterogeneous, and anisotropic domains.

6.12 Exercises and Conceptual Applications

1. Derive the kinematic relation

$$V \approx V_0 - D\kappa \tag{6.64}$$

for a circular wavefront propagating in an isotropic medium. Clearly explain the assumptions involved and interpret the physical meaning of V_0, D, and κ.

2. Prove the geodesic deviation equation:

$$\nabla_{\mathbf{t}}\nabla_{\mathbf{t}}\mathbf{n} + \mathscr{R}(\mathbf{n}, \mathbf{t}) = 0 \tag{6.65}$$

from first principles using the definition of the Lie bracket and properties of covariant derivatives. Discuss its physical meaning in the context of wavefront divergence.

3. Given orthonormal moving frames $\mathbf{e}_1$ and $\mathbf{e}_2$ on a 2D surface, derive the connection 1-form $\omega_{212} = \omega_{21}(\mathbf{e}_2)$ using the formula:

$$\omega_{ij}(\mathbf{e}_k) = \mathbf{e}_j \cdot (\nabla \mathbf{e}_i \cdot \mathbf{e}_k). \tag{6.66}$$

Apply this to compute ω_{212} for polar coordinates (r, θ) on a flat plane.

4. Implement a numerical scheme to compute the relative acceleration:

$$\nabla_{\mathbf{e}_1}\nabla_{\mathbf{e}_1}\mathbf{e}_2, \tag{6.67}$$

using the expression:

$$\nabla_{\mathbf{e}_1}\nabla_{\mathbf{e}_1}\mathbf{e}_2 = \left(\omega_{211}\omega_{212} - \nabla_{\mathbf{e}_1}\omega_{211}\right)\mathbf{e}_1 + \left(\omega_{212}^2 - \nabla_{\mathbf{e}_1}\omega_{212}\right)\mathbf{e}_2. \tag{6.68}$$

Test your code on a domain with a circular hole and verify that a conduction block forms behind the hole.

5. Consider a pseudosphere defined by the parametrization:

$$\begin{aligned} x &= r\,\mathrm{sech}\,\theta\cos\phi, \\ y &= r\,\mathrm{sech}\,\theta\sin\phi, \\ z &= r(\theta - \tanh\theta). \end{aligned} \tag{6.69}$$

Compute the unit tangent vectors $\mathbf{e}_\theta$ and $\mathbf{e}_\phi$, and evaluate the connection 1-form component $\omega_{\phi\theta\phi}$. Explain why relative acceleration on the pseudosphere differs in sign from that on a sphere.

6. Design a simulation for electric wave propagation on an anisotropic 2D atrium model using manually defined fiber orientations. Analyze how the fiber orientation affects ω_{212} and the emergence of conduction blocks. Discuss whether your results align with known electrophysiological behavior of the atrium.

7. Using the Aliev–Panfilov model or a similar reaction—diffusion system, simulate wave propagation across gaps of decreasing width. Determine the critical width below which propagation fails, and relate this threshold to the magnitude of ω_{212} and relative acceleration.

Appendix A
MATLAB: Spectral Element Codes

A.1 SetupParameter.m

The MATLAB script `SetupParameter.m` initializes fundamental parameters necessary for numerical computations on spherical geometries. It defines critical geometric parameters such as the radius of the sphere (R), angular ranges (`Lphi` and `Lth`), and discretization adjustments to mitigate singularities near the poles. Parameters also include dimensions for moving frames and spatial discretization.

```
%% SetupParameter.m
% This script sets global parameters for the spherical grid generation
% and moving frame computations.

%% Sphere Parameters
R = 1.0;                % Radius of the sphere

gap = 0.1;              % Small gap at polar angle boundaries to avoid
     singularities
Lphi = 2 * pi;          % Full azimuthal angle range, from 0 to 2\pi
Lth = pi - gap;         % Polar angle range, slightly less than full range (\
     pi)

%% Moving Frame and Spatial Dimensions
mfdim = 3;              % Dimension of moving frames (3D orthonormal basis)
spacedim = 3;           % Dimension of the ambient space (3D Cartesian
     coordinates)
expdim = 2;             % Dimension of the domain

%% Adjusted Grid Dimensions
Nphip = Nphi + 1;       % Adjusted grid points along azimuthal direction (
     includes endpoint)
Nthetap = Ntheta + 1; % Adjusted grid points along polar direction (includes
      endpoint)

```


S. Chun, *Moving Frames for the Numerical Solution of Partial Differential Equations in Complex Domains*, Springer Asia Pacific Mathematics Series 7,
https://doi.org/10.1007/978-981-95-4495-0

```
% Usage:
% Include this script in other functions to ensure consistency of parameters
% Example: call this script at the start of visualization or computation routines.
```

Listing A.1 Initialization of numerical parameters (MATLAB script: `SetupParameter.m`)

A.2 Generategrid.m

The script `Generategrid.m` creates spectral grids and differentiation matrices for discretizing angular coordinates on a spherical domain. It employs the spectral element method (SEM) using Gauss-Lobatto-Legendre (GLL) nodes, mapping these nodes from a reference interval $[-1, 1]$ to the physical domain. It computes differentiation matrices essential for numerical differentiation in both the azimuthal (ϕ) and polar (θ) directions.

```
function [phigrid, thetagrid, Dbphi, Dbtheta] = Generategrid(Nphi, Ntheta)
% Generategrid.m
%
% This function generates the spherical grid points and spectral
% differentiation matrices using the Spectral Element Method (SEM).
%
% Inputs:
%   Nphi   - Number of discretization points in the azimuthal direction
%   Ntheta - Number of discretization points in the polar direction
%
% Outputs:
%   phigrid    - Grid points in the azimuthal direction (0 to 2\pi)
%   thetagrid  - Grid points in the polar direction (gap/2 to \pi - gap/2)
%   Dbphi      - SEM differentiation matrix for the azimuthal direction
%   Dbtheta    - SEM differentiation matrix for the polar direction
%
% Usage example:
%   [phigrid, thetagrid, Dbphi, Dbtheta] = Generategrid(20, 20);

%% --- Load Parameters ---
SetupParameter;  % Ensure global parameters (e.g., Lphi, Lth, gap) are consistent

%% --- Generate Grid and Differentiation Matrices ---

% Azimuthal direction (phi) using Gauss-Lobatto-Legendre (GLL) nodes
% Generate SEM nodes (zh_phi) and differentiation matrix (Dphi) in [-1,1]
[~, ~, ~, Dphi, zh_phi, ~] = semhat(Nphi);

% Map nodes from reference domain [-1, 1] to physical domain [0, 2\pi]
phigrid = Lphi * (zh_phi + 1) / 2;

% Scale differentiation matrix for the azimuthal direction
hphi = 1 / Lphi;
Dbphi = 2.0 * hphi * Dphi;

% Polar direction (theta) using Gauss-Lobatto-Legendre (GLL) nodes
% Generate SEM nodes (zh_theta) and differentiation matrix (Dtheta) in [-1,1]
[~, ~, ~, Dtheta, zh_theta, ~] = semhat(Ntheta);

```

```
% Map nodes from reference domain [-1, 1] to physical domain [gap/2, \pi -
    gap/2]
thetagrid = 0.5 * gap + Lth * (zh_theta + 1) / 2;

% Scale differentiation matrix for the polar direction
htheta = 1 / Lth;
Dbtheta = 2.0 * htheta * Dtheta;

end
```

Listing A.2 Generation of spectral grids and differentiation matrices (MATLAB script: `Generategrid.m`)

A.3 CartesianToMovingframes.m

The MATLAB function `CartesianToMovingframes.m` transforms vector fields expressed in global Cartesian coordinates into a local moving frame representation. This conversion is performed via the projection of the vector field onto local orthonormal bases defined at each grid point. This step is crucial when dealing with numerical problems defined on curved surfaces, ensuring that computations accurately reflect local geometry.

```
function inarray = CartesianToMovingframes(movingframes, velocity)
% CartesianToMovingframes - Project a Cartesian vector field into a local
    moving frame
%
% Syntax:
%   inarray = CartesianToMovingframes(movingframes, velocity)
%
% Description:
%   This function transforms a 3D vector field from the global Cartesian
    basis
%   into a local orthonormal basis defined by the moving frames at each grid
     point.
%   The projection is performed by computing the dot product between the
    velocity
%   vector and each of the moving frame basis vectors at every grid location
    .
%
% Inputs:
%   movingframes - 4D array [Nphi x Ntheta x mfdim x 3], where each (i,j,:)
    is an orthonormal basis
%   velocity     - 3D array [Nphi x Ntheta x 3], Cartesian vector field to
    be projected
%
% Output:
%   inarray - 3D array [Nphi x Ntheta x mfdim], velocity components in the
    local moving frame

% Get grid and frame dimensions
[Nphip, Nthetap, mfdim, ~] = size(movingframes);  % mfdim usually 3 (e1, e2,
     e3)

% Initialize output: one scalar projection per frame direction at each grid
    point
inarray = zeros(Nphip, Nthetap, mfdim);

% Loop through each grid point
```

```
for i = 1:Nphip
    for j = 1:Nthetap

        % Loop over each basis direction in the moving frame
        for k = 1:mfdim
            vel = velocity(i,j,:);      % Cartesian vector at (i,j)
            vel = vel(:);               % Convert to column vector

            mf = movingframes(i,j,k,:); % Moving frame basis vector e_k at (
                i,j)
            mf = mf(:);                 % Convert to column vector

            % Project velocity onto the k-th moving frame vector
            inarray(i,j,k) = dot(vel, mf);
        end
    end
end

end
```

Listing A.3 Transformation of Cartesian vectors into local moving frames (MATLAB script: `CartesianToMovingframes.m`)

A.4 zwgll.m

The `zwgll.m` script, authored by Luca Pavarino (University of Pavia) Fischer (2023), computes Gauss-Lobatto-Legendre (GLL) quadrature nodes and weights on the reference interval $[-1, 1]$. These nodes include endpoints and internal roots of the first derivative of Legendre polynomials. GLL quadrature is a cornerstone of spectral element methods due to its accuracy and stability for numerical integration.

```
function [z, w] = zwgll(p)
% zwgll  Computes Gauss-Lobatto-Legendre nodes and weights on [-1, 1]
%
%   [z, w] = zwgll(p) returns the p+1 GLL nodes z and weights w.
%   These are the endpoints -1 and 1, and the roots of the first
%   derivative of the Legendre polynomial of degree p.
%
%   Inputs:
%     p - Polynomial degree (number of nodes is p+1)
%
%   Outputs:
%     z - GLL nodes (column vector)
%     w - GLL weights (column vector)

n = p + 1;
z = zeros(n, 1); z(1) = -1; z(n) = 1;  % Initialize and assign endpoints
w = zeros(n, 1);                       % Initialize weights

% Interior nodes: roots of P'_p(x)
if p > 1
    if p == 2
        z(2) = 0;  % Special case: P'_2(x) = 3x
    else
        % Tridiagonal Jacobi matrix for P'_p roots (Golub-Welsch)
        i = (1:p-2)';
        off_diag = 0.5 * sqrt((i .* (i + 2)) ./ ((i + 0.5) .* (i + 1.5)));
        M = diag(off_diag, 1) + diag(off_diag, -1);
        z(2:p) = sort(eig(M));
```

```
    end
end

% Compute weights using Legendre polynomial recurrence
w([1, n]) = 2 / (p * n);  % Endpoints have the same weight
for i = 2:p
    x = z(i);
    % Recurrence to compute P_p(x)
    P0 = 1; P1 = x;
    for j = 1:p-1
        P2 = ((2*j + 1)*x*P1 - j*P0) / (j + 1);
        P0 = P1; P1 = P2;
    end
    w(i) = 2 / (p * n * P2^2);  % GLL weight formula
end
end
```

Listing A.4 Computation of GLL nodes and weights on the interval $[-1, 1]$ (MATLAB script: `zwgll.m`)

A.5 semhat.m

The MATLAB script `semhat.m`, authored by Paul Fisher (University of Urbana Champagne and Argonne National Laboratory) Fischer (2023), calculates one-dimensional spectral element matrices essential for discretizing PDEs on the reference element $[-1, 1]$. It computes the stiffness, mass, convection, and derivative matrices based on GLL nodes and weights. These matrices facilitate accurate and efficient numerical solutions in spectral element discretizations.

```
function [Ah, Bh, Ch, Dh, z, w] = semhat(N)
% semhat  Computes 1D spectral element matrices on reference element [-1, 1]
%
%   [Ah, Bh, Ch, Dh, z, w] = semhat(N) returns the stiffness matrix (Ah),
%   mass matrix (Bh), convection matrix (Ch), derivative matrix (Dh),
%   Gauss-Lobatto-Legendre nodes (z), and weights (w) for polynomial degree N.
%
%   Inputs:
%     N - Polynomial degree
%
%   Outputs:
%     Ah - Stiffness matrix      (A = D' * B * D)
%     Bh - Mass matrix           (B = diag(w))
%     Ch - Convection matrix     (C = B * D)
%     Dh - Derivative matrix
%     z  - GLL nodes
%     w  - GLL weights

[z, w] = zwgll(N);           % Gauss-Lobatto-Legendre nodes and weights
Bh     = diag(w);            % Mass matrix (diagonal with quadrature weights)
Dh     = deriv_mat(z);       % Derivative matrix using GLL nodes

Ah     = Dh' * Bh * Dh;      % Stiffness matrix via Galerkin inner product
Ch     = Bh * Dh;            % Convection matrix
end
```

Listing A.5 Computation of 1D spectral element matrices on reference interval $[-1, 1]$ (MATLAB script: `semhat.m`)

Glossary

∇ **(nabla)** Differential operator; acting on scalars gives the gradient; components and contractions produce divergence and curl on surfaces.

Δ **or** ∇^2 **(Laplacian)** Divergence of the gradient; on surfaces, the intrinsic Laplace–Beltrami operator.

Γ^k_{ij} **(Christoffel symbols)** Connection coefficients of a coordinate basis; encode how coordinate tangents change across space.

ω_{ij} **(connection 1-forms)** Components that encode infinitesimal rotation of an orthonormal moving frame along directions in the tangent plane.

g_{ij} **(metric tensor)** Inner products of basis tangent vectors; determines distances, angles, areas, and the Christoffel symbols.

$\mathbf{e}_i$ **(moving-frame vectors)** Local orthonormal tangent basis vectors at a surface point; $\mathbf{e}_3$ aligns with the surface normal.

$\mathbf{k}$ **(surface normal)** Unit vector orthogonal to the tangent plane at a point.

$\mathbf{n}$ **(boundary normal)** Outward unit normal along an element boundary; used in numerical fluxes.

D **(diffusion tensor)** Symmetric positive-definite tensor that scales gradients anisotropically; $D = Q\Lambda Q^T$.

Q**,** Λ **(eigendecomposition)** Q has eigenvectors (columns) and Λ holds eigenvalues used to align/scale transport with material axes.

Ω**,** $\partial\Omega$ **(domain, boundary)** Surface or volume where the PDE is posed and its boundary curve/surface.

Adapted frame Moving frame aligned to the tangent plane (two tangents) and surface normal (one normal).

Anisotropic diffusion Diffusion governed by a tensor with distinct eigenvalues/eigenvectors; aligns transport with material axes.

Anisotropy/Isotropy Direction-dependent/direction-independent material or propagation properties.

Bidomain model Continuum model separating intra- and extracellular media coupled by transmembrane currents; used in cardiac/neural contexts.

S. Chun, *Moving Frames for the Numerical Solution of Partial Differential Equations in Complex Domains*, Springer Asia Pacific Mathematics Series 7,
https://doi.org/10.1007/978-981-95-4495-0

Connection (Levi–Civita) Rule for differentiating and transporting vectors consistently; defined by the metric and torsion-free compatibility.

Conservation law Equation expressing local conservation via a divergence of a flux, e.g., $\partial_t u + \nabla \cdot (uv) = 0$.

Covariant derivative Derivative operator that adjusts for spatial variation of the basis and curvature (via the connection).

Curved vs. regular elements Curved elements have varying Jacobian and better geometric fidelity than straight-edged (regular) elements.

Diffusion equation Elliptic/parabolic equation with Laplacian or tensor $D\nabla$ acting on scalar fields.

Diffusion–reaction system Coupled PDEs combining diffusion with local reaction kinetics (chemistry or bioelectric).

Diffusion tensor imaging (DTI) Tensor field describing anisotropy of conduction/diffusion; eigenvectors indicate preferred fiber directions.

Discontinuous Galerkin (DG) Weak formulation with discontinuous trial spaces and numerical fluxes for inter-element coupling.

Drude model Dispersion model for permittivity used in electromagnetic cloak simulations.

Electromagnetic cloak Metamaterial structure guiding waves around a region to reduce scattering; modeled on curved geometries.

Element mapping Smooth mapping from a standard element to a curved element; its derivatives define curved tangents on the surface.

Ephaptic coupling Interaction between excitable cells mediated by extracellular electric fields rather than synapses or gap junctions.

GLL nodes Gauss–Lobatto–Legendre nodes used in spectral/*hp* element discretizations.

Gmsh Open-source 3D mesh generator used to construct curved surface meshes.

h-/p-convergence Error decay under mesh refinement (h) and polynomial order refinement (p); exponential in smooth settings.

Hairy Ball Theorem Any continuous tangent vector field on the 2-sphere must vanish somewhere; motivates piecewise frames.

Helmholtz–Hodge decomposition Decomposition of a tangent vector field into curl-free, divergence-free, and harmonic parts.

Holonomic vs. non-holonomic Coordinate-induced (holonomic) frames arise from coordinates; general orthonormal moving frames are typically non-holonomic on curved surfaces.

Internal point error Deviation of interior nodes from exact geometry; can degrade accuracy of covariant operators.

Laplace–Beltrami operator Intrinsic Laplacian induced by the metric on a surface.

Local DG (LDG) DG variant for diffusion/Laplacian using mixed formulations with auxiliary variables.

Maxwell (TM/TE) Surface formulations of Maxwell's equations often decomposed into transverse magnetic (TM) and electric (TE) modes in moving frames.

Metric tensor Positive-definite bilinear form providing inner products and distances on the surface.

Moving frame A local orthonormal basis attached to each point of a surface; adapts to geometry and material anisotropy.

NekMesh High-order mesh generator associated with Nektar++; produces curvilinear elements with high geometric fidelity.

Nektar++ Open-source spectral/*hp* element library used for the implementations in the book.

Normal coordinates Locally Euclidean-like coordinates minimizing connection coefficients in a neighborhood.

Numerical flux Interface term (e.g., Lax–Friedrichs) that stabilizes and couples neighboring elements in DG.

Parallel transport Transport of a vector along a curve while preserving inner products; realizes the connection geometrically.

Projection mesh (ProjMesh) Sphere meshes obtained by projecting subdivided cube vertices; may introduce internal point error.

Relative acceleration Second-order variation of frame angles/transport along trajectories; linked to curvature diagnostics.

Riemann curvature tensor Tensor measuring the non-commutativity of covariant derivatives; derived from the connection.

Rossby–Haurwitz wave Standard SWE benchmark; non-dispersive wave pattern rotating about the polar axis.

SEM differentiation matrices Precomputed derivative operators on reference elements used by spectral element codes.

Shallow Water Equations (SWE) Depth-averaged balance laws for large-scale flows on the sphere, including Coriolis and topography effects.

Spectral/*hp* methods High-order methods with polynomial expansions (p-refinement) on meshes (h-refinement).

Tangent plane The plane spanned by the two tangent frame vectors at a point on a surface.

Topography (bathymetry) Bottom height H_b (or H_0 for still water depth) influencing momentum through ∇H_0.

Wavefront Locus of equal phase in a propagating signal; curvature can be inferred from frame connection derivatives.

Weak formulation Integral form obtained by testing and integration by parts, reducing differentiability requirements of the solution.

Zonal flow Longitude-directed flow aligned with parallels on the sphere; used in steady and unsteady tests.

References

Abraham, R., Marsden, J.E., Ratiu, T.: Manifolds, Tensor Analysis, and Applications. Springer, New York (1988)

Akivis, M.A., Rosenfeld, B.A.: Élie Cartan (1869–1951). Translations of Mathematical Monographs. American Mathematical Society, Providence (2011)

Aliev, R.R., Panfilov, A.V.: A simple two-variable model of cardiac excitation. Chaos Solitons Fractals **7**(3), 293–301 (1996)

Anderson, R.H., Cook, A.C.: The structure and components of the atrial chambers. Europace **9**(Suppl. 6), vi3–vi9 (2007)

Arfken, G.B., Weber, H.J., Harris, F.E.: Mathematical Methods for Physicists, 7th edn. Elsevier, India (2012)

Arnold, D.N., Falk, R.S., Winther, R.: Finite element exterior calculus from Hodge theory to numerical stability. Bull. Am. Math. Soc. **47**(2), 281–354 (2010)

Beffa, G.M., Mansfield, E.L., Wang, J.P.: Discrete moving frames and discrete integrable systems. Found. Comput. Math. **13**, 545–582 (2013)

Beinart, R., Abbara, S., Blum, A., Ferencik, M., Heist, K., Ruskin, J., Mansour, M.: Left atrial wall thinness variability measured by CT scans in patients undergoing pulmonary vein isolation. J. Cardiovasc. Electrophysiol. **22**, 1232–1236 (2011)

Bernard, P.-E., Remacle, J.-F., Comblen, R., Legat, V., Hillewaert, K.: High-order discontinuous Galerkin schemes on general 2D manifolds applied to shallow water equations. J. Comput. Phys. **228**(17), 6514–6535 (2009)

Bhatia, H., Norgard, G., Pascucci, V., Bremer, P.-T.: The Helmholtz–Hodge decomposition–a survey. IEEE Trans. Vis. Comput. Graph. **19**(8), 1386–1404 (2013)

Bishop, R.L.: There is more than one way to frame a curve. Am. Math. Mon. **82**(3), 246–251 (1975)

Burke, W.L.: Applied Differential Geometry. Cambridge University Press, Cambridge (1985)

Cai, W., Chettiar, U.K., Kildishev, A.V., Shalaev, V.M.: Optical cloaking with metamaterials. Nat. Photon. **1**, 224–227 (2007)

Cantwell, C.D., Yakovlev, S., Kirby, R.M., Peters, N.S., Sherwin, S.J.: High-order spectral/*hp* element discretisation for reaction–diffusion problems on surfaces: application to cardiac electrophysiology. J. Comput. Phys. **257**, 813–829 (2014)

Cantwell, C.D., Moxey, D., Comerford, A., Bolis, A., Rocco, G., Mengaldo, G., de Grazia, D., Yakovlev, S., Lombard, J.-E., Ekelschot, D., Jordi, B., Mohamied, Y., Eskilsson, C., Nelson, B., Vos, P., Biotto, C., Kirby, R.M., Sherwin, S.J.: Nektar++: an open-source spectral/hp element framework. Comput. Phys. Commun. **192**, 205–219 (2015)

S. Chun, *Moving Frames for the Numerical Solution of Partial Differential Equations in Complex Domains*, Springer Asia Pacific Mathematics Series 7,
https://doi.org/10.1007/978-981-95-4495-0

Cartan, É.: Leçons sur la géométrie des espaces de Riemann. Gauthier-Villars, Paris (1922)

Cartan, É.: Leçons sur la théorie des espaces à connexion projective. Gauthier-Villars, Paris (1924)

Cartan, É.: La théorie des groupes finis et continus et la géométrie différentielle traitées par la méthode du repère mobile. Gauthier-Villars, Paris (1937)

Cartan, É.: Riemannian Geometry in an Orthogonal Frame (Vladislav V. Goldberg, Trans.). World Scientific, Singapore (2001a)

Cartan, É.: Geometry of Riemannian Spaces (R. Hermann, Trans.). Mathematical Science Press, Brookline (2001b)

Chun, S.: Method of moving frames to solve conservational laws on curved surfaces. J. Sci. Comput. **53**(2), 268–294 (2012)

Chun, S.: Method of moving frames to solve (an)isotropic diffusion equations on curved surfaces. J. Sci. Comput. **59**(3), 626–666 (2014a)

Chun, S.: A mathematical model of the unidirectional block caused by the pulmonary veins for anatomically-induced atrial reentry. J. Biol. Phys. **40**(3), 219–258 (2014b)

Chun, S.: Method of moving frames to solve the time-dependent Maxwell's equations on anisotropic curved surfaces: applications to invisible cloak and ELF propagation. J. Comput. Phys. **340**, 85–104 (2017)

Chun, S.: High-order method with moving frames to compute the covariant derivatives of vectors on general 2D curved surfaces. Commun. Appl. Math. Comput. **5**, 1534–1563 (2023a)

Chun, S.: Relative acceleration for conduction failure of cardiac electric propagation on a 2D general curved surface. J. Compt. Phys. X **17**, 100135 (2023b)

Chun, S., Eskilsson, C.: Method of moving frames to solve the shallow water equations on arbitrary rotating curved surfaces. J. Comput. Phys. **333**, 1–23 (2017)

Chun, S., Oh, T.: Divergence/connection preservation scheme in the curvilinear domain with a small geometric approximation error. J. Sci. Comput. **92**(15) (2022). https://doi.org/10.1007/s10915-022-01865-0

Cummer, S.A.: Modeling electromagnetic propagation in the Earth-Ionosphere waveguide. IEEE Trans. Antennas Propag. **48**(9), 1420–1429 (2000)

Davydov, V.A., Zykov, V.S., Mikhailov, A.S.: Kinematics of autowave structures in excitable media. Sov. Phys. Usp. **34**, 665–684 (1991)

do Carmo, M.P.: Differential Geometry of Curves and Surfaces. Prentice Hall, Upper Saddle River (1976)

Dray, T.: Differential Forms and the Geometry of General Relativity. CRC Press, Boca Raton (2015)

Eigenberg, M.: A proof of the Hairy Ball theorem. Am. Math. Mon. **86**(7), 571–574 (1979)

Fels, M., Olver, P.J.: Moving coframes I: a practical algorithm. Acta Appl. Math. **51**, 161–213 (1998)

Fischer, P.: TAM570: Spectral Methods for CFD. University of Illinois Urbana-Champaign, Champaign (2023)

Fischer, P.F., Lottes, J.W., Pointer, D.A.; Petascale algorithms for reactor hydrodynamics. J. Phys.: Conf. Ser. **16**, 254–265 (2005)

FitzHugh, R.: Impulses and physiological states in theoretical models of nerve membrane. Biophys. J. **1**(6), 445–466 (1961)

Frankel, T.: The Geometry of Physics: An Introduction, 3rd edn. Cambridge University Press, Cambridge (2011)

Galewsky, J., Scott, R.K., Polvani, L.M.: An initial-value problem for testing numerical methods of the global shallow-water equations. Tellus A, **56**, 429–440 (2004)

Geuzaine, C., Remacle, J.-F.: Gmsh: a three-dimensional finite element mesh generator with built-in pre- and post-processing facilities. Int. J. Numer. Meth. Eng. **79**(11), 1309–1331 (2009)

Giraldo, F.X., Warburton, T.: A nodal triangle-based spectral element method for the shallow water equations on the sphere. J. Comput. Phys. **207**, 129–150 (2005)

Giraldo, F.X., Hesthaven, J.S., Warburton, T.: Nodal high-order discontinuous Galerkin methods for the spherical shallow water equations. J. Comput. Phys. **181**, 499–525 (2002)

Green, M.D., Kirilov, K.S., Turner, M., Marcon, J., Eichstädt, J., Laughton, E., Cantwell, C.D., Sherwin, S.J., Peiró, J., Moxey, D.: NekMesh: an open-source high-order mesh generation framework. Comput. Phys. Commun. **298**, 109088 (2024)

Griffiths, David J.: Introduction to Electrodynamics, 4th edn. Cambridge University Press, Cambridge (2017)

Hesthaven, J.S., Warburton, T.: Nodal Discontinuous Galerkin Methods: Algorithms, Analysis, and Applications. Springer, New York (2008)

Hocini, M., Loh, P., Ho, S.Y., Sánchez-Quintana, D., Thibault, B., de Bakker, J.M., Janse, M.J.: Anisotropic conduction in the triangle of koch of Mammalian hearts: electrophysiologic and anatomic correlations. J. Am. Coll. Cardiol. **31**(3), 629–636 (1998)

Hodgkin, A.L., Huxley, A.F.: A quantitative description of membrane current and its application to conduction and excitation in nerve. J. Physiol. **117**(4), 500–544 (1952)

Ivey, T.A., Landsberg, J.M.: Cartan for Beginners: Differential Geometry via Moving Frames and Exterior Differential Systems. Graduate Studies in Mathematics, vol. 61. American Mathematical Society, Providence (2003)

Jackson, J.D.: Classical Electrodynamics, 3rd edn. Wiley, New York (1999)

Jones, D.K.: Diffusion MRI: Theory, Methods, and Applications. Oxford University Press, New York (2011)

Karniadakis, G., Sherwin, S.: Spectral/hp Element Methods for Computational Fluid Dynamics, 2nd edn. Oxford University Press, New York (2013)

Kenmotsu, K.: Surfaces with Constant Mean Curvature. American Mathematical Society, Providence (2003)

Kobayashi, S., Nomizu, K.: Foundations of Differential Geometry, vols. I and II. Wiley-Interscience, New York (1963/1969)

König, M., Busch, K., Niegemann, J.: The discontinuous Galerkin time-domain method for Maxwell's equations with anisotropic materials. Photon. Nanostruct. **8**, 303–309 (2010)

Läuter, M., Handorf, D., Dethloff, K.: Unsteady analytical solutions of the spherical shallow water equations. J. Comput. Phys. **201**, 535–553 (2005a)

Läuter, M., Giraldo, F.X., Handorf, D., Dethloff, K.: A discontinuous Galerkin method for the shallow water equations in spherical triangular coordinates. J. Comput. Phys. **201**, 535–553 (2005b)

Lee, J.M.: Introduction to Riemannian Manifolds, 2nd edn. Springer, Cham (2018)

Li, J., Pendry, J.B.: Hiding under the carpet: a new strategy for cloaking. Phys. Rev. Lett. **101**(20), 203901 (2008)

Li, S., Xiao, F.: A global shallow water model using high order multi-moment constrained finite volume method and icosahedral grid. J. Comput. Phys. **229**, 1774–1796 (2010)

Mansfield, E.L.: A Practical Guide to the Invariant Calculus. Cambridge University Press, Cambridge (2010)

Mansfield, E.L., Rojo-echeburúa, A., Hydon, P.E., Peng, L.: Moving frames and Noether's finite difference conservation laws I. Trans. Math. Appl. **3**, 1–47 (2019)

Mikhailov, A.S.: Kinematics of wave patterns in excitable media. In: Holden, A.V., Markus, M., Othmer, H.G. (eds.) Nonlinear Wave Processes in Excitable Media. Springer, Boston (1991)

Misner, C.W., Thorne, K.S., Wheeler, J.A.: Gravitation. Princeton University Press, Princeton (2017)

Mohammadian, A.H., Shankar, V., Hall, W.F.: Computation of electromagnetic scattering and radiating using a time-domain finite-volume discretization procedure. Comput. Phys. Commun. **68**, 175–196 (1991)

Moore, T.A.: A General Relativity Workbook. University Science Books, Mill Valley (2013)

Nair, R.D., Thomas, S.J., Loft, R.D.: A discontinuous Galerkin global shallow water model. Mon. Weather Rev. **133**, 876–888 (2005)

Needham, T.: Visual Differential Geometry and Forms. Princeton University Press, Princeton (2021)

Olver, P.J.: Equivalence, Invariants and Symmetry. Cambridge University Press, Cambridge (1995)

Olver, P.J., Valiquette, F.: On the construction of moving frames. Found. Comput. Math. **8**(3), 267–308 (2008)

O'Neill, B.: Elementary Differential Geometry, 2nd edn. Academic, San Diego (1997)

Piuze, E., Sporring, J., Siddiqi, K.: Maurer–Cartan forms for fields on surfaces: application to heart fiber geometry. IEEE Trans. Pattern Anal. Mach. Intell. **37**(12), 2492–2504 (2015)

Post, E.J.: Sagnac effect. Rev. Mod. Phys. **39**(2), 475–493 (1967)

Pudykiewicz, J.A.: Numerical solution of the reaction-advection-diffusion equation on the sphere. J. Comput. Phys. **213**, 358–390 (2006)

Pudykiewicz, J.A.: On numerical solution of the shallow water equations with chemical reactions on icosahedral geodesic grid. J. Comput. Phys. **230**, 1956–1991 (2011)

Renteln, P.: Manifolds, Tensors, and Forms: An Introduction for Mathematicians and Physicists. Cambridge University Press, Cambridge (2013)

Rohr, S., Kucera, J.P., Fast, V.G., Kléber, A.G.: Paradoxical improvement of impulse conduction in cardiac tissue by partial cellular uncoupling. Science **275**(5301), 841–844 (1997)

Schurig, D., Mock, J.J., Justice, B.J., Cummer, S.A., Pendry, J.B., Starr, A.F., Smith, D.R.: Metamaterial electromagnetic cloak at microwave frequencies. Science **314**, 977–979 (2006)

Sharpe, R.W.: Differential geometry: Cartan's generalization of Klein's Erlangen program. Springer, New York (1997)

Sherwin, S.J., Peiró, J.: Mesh generation in curvilinear domains using high-order elements. Int. J. Numer. Meth. Eng. **53**, 207–223 (2002)

Simpson, J.J., Taflove, A.: Two-dimensional FDTD model of antipodal ELF propagation and Schumann resonance of the Earth. IEEE Antennas Wirel. Propag. Lett. **1**, 53–54 (2002)

Smith, R.K., Dritschel, D.G.: Revisiting the Rossby–Haurwitz wave test case with contour advection. J. Comput. Phys. **217**, 473–484 (2006)

Spivak, M.: A Comprehensive Introduction to Differential Geometry, vol. II. Publish or Perish, Houston (1999)

Sternberg, S.: Curvature in Mathematics and Physics. Dover Publications, Mineola (2012)

Takacs, L.L.: Effects of using a posteriori methods for the conservation of integral invariants. Mon. Weather Rev. **116**, 525–545 (1988)

Toro, E.F.: Riemann Solvers and Numerical Methods for Fluid Dynamics. Springer, Dordrecht (2009)

Treibergs, A.: Notes on Cartan's Method of Moving Frames, June 2019

Turing, A.M.: The chemical basis of morphogenesis. Philos. Trans. R. Soc. Lond. Ser. B, Biol. Sci. **237**, 37–72 (1953)

Turner, M.: High-Order Mesh Generation for CFD Solvers, Ph.D. Thesis, Imperial College London (2017)

Turner, M., Moxey, D., Sherwin, S.J., Peiró, J.: ECCOMAS congress 2016. Int. J. Numer. Meth. Eng. **53**, ECCOMAS Proceedings, 428–443 (2016)

Ward, A.J., Pendry, J.B.: Refraction and geometry in Maxwell's equations. J. Mod. Opt. **43**, 773–793 (1996)

Weatherburn, C.E.: Introduction to Riemannian geometry and the tensor calculus. Cambridge University Press, Cambridge (1957)

Williamson, D.L., Drake, J.B., Hack, J.J., Jakob, R., Swarztrauber, P.N.: A standard test set for numerical approximations to the shallow water equations in spherical geometry. J. Comput. Phys. **102**(1), 211–224 (1992)

Xie, Y., Sato, D., Garfinkel, A., Qu, Z., Weiss, J.N.: So little source, so much sink: requirements for afterdepolarizations to propagate in tissue. Biophys. J. **99**(5), 1408–1415 (2010)

Xu, G.: Convergence of discrete laplace–Beltrami operators over surfaces. Comput. Math. Appl. **48**, 347–360 (2004)

Zhao, Y., Argyropoulos, C., Hao, Y.: Full-wave finite-difference time-domain simulation of electromagnetic cloaking structure. Opt. Exp. **16**, 6717–6730 (2008)

Zykov, V.S., Winfree, A.T.: Simulation of wave processes in excitable media. Wiley, New York (1992)

MIX
Papier aus verantwortungsvollen Quellen
Paper from responsible sources
FSC® C105338
FSC
www.fsc.org

If you have any concerns about our products,
you can contact us on
ProductSafety@springernature.com

In case Publisher is established outside the EU,
the EU authorized representative is:
Springer Nature Customer Service Center GmbH
Europaplatz 3, 69115 Heidelberg, Germany

Printed by Libri Plureos GmbH
in Hamburg, Germany